T0260857

FOSSIL SALAMANDERS OF NORTH AMERICA

Life of the Past *James O. Farlow, editor*

FOSSIL SALAMANDERS OF NORTH AMERICA

BY

J. ALAN HOLMAN

Indiana University Press

BLOOMINGTON AND INDIANAPOLIS

This book is a publication of

Indiana University Press
601 North Morton Street
Bloomington, IN 47404-3797 USA

http://iupress.indiana.edu

Telephone orders 800-842-6796
Fax orders 812-855-7931
Orders by e-mail iuporder@indiana.edu

The paper used in this publication meets the minimum requirements of American National Standard for Information Sciences—Permanence of Paper for Printed Library Materials, ANSI
Z39.48-1984.

Manufactured in the United States of America

Library of Congress Cataloging-in-Publication Data

Holman, J. Alan, date
Fossil salamanders of North America / J. Alan Holman.
p. cm.—(Life of the past)
Includes bibliographical references and index.
ISBN 0-253-34732-7 (alk. paper)
1. Salamanders, Fossil—North America. 2. Paleontology—North America. I. Title. II. Series.
QE868.C2H655 2006
567'.8—dc22 2005036842
1 2 3 4 5 11 10 09 08 07 06

Dedicated to my wife and favorite traveling companion,
Peggy (Margaret B.) Holman

"I think I just caught a bullhead [catfish] with legs!"

—Anonymous northern Michigan fisherman in reference to a Mudpuppy (*Necturus maculosus*), a large, aquatic salamander

CONTENTS

PREFACE

Salamanders are puzzling animals to most people, as they may look like flattened fish with legs, eels, slimy lizards, or lizards with toad-like skins. In general, salamanders have a life history that mirrors the ancient evolutionary transition from aquatic to terrestrial vertebrates, but several important groups remain permanently aquatic. Many salamanders are colorful, and some ooze or jet-spray poisonous secretions from their skin as protection from would-be predators.

Because of their body forms, some salamanders have ended up with odd common names that have been deemed "standard" by committees of herpetologists. For instance, the Mudpuppy is a dull-colored, flattened, permanently aquatic salamander. The Hellbender is a very big, flat, roughly shaped aquatic species. The name is thought to have originated from the idea that anything that ugly is bent for Hell! Large and small elongate aquatic salamanders with tiny eyes and limbs have "siren" as both the common and the scientific names, but local people often call the large ones mud eels and the small ones eel worms. The terrestrial form of the salamander genus *Notophthalmus* has a dry, toad-like skin and is called an eft, but the aquatic form of this genus is called a newt.

Several books have been written on modern North American salamanders—the latest comprehensive one by Petranka (1998)—but oddly, there is no book available on North American fossil salamanders. Thus, I have eagerly accepted the challenge of writing one, although identifying and interpreting salamander fossils is difficult, for a number of reasons. Fossil salamander remains may represent metamorphosed adults, various stages of larval development, or one of those strange flattened or eel-like aquatic creatures that retain larval characteristics. To add to the confusion is the fact that salamander larvae of the same species may turn into large aquatic creatures in one habitat but may metamorphose into a smaller terrestrial animal in another! That is why I have included sections on the life histories of salamanders and on the process of identifying and interpreting fossil ones. Hopefully, this will inspire the next generation of paleoherpetologists to give salamanders a try.

In writing this book, I have attempted to root out all the pertinent literature on the subject; I also re-examined my own work, which began in 1955. The heart of the book (chapter 2) consists of systematic accounts of fossil salamanders of North America (exclusive of Mexico), as well as the geologic age of the rocks and a listing of the localities where each taxon occurs. Extinct and currently living salamanders are presented in a somewhat different format. The extinct ones are diagnosed, described, discussed, and usually figured.

Fossil taxa that are also currently living are first presented as fossil occurrences. Then their modern characteristics, ecological attributes, and modern ranges are given. The reader will also find criteria for identifying most of them as fossils, as well as illustrations of diagnostic skeletal elements.

The book begins (chapter 1) with a definition and discussion of the Amphibia, Lissamphibia, Anura (frogs), and Caudata (salamanders). This is followed by a short discussion of the ancient Paleozoic amphibians. Comments on the cranial and postcranial morphology (form and structure) of lissamphibians (caecilians, frogs, and salamanders) are given, with special emphasis on pedicellate teeth. The primary morphological characters of the salamanders are then detailed.

Next, after a general discussion of salamander life histories, caudate specializations are detailed; included are such subjects as the skin, food procurement, vocalization, courtship and mating, eggs, larvae, and neoteny (explained in text). After this is a detailed account of the salamander skeleton, including individual bones that are useful in fossil studies. Finally, definitions of chronological terms are given, and geological time scales are presented.

Chapter 3 gives an epoch-by-epoch discussion of Mesozoic, Tertiary, and Pleistocene salamanders. Here the differences in the salamander groups of each epoch are stressed, with emphasis on the extinction of old groups and the appearance of new ones. A detailed overview of Pleistocene (Ice Age) salamanders ends the chapter.

This book is meant to be an introduction to the study of North American fossil salamanders; thus, I am writing it as simply as possible. To make the book more readable, I have mixed in a little levity, as well as some whimsical accounts of research in the field and laboratory. I have tried to avoid convoluted arguments and hazy scientific jargon. The book is intended for neoherpetologists, paleoherpetologists, general paleontologists, biologists, zoologists, and of course, anybody who likes salamanders.

ACKNOWLEDGMENTS

I very gratefully acknowledge those fine people that have shared their special interest in North American fossil salamanders with me. The book could not have been written without them. These people are the late Walter Auffenberg, Bayard Brattstrom, Charles Chantell, the late Richard Estes, Leslie Fay, Kenneth Ford, James Gardner, Frederic Grady, James Harding, the late Max Hecht, the late Claude Hibbard, Jim I. Mead, Charles Meszoely, Andrew Milner, Bruce Naylor, Dennis Parmley, J.-C. Rage, Ronald Richards, Zbyněk Roček, Karel Rogers, Elise Schroeder, Anthony Stuart, Robert Sullivan, Joseph Tihen, George Van Dam, Thomas Van Devender, Michael Voorhies, Richard Wilson, the late Vincent Wilson, and Alisa Winkler.

I thank Robert Sloan and the other staff members of Indiana University Press for their effort in the production of this book. James Farlow was editor of this volume, and I thank him for his input, which has greatly improved the text. Sharon Stewart copyedited the book. Providers of illustrations are acknowledged in the Figure Credits section. James Gardner was especially helpful in providing fossil salamander literature. The Michigan State University Museum has provided office and research space for my work.

ABBREVIATIONS

AMNH	American Museum of Natural History
CM	Carnegie Museum of Natural History
FGS	Florida Geological Survey
FMNH	Field Museum of Natural History
KU	Museum of Natural History, University of Kansas
LACMVP	Natural History Museum of Los Angeles County, Vertebrate Paleontology
MAMCT	Museum, Texas A&M University, College Station
MCZ	Museum of Comparative Zoology, Harvard University
MPUM	Museum of Paleontology, University of Montana
MSUMP	Montana State University Museum of Paleontology
MSUVP	Michigan State University Museum, Vertebrate Paleontology Collection
NJSM	New Jersey State Museum
NMC	National Museum of Canada
OMNH	Oklahoma Museum of Natural History
PU	Princeton University (its vertebrate fossils now at the YPM)
ROM	Royal Ontario Museum
RTMP	Royal Tyrrell Museum of Palaeontology
SMNH	Royal Saskatchewan Museum of Natural History
SMPSMU	Shuler Museum of Paleontology, Southern Methodist University
UALVP	University of Alberta Laboratory for Vertebrate Palaeontology
UCM	University of Colorado Museum of Natural History
UCMP	University of California Museum of Paleontology
UF	Florida Museum of Natural History
UM	University of Minnesota
UMMPV	University of Michigan Museum of Paleontology, Vertebrate Collection
UNSM	University of Nebraska State Museum
UOMNH	University of Oregon Museum of Natural History
USNM	United States National Museum
UT	University of Texas
YPM	Yale Peabody Museum, Yale University

FOSSIL SALAMANDERS OF NORTH AMERICA

CHAPTER

1 Introduction

An Overview of Salamanders

Our understanding of the evolutionary relationships of the major groups of salamanders is incomplete because of major gaps in the fossil record. Therefore, the "informal classification" of salamanders in Heatwole and Carroll (2000) is presented here in a modified form, with only the families and genera included. Extinct families and genera are prefixed by #. Later in the book, extinct species of modern genera will be designated by *. Genera found in North America are followed by **NA**. Not all these North American genera have been found as fossils.

Order Caudata

#Family Karauridae
- #*Karaurus*
- #*Kokartus*
- #*Marmorerpeton*

Family Sirenidae
- *Siren* **NA**
- *Pseudobranchus* **NA**
- #*Habrosaurus* **NA**
- #*Noterpeton*
- #*Kababaisha*

Family Hynobiidae
- *Onychodactylus*
- *Salamandrella*
- *Batrachuperus*
- #*Paradactylodon*
- *Hynobius*
- *Ranodon*

Family Cryptobranchidae

Andrias **NA**

Cryptobranchus **NA**

#*Aviturus*

#*Ulanurus*

#*Zaissanurus*

Family not designated ("stem salamandriform salamanders")

#*Valdotriton*

#*Galverpeton*

Family Proteidae

Necturus **NA**

Proteus

#*Mioproteus*

#*Orthophylia*

Family Rhyacotritonidae

Rhyacotriton **NA**

Family Plethodontidae

Desmognathus **NA**

Leurognathus **NA**

Phaeognathus **NA**

Aneides **NA**

Ensatina **NA**

Plethodon **NA**

Gyrinophilus **NA**

Pseudotriton **NA**

Batrachoseps **NA**

Family Amphiumidae

Amphiuma **NA**

Family Dicamptodontidae

Dicamptodon **NA**

#*Chrysotriton* **NA**

Family Ambystomatidae

Ambystoma **NA**

Rhyacosiredon **NA** (Mexico only)

Family Salamandridae

#*Palaeopleurodeles*

Salamandra

Mertensiella

Chioglossa

Pleurodeles

Tylototriton

Echinotriton

Triturus

Euproctus

Neurergus

#*Koaliella*

#*Brachycormus*

Cynops

Pachytriton

Paramesotriton

#Procynops
Notophthalmus **NA**
Taricha **NA**
#Family Batrachosauroididae (an enigmatic group)
#Opisthotriton **NA**
#Palaeoproteus
#Mynbulakia
#Parrisia **NA**
#Prodesmodon **NA**
#Batrachosauroides **NA**
#Peratosauroides **NA**
#Hylaeobatrachus
#Prosiren **NA**
#Family Scapherpetontidae (an enigmatic group)
#Scapherpeton **NA**
#Lisserpeton **NA**
#Piceoerpeton **NA**
#Geyeriella
#Woltersdorfiella
#Bargmannia
#Family Albanerpetontidae (salamander-like lissamphibians)
#Albanerpeton **NA**
#Celtedens
#Ramonellus (a questionable member of this family)

One must remember that taxonomic names like the ones above are anything but written in stone. In fact, with every new publication that appears on the subject of vertebrates (including those on fossil amphibians and reptiles), existing names may be verified or scrapped, entirely new names may be proposed, and taxa may be rearranged at any level. This is to be expected in a science where new methods of systematic study, including molecular ones, are constantly emerging.

The salamanders are recognized as a monophyletic group. In other words, they are thought to be a natural group that arose from a single ancestral species. The evolutionary relationships of the various taxonomic groups above are currently being hotly debated. Here I present two analyses, by Milner (1983) and by Larson and Dimmick (1993), of the evolutionary relationships of the major groups of salamanders. The analyses are in the form of cladograms (Fig. 1), which are defined by Lincoln et al. (1982, p. 48) as a "branching diagram representing the relationships between characters or character states [in organisms], from which phylogenetic [evolutionary] inferences can be made."

Definitions of the terms *Amphibia, Lissamphibia,* and *Caudata* are provided below, as are details of some of the specializations found in fossil and living salamanders.

Amphibia. The members of the class Amphibia are tetrapods (four legged animals) or derivatives of tetrapods that are characterized by a non-amniote egg (often referred to as an aquatic egg) and usually by aquatic larvae that metamorphose into at least partly terrestrial adults. The mod-

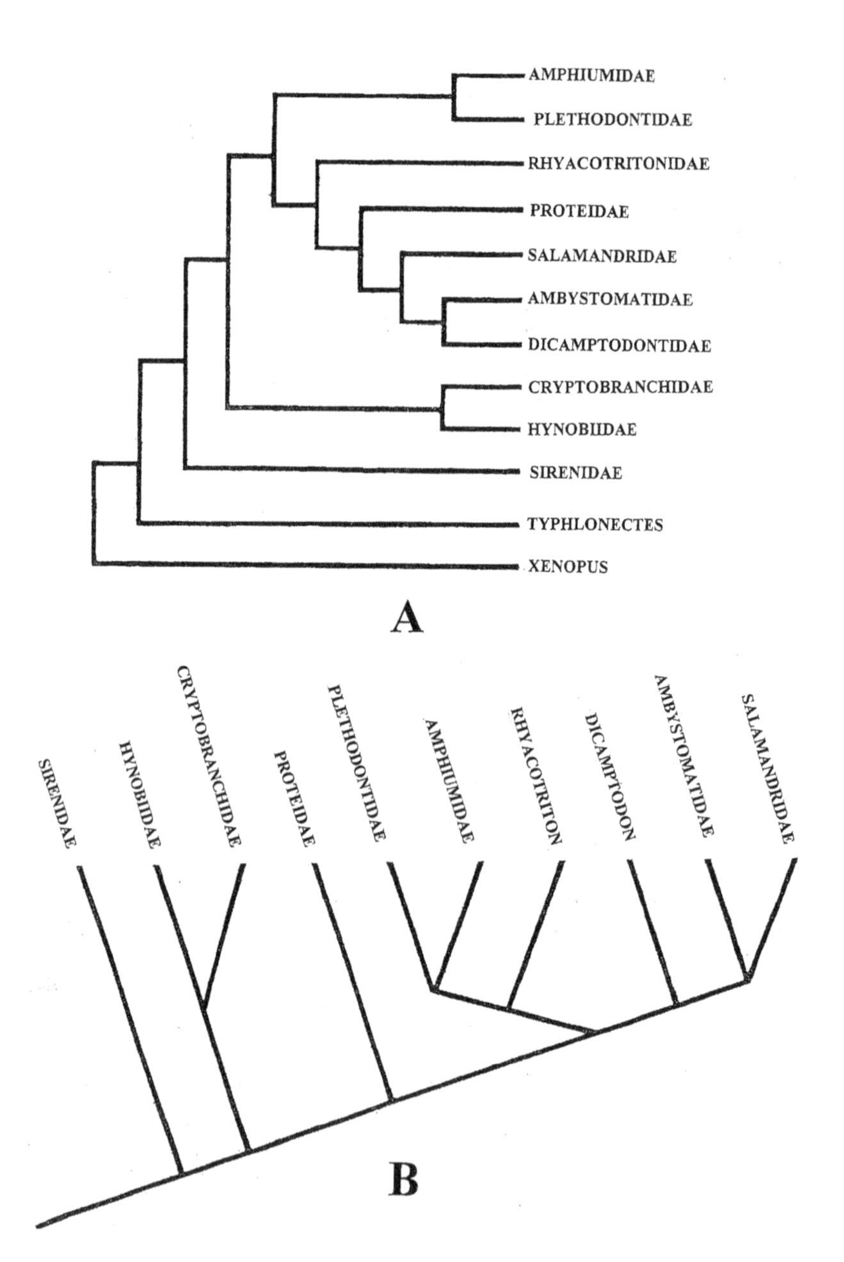

FIGURE 1. Two cladograms depicting suggested evolutionary relationships of modern salamander groups. (A) Cladogram adapted from Milner (1983). (B) Cladogram adapted from Larson and Dimmick (1993).

ern members of this class are ectotherms ("cold-blooded" animals) that all have a glandular skin lacking scales, feathers, or other epidermal (outer skin) structures (Duellman and Trueb, 1986).

Many of the ancient amphibians, however, were large or even giant forms (e.g., DeFauw, 1989) and had highly ossified, flat bodies that were very different from those of the amphibians of today. Some of these ancient amphibians, the temnospondyls, were adapted to a variety of terrestrial, semiaquatic, and even fully aquatic modes of life (Figs. 2–4). One has been described by S. L. DeFauw (personal communication, March 1988) as being the size of a Volkswagen Microbus—truly an amazing amphibian specimen. These ancient groups are placed in the class Am-

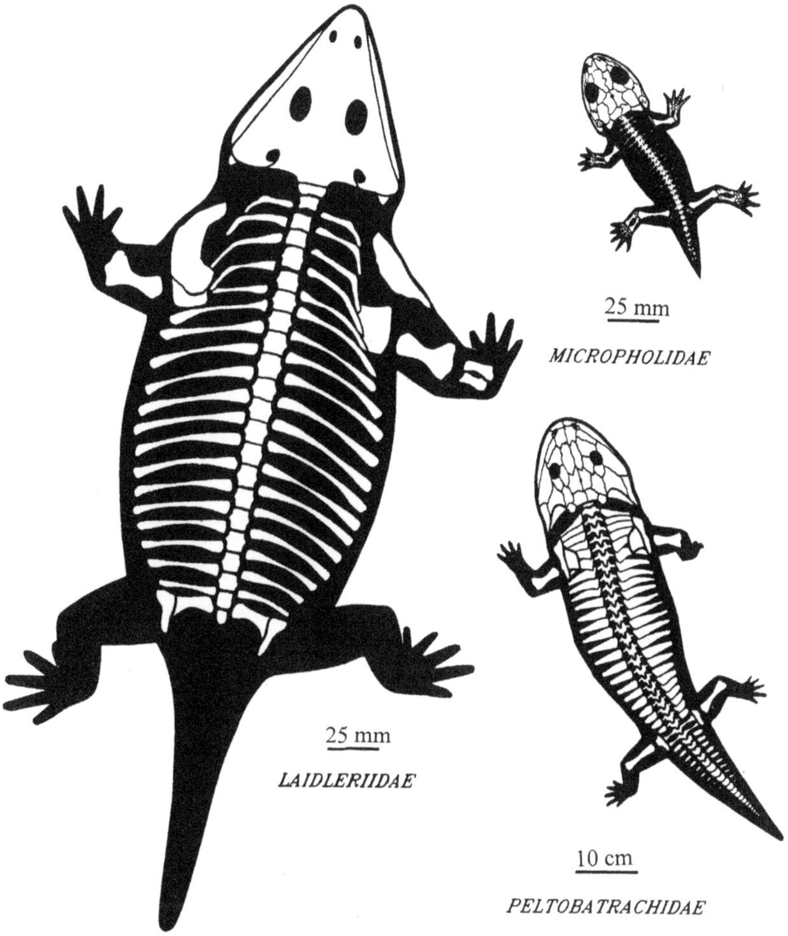

25 mm

MICROPHOLIDAE

25 mm

LAIDLERIIDAE

10 cm

PELTOBATRACHIDAE

FIGURE 2. Representatives of three terrestrial families of the Temnospondyli.

phibia with the modern amphibians, mainly because the fossil evidence shows that they laid their eggs in water, had an aquatic larval stage, and were at least partially terrestrial as adults or had evolved from lineages that were terrestrial as adults.

Lissamphibia. Modern amphibians are placed in three orders (Fig. 5): Anura (frogs and toads), Caudata (salamanders), and Gymnophiona (caecilians). These orders are grouped together as the Lissamphibia, a term that has been widely discussed and usually accepted by recent authors (Trueb and Cloutier, 1991; Cannatella and Hillis, 1993; Trueb, 1993; Pough et al., 1998; Sanchiz, 1998; Carroll, 2000a; Holman, 2003). The term *Lissamphibia* grew out of the pioneering work of Parsons and Williams (1963), who studied the morphology of the living species of amphibians and suggested that they must have had a common ancestry from a specific lineage of ancient amphibians.

Skeletal characters used by Parsons and Williams (1963) to separate the lissamphibians from the ancient amphibian groups include (1) the unique pedicellate teeth (Fig. 6), which consist of a pedicel and a crown; (2) the operculum–plectrum complex of the inner ear; (3) the specialized fenestration of the posterolateral part of the skull roof; (4) the loss of

FIGURE 3. Representatives of semiaquatic freshwater or saltwater families of the Temnospondyli. The Lydekkerinidae are typical examples of semiaquatic freshwater temnospondyls; the Mastodonsauridae are typical examples of semiaquatic saltwater temnospondyls.

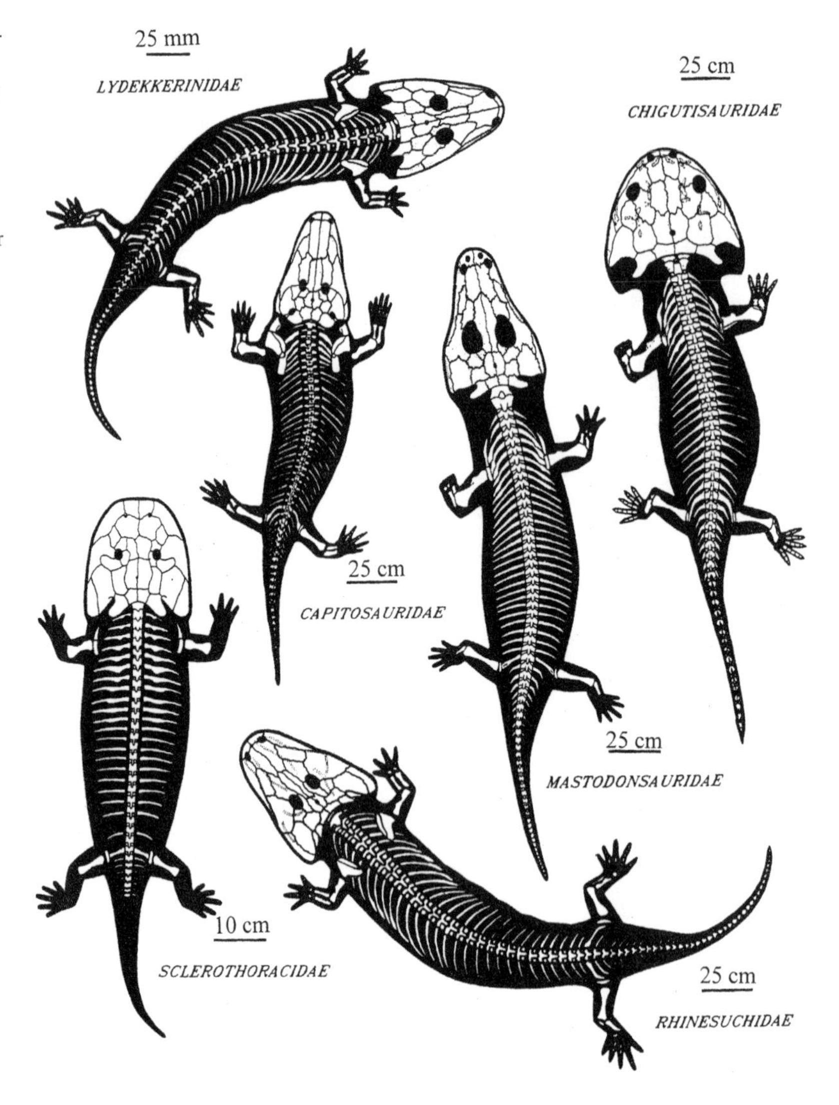

posterior bones on the skull roof; (5) the open palate; (6) the presence of two occipital condyles; and (7) the characteristically shaped atlas. However, documenting the evolutionary relationships between the ancient and modern amphibians is very difficult because of the lack of fossils that bridge the gap.

The presence of pedicellate teeth is undoubtedly the strongest character that unifies the salamanders, frogs, and caecilians. These teeth have two main portions: a basal pedicel and a distal pointed crown (Fig. 6). Both the pedicel and the crown are composed mainly of dentine. These two dentine components are divided by uncalcified dentine in some frogs. In salamanders, caecilians, and some other frogs, a ring of fibrous connective tissue divides the pedicel and the crown (Duellman and Trueb, 1986). Usually, the crown is capped by enamel-like material and is not as long as the pedicel.

A few lissamphibians, such as the aquatic salamanders of the genus

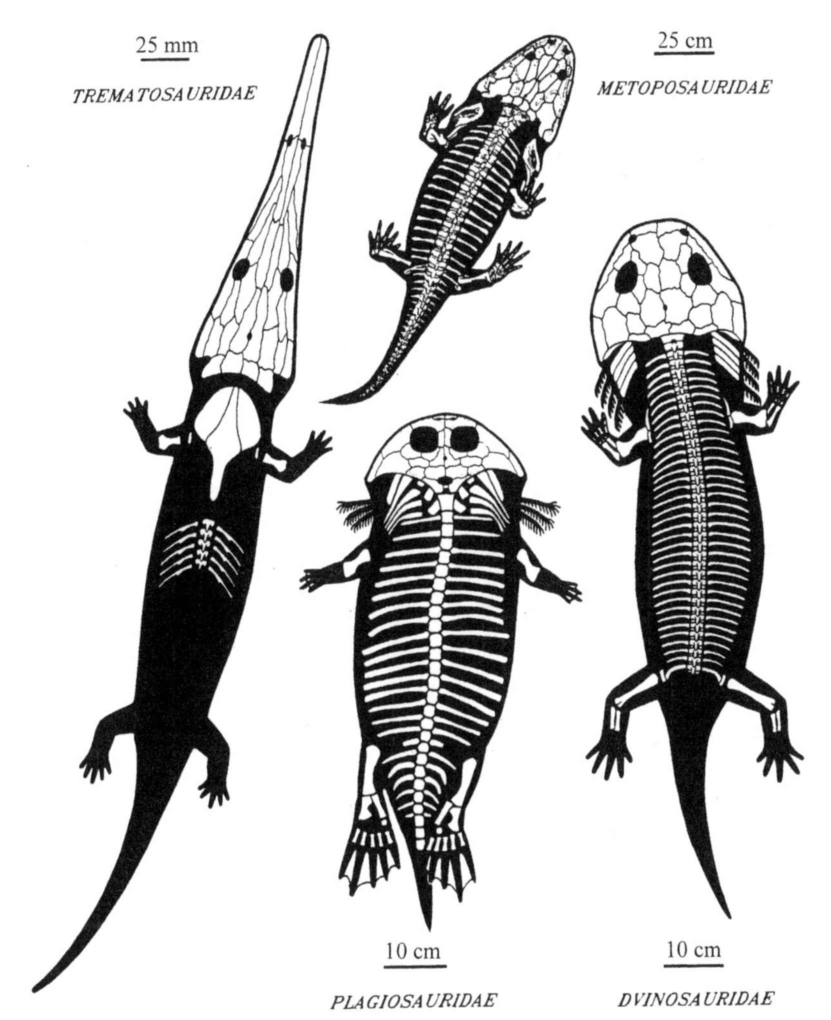

Figure 4. Representatives of fully aquatic freshwater or saltwater families of the Temnospondyli. All the families depicted are fully aquatic saltwater temnospondyls except for the Dvinosauridae, which are fully aquatic freshwater temnospondyls. Notice the external gills in the Dvinosauridae and Plagiosauridae.

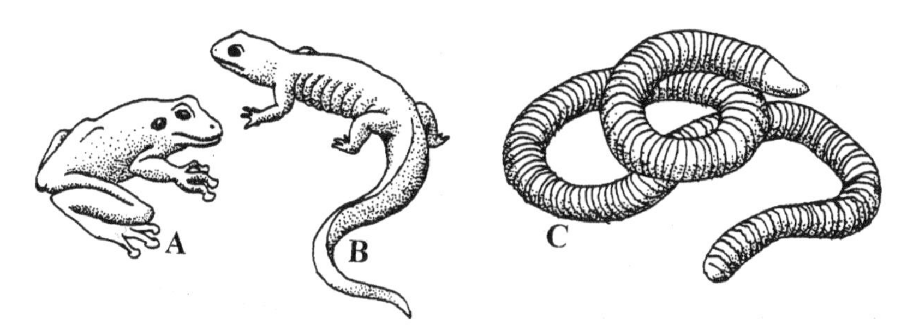

Figure 5. The three lissamphibian orders. (A) Anura, represented by a Treefrog (Hylidae). (B) Caudata, represented by a Mole Salamander (Ambystomatidae). (C). Gymnophiona, represented by a terrestrial caecilian (Caeciliidae).

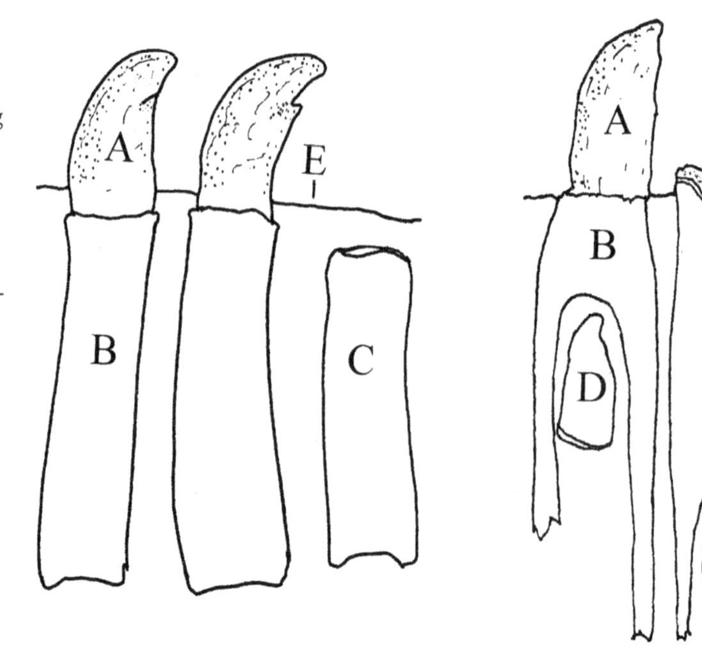

FIGURE 6. Pedicellate teeth in the salamander *Amphiuma* (left group) and in the frog *Thaumastosaurus* (right group). (A) Functional erupted tooth crowns. (B) Pedicels with functional tooth crowns attached. (C) Pedicels with tooth crowns missing. (D) Replacement crowns. (E) Gum line.

Siren, lack pedicellate teeth. Moreover, the separation between the crown and the pedicel can be weak or obscured in some salamanders in the family Proteidae, which includes the genus *Necturus*, the aquatic Mudpuppy that has been dissected in biology labs for years as a "typical" lower tetrapod. Some frogs, such as the frog-eating *Ceratophrys* (now in the pet trade), have undivided teeth in the adult that develop secondarily from divided teeth. In contrast, some salamander larvae have undivided teeth that do not become divided until the larvae transform.

Most lissamphibians have two small bones that function to transmit sounds to the inner ear. The operculum is the proximal (toward the body) element of the two, and the plectrum, or columella, is the distal (away from the body) one. In some lissamphibians the two bones are fused. The operculum develops in association with the fenestra ovalis (oval window). In all but a few lissamphibians the plectrum is joined to the tympanum (ear drum).

In the ancient amphibians the posterolateral (toward the back and the side) part of the skull is bounded by the jugal and postorbital bones, which separate the orbit (eye socket) from the temporal region. But in the lissamphibian orders both the jugal and the postorbital bones are missing, thus forming an odd fenestration (opening) in the region. All the living amphibians lack bones in the posterior part of the skull roof that were present in the ancient amphibians. These missing elements include the supratemporal, intertemporal, tabular, and postparietal bones.

An open palate is characteristic of lissamphibians; the ancient groups have palates that are much more closed. Relative to the palatal region, the parasphenoid in the lissamphibians tends to be broad and to form a significant part of the palate. The modern amphibian taxa characteristically have pterygoid bones that are reduced and widely separated and usually articulate with the braincase.

In lissamphibians, two occipital condyles on the back of the skull articulate with the vertebral column; in contrast, the ancient amphibians had only one occipital condyle. Moreover, in the modern amphibians the atlas lacks ribs and bears two cotyles (cuplike structures) for receiving the two occipital condyles (convex knobs) of the skull.

Caudata. Salamanders differ from anurans (frogs) in having tails and in lacking bodies that are specially adapted for jumping (see Fig. 5). They differ from caecilians in lacking ringed grooves around the body, and they also lack the tiny eyes of those wormlike, burrowing animals. Modern salamanders are not familiar to most people and are often given folk names that relate to other widely different vertebrate groups. Salamanders may be named after fishes ("mud eels" for sirenids), lizards ("spring lizards" for several plethodontid species), and mammals (Mudpuppies [also an official name] for *Necturus*), or they may have uncomplimentary names (Hellbenders [also an official name] for *Cryptobranchus*). About 400 species of salamander, worldwide, are currently recognized, about 130 of them living in the United States and Canada. Of the more than 4600 species of amphibians that have been described in the world, only about 8.5% are salamanders (Petranka, 1998).

Today, the salamanders are recognized as a monophyletic group (a group of taxa descended from a single taxon) by many modern authors (e.g., Larson, 1991; Trueb and Cloutier, 1991; Hay et al., 1995; and Milner, 2000). On the other hand, the relationships between the various taxonomic salamander groups are still poorly understood because of major gaps in the fossil record (Heatwole and Carroll, 2000).

Many of the approaches used in salamander study concern molecular or nonskeletal material of these animals and are thus of not much use to the paleontologist. Here we provide skeletal criteria for the recognition of fossil salamanders, mainly following Schoch (1998) and Milner (2000).

1. Almost all anurans lack toothed mandibles (lower jaws), and caecilian teeth are characteristically recurved. Thus, one can almost certainly identify a post-Triassic mandible with pedicellate, bicuspid, non-recurved teeth as a salamander.
2. The presence of Y-shaped trunk ribs (Fig. 7A) and two-headed (bicapitate) rib bearers on the trunk vertebrae (Fig. 7B) are important features in most salamanders. In the aquatic cryptobranchid salamanders these bones often fuse, but they are still recognizable. In the elongate, aquatic sirenid and amphiumid salamanders, ribs and rib-bearers occur only on the first few trunk vertebrae.
3. There are no sutures or other divisions within salamander vertebrae. Moreover, the way intervertebral foramina (small openings) are distributed along the vertebral column is important in the study of the relationships between various salamander groups (Edwards, 1976).
4. Salamander vertebrae may have basapophyses on the anteroventral ends of the centra (Fig. 7C).

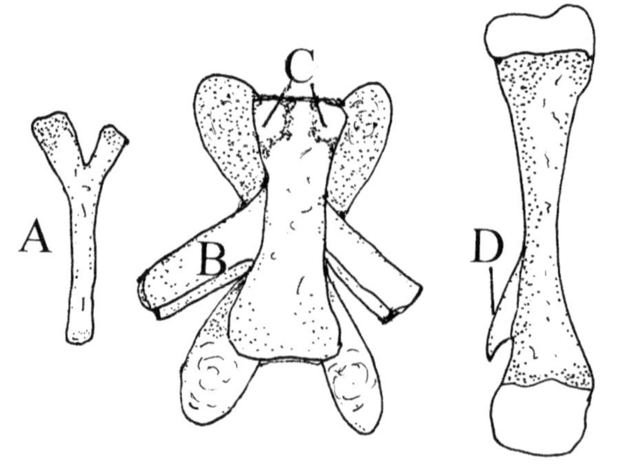

FIGURE 7. (A) Y-shaped trunk rib of the salamander *Ambystoma*. (B) Bicapitate rib-bearers of *Ambystoma*. (C) Basapophyses of trunk vertebra of *Ambystoma*. (D) Hook-like trochanter crest of the right femur of *Ambystoma*.

5. The salamander femur is short compared with that of the frog. It also has a pronounced, sometimes hook-like trochanter crest (Fig. 7D).
6. Tooth-bearing vomers are important in the feeding process of larval salamanders, but they are restructured at metamorphosis, at about the time the maxillary teeth belatedly appear. Some salamanders that are neotenic (retaining larval characteristics in sexually mature individuals) can be recognized by their reduced, obsolete, or absent maxillaries, together with the presence of well-developed vomers with anteroposterior tooth rows.
7. Unlike frogs, salamanders have separate frontals and parietals, a primitive condition. In the palate, the palatines are absent in salamanders but are retained in frogs.

Non-skeletal Specializations in Salamanders. Like frogs, salamanders are not "typical" vertebrates at all, and the dissection of Mudpuppies (*Necturus maculosus*) by the millions in biology labs as examples of vertebrates intermediate between sharks and cats was ludicrous. This misuse was based on the former abundance and the low cost of these animals. The following sections deal with the skin, food procurement, vocalization, enemies and defense, types of reproduction, courtship and mating, eggs and larvae, neoteny in living salamanders, and habitats and habits.

THE SKIN

The skin of lissamphibians, including salamanders, is "naked" in that it lacks the scales, feathers, or hair found in other classes of vertebrates. Modern amphibian skin in both the larval and the adult forms is permeable to water, which is not the case for almost all other adult vertebrates. This factor in itself is the most important specialization of this organ. Salamander skin, like that of other vertebrates, is composed of an outer epidermis, which is relatively thin, and a thicker inner dermis. But in salamanders there is an almost imperceptible gradation from the compact part of the dermis to the connective tissue covering the underlying muscles and bones (Duellman and Trueb, 1986). Caecilians have this character even more developed than salamanders. Frogs, on the other hand, have a very loose skin that is attached to the body wall only in certain places. In other words, it is easy to skin frogs in one piece, but that is difficult to do in salamanders or caecilians.

Salamanders, as well as the other living amphibian groups, have glands that are derived from the epidermis. These glands become embedded in the dermis. All salamanders have three types of glands that together occur widely over the surface of body. These are termed mucous glands, granular glands, and mixed glands. Mucous glands and the mucoid component of mixed glands produce homogeneous, sticky secretions. The main function of these secretions on land is to keep the body of the salamander from drying out so that important respiratory exchanges can occur. A dried out salamander is quickly a "gone goose." This mucous helps maintain a water–salt balance in the body fluids of the salamander and acts as a lubricant when the animal is swimming or crawling through the water.

Granular glands and the granular part of mixed glands produce poisons, some of which may give off characteristic smells. Granular poison-producing glands are on the top of the head, on the back of the eyes, on the sides of the back, and on the tail. Mixed glands are located just about everywhere in the skin. Some species of the families Ambystomatidae, Plethodontidae, and Salamandridae have a fourth gland type, hedonic glands, which are located in restricted zones, rather than being spread out over much or all of the body. These glands secrete pheromones, which are substances associated with courtship and mating. The chin gland in male plethodontids (lungless salamanders) is a well-known type of hedonic gland.

The skin of salamanders is also is replete with color-producing pigment cells. Several species of salamanders are generally brightly colored or have bright spots or other markings on their bodies (e.g., some plethodontids and salamandrids); many other species show mixtures of brownish, grayish, or yellowish patterns. North American salamanders that live underground in cave streams tend to lack pigment and appear pink or white (e.g., some plethodontids). In many salamanders the skin is smooth (e.g., most plethodontids and ambystomatids), whereas in others it is rumpled or bumpy (e.g., cryptobranchids and proteids), or even toad-like (salamandrids). As far as I am aware, no salamanders in the world have true nails on their toes, although some claw-like structures appear on the toes in various families of salamanders (Duellman and Trueb, 1986).

Food Procurement

Available information indicates that almost all adult amphibians are carnivores and that most are insectivorous. On the other hand, it was reported that North American aquatic salamanders of the genus *Siren* (all true filter feeders) were found to have had large amounts of plant material in their digestive tracts and that they eat the water plant *Elodea* (Ultsch, 1973). Some large North American salamanders eat relatively large prey. I have seen accidentally hooked Mudpuppies (*Necturus maculosus*) disgorge small crayfish in northern Michigan, and Bury (1972) reported that the California Giant Salamander (*Dicamptodon ensatus*) eats other salamanders, frogs, snakes, mice, and shrews. Hellbenders (*Cryptobranchus alleganiensis*) are known to eat crayfish, fish, lampreys, a toad, and even a small mammal (see Petranka, 1998).

Analysis of the stomach contents of several species of salamanders in New Hampshire (Burton, 1976), as well as of the terrestrial stage of the Eastern Newt (*Notophthalmus viridescens*) in New York (MacNamara, 1977), showed that the abundance of food items in the stomachs of these salamanders was correlated with the relative abundance of the various prey species.

Salamanders may locate prey by merely sitting and waiting for it to come by. But many caudates actively hunt for their food—in fact, more so than do frogs. Field observations of some salamanders, as well as laboratory experiments, indicate that vision is most important for locating prey. On the other hand, several salamanders may locate prey on the basis of the sense of smell (olfaction). Experiments on the Tiger Salamander (*Ambystoma tigrinum*) indicate that it has the greatest success in detecting, finding, and catching prey when both sight and smell are used (Lindquist and Bachmann, 1982). Movement of the prey is not necessary. I have seen Tiger Salamanders crawl out from under pieces of flower pot and move across a terrarium to pick up and swallow bits of semi-frozen beef. Wood and Goodwin (1954) indicated that the Eastern Newt can locate pill clams by olfaction alone.

All anurans and salamanders use the tongue to capture prey on land. When in water, both larval and adult salamanders basically procure food by opening the jaws and expanding the buccal (mouth) cavity. This basic process has several variations. In terrestrial salamanders with lungs, the tongue is fairly short and is attached to the anterior part of the mouth. The salamander protrudes its tongue, and the sticky surface initially contacts the prey, which is drawn into the jaws. Some lungless salamanders (family Plethodontidae) have a very long tongue that may be flipped out to capture prey in the manner of chameleon lizards.

Vocalization

Vocalization in all animals is mainly a method of advertising their presence to other animals of the same species (conspecific). Sound production is very important and almost universal in frogs, but it is uncommon in salamanders and caecilians. Various muted squeaks are made by some plethodontid salamanders. The aquatic salamander *Siren* yelps when it is bitten by its own species during territorial disputes, and the aquatic *Amphiuma* makes low whistling sounds. Species of the salamander genera *Dicamptodon* and *Ambystoma* produce several kinds of sounds, including barks, clicks, whistles, and squeaks. These sounds have variously been interpreted as an aid to orientation or as a defense mechanism, but the consensus appears to be that it is mainly a defense mechanism. Unexpected sounds might startle predators or warn other conspecific salamanders of the presence of predators.

Enemies and Defense

All amphibians may be infected with several kinds of diseases, including those of viral, bacterial, and fungal origin. These diseases occur in specimens in the wild, sometimes in what would appear to be fairly pristine situations. Such diseases are prevalent in captive salamanders, especially aquatic species. Parasites also are detrimental to the health of all amphibians. Such parasites include intestinal worms (helminths); annelids, especially leeches; arthropods, including mites, chiggers, and copepods; and fly larvae. Many salamanders are highly susceptible to the toxic secretions of the skin glands of other salamanders; in fact, different species of salamanders from the same forest floor may quickly die when put in the same collecting jar or bag!

Salamanders are prey for a large variety of predators, ranging from humans to Venus flytrap plants. Aquatic salamander eggs are eaten mainly by aquatic invertebrates and fishes. These aquatic invertebrates include leeches, which eat the eggs of the Spotted Salamander (*Ambystoma maculatum*); and caddis fly larvae, which feed on both Spotted Salamander and Tiger Salamander (*Ambystoma tigrinum*) eggs. Moreover, larval and adult Eastern Newts (*Notophthalmus viridescens*) eat eggs of several other species of salamanders.

Some plethodontid salamanders lay terrestrial eggs, and these are eaten by several animal species. Not surprising is the fact that some plethodontids eat the eggs of other plethodontids. These terrestrial eggs are also eaten by insects, especially carabid and tenebrionid beetles.

Larval salamanders are eaten by a fairly wide variety of predators. The widespread Common Gartersnake (*Thamnophis sirtalis*) is known to eat salamander larvae in the United States and Canada, and North American fishes are especially fond of caudate larvae. Some Tiger Salamander larvae are actually specialized for eating other Tiger Salamander larvae. Adult salamanders are also the food of many predators. The Ring-necked Snake (*Diadophis punctatus*) and several species of the gartersnake genus *Thamnophis* are fond of various small plethodontid salamanders. The large aquatic Mudsnake (*Farancia abacura*) is such a feeding specialist on the large eel-like salamander amphiuma (*Amphiuma*) that this snake is difficult or impossible to keep in captivity if amphiumas are not provided as food. Foraging passerine birds look for plethodontid salamanders in the woods by scratching in leaf litter.

Salamanders have evolved several antipredator mechanisms. Most salamanders appear to move very slowly (especially when compared with small lizards). But some salamanders may startle the observer by (1) moving rapidly along by quickly coiling and uncoiling the body, (2) flipping the tail while running, or (3) laterally writhing in a somewhat snakelike manner. Some salamanders of the plethodontid genus *Desmognathus* can become airborne by flipping the body. I once observed a herpetologist friend of mine become frustrated by a small *Desmognathus* he posed on a piece of woodland litter to photograph. A millisecond before each camera click, the little creature would flip off the prop onto the floor or onto another prop.

Many salamanders have coloration that conceals them from would be predators. A striking example of this is the plethodontid Green Salamander (*Aneides aeneus*), which climbs on trees and has a greenish background color and a lichen-like pattern that makes it almost invisible to humans. A plethora of woodland plethodontid salamanders have subdued colors that match the woodland floors on which they live. Adult newts of the genus *Notophthalmus* have counter-shading, which helps conceal them in the aquatic situations in which they live. Snakes and other elongate animals often have linear color patterns, such as stripes, that create an optical illusion when they are moving. Now you see what you think is a stationary gartersnake in the grass, and now you don't. A few plethodontid salamanders have linear color patterns, and so does the elongate aquatic sirenid salamander *Pseudobranchus*.

Some salamanders have defense mechanisms that are termed encounter behavior. This type of behavior may allow the salamander to be injured in the process, but it may also allow it to escape. The so-called unken reflex, oddly enough, occurs in both frogs and salamanders. The term is named for the behavior of the European frog *Bombina* as it makes its "unk, unk, unk" call. The behavior consists of a rigid posture, with the chin and the tail raised up to expose the bright colors on the lower surfaces of the animal. North American salamanders of the genera *Notophthalmus* and *Taricha*, both of the family Salamandridae, display the unken reflex. These salamanders have poisonous skin glands on the back. Both the bright colors and the posture of these salamanders are cues associated with noxiousness by the would-be predator. Salamanders that have the unken reflex have expanded bony frontosquamosal arches on the head skeleton and expanded neural arches on their vertebrae that are said to reduce injury by predators (Naylor, 1978a). Incidentally, these bony frontosquamosal arches and expanded neural arches are useful characters in the identification of fossil salamandrid salamanders.

Other salamander defenses against predators involve the use of the tail. Some salamanders that lash their tails at predators have especially well developed musculature in the tail region and have poison glands on the top of the tail. In North America, salamanders with this tail-lashing mechanism are usually large or moderately large. Some North American salamander genera that have species with this active antipredator behavior include *Dicamptodon*, *Ensatina*, *Rhyacotriton*, *Ambystoma*, and *Plethodon*.

Tail undulation is a passive antipredator behavior. This involves a sinuous movement of the salamander's tail while the body remains motionless. Usually, the tail extends straight up, with the body coiled and the head tucked under the tail base. Undulatory-type tails are long and slender and have poison glands on their upper surface. Most species with these kinds of tail can shed them at will. Salamanders with this type of antipredator behavior are usually small and slender. Most are in the family Plethodontidae. North American genera of salamanders that have the tail-undulation mechanism include *Ambystoma* (the only non-plethodontid), *Aneides*, *Eurycea*, *Gyrinophilus*, *Hemidactylium*, *Hydromantes*, *Plethodon*, *Pseudotriton*,and *Typhlotriton*.

Head butting is an antipredator behavior occurring in some heavy-bodied salamanders that have concentrations of poison glands near the top of the head. These animals move the head into a downward position and either swing the head toward or lunge at predators. During this process the body is raised up, and the head and the front part of the body are bent toward the predator. Some species vocalize during this behavior. The only North American genus that I am aware of that has species with this antipredator mechanism is *Ambystoma*.

The dicamptodontid *Dicamptodon*, the plethodontid *Desmognathus*, and some species of *Ambystoma* bite vigorously. The large eel-shaped *Amphiuma* has a painful bite, sometimes accompanied by a vocal squeak.

"Playing dead" may help certain species of salamanders evade death by predation, as some predators eat only living animals. Arnold (1982) showed that the skin secretions of large plethodontids can immobilize would-be snake predators by gluing their mouths or coils together.

Types of Reproduction

With the exception of the Hynobiidae, the Cryptobranchidae, and the Sirenidae (of which only the latter two families occur in North America), salamanders have a unique way of internal fertilization. This involves sperm-containing structures, spermatophores, which are produced and deposited by the males and picked up by the females, which results in fertilization. The spermatophore is a conical, gelatinous structure capped with a mass of sperm.

Frogs are famous for their variety of reproductive strategies, but caudates also have interesting differences in their types of reproduction. Salamanders have two major types: fish-like external fertilization, where the eggs and the larvae are aquatic; and internal fertilization (discussed above), which itself has six modes of reproduction (Duellman and Trueb, 1986).

In external fertilization, sperm released into the water by males fertilizes the eggs released by the females in the vicinity. In internal fertilization, the male deposits spermatophores on the substrate, which are picked up by the cloacal lips (lips of the vent opening) of the female. External fertilization is known to occur in the Hynobiidae (an Old World genus) and the Cryptobranchidae, and it probably occurs in the Sirenidae, although I am unaware that anyone has ever seen sirenids mate. External fertilization is considered to be the primitive state in salamanders.

Modes of reproduction in salamanders when internal fertilization takes places include situations where (1) both the eggs and the larvae are aquatic; (2) the eggs are terrestrial, but the larvae are aquatic; (3) the eggs are terrestrial, and the larvae are terrestrial but non-feeding; (4) the eggs are terrestrial and undergo direct development; (5) the eggs are retained in the oviducts, where they develop until aquatic larvae hatch; and (6) the eggs are retained in the oviducts until fully developed young hatch.

Internal fertilization occurs in approximately 90% of salamander species. In North America, type 1 reproduction occurs in the Proteidae, Amphiumidae, most Ambystomatidae, and aquatic Plethodontidae. Type 2 reproduction is found in the North American ambystomatid salamanders *Ambystoma cingulatum* and *Ambystoma opacum*. These salamanders deposit terrestrial eggs in depressions. When rain comes the depressions fill with water and the eggs hatch into aquatic larvae. The type 2 situation also occurs in the North American plethodontids *Hemidactylium scutatum* and *Stereochilus marginatus*. These animals have terrestrial nests in rotting wood or clumps of moss, and when hatching occurs the hatchlings squirm into the water below.

Desmognathus aeneus, a plethodontid, is in the type 3 situation, in which the eggs hatch into larvae in a terrestrial nest, but these larvae do not feed until they metamorphose. *Plethodon cinereus*, a widespread small plethodontid of the woodlands, is a type 4 form. It lays its eggs under moist leaves or in rotting wood. These eggs develop directly into the adult form. The type 5 and type 6 situations have not been recognized in North America.

Courtship and Mating

Location of breeding sites is of primary importance in both frogs and salamanders. As far as we know, vocalization, which plays such a vital part of the reproductive process in frogs, including location of breeding sites, has little or no significance in the breeding processes in salamanders. Odor may be the most important way that salamanders find breeding sites, but visual cues and other factors are important as well. Sexual differences (sexual dimorphism) are often related to courtship and breeding. Size differences in male and female salamanders are usually not as pronounced as they are in some other vertebrates, but some other sexual differences are prominent in caudates.

During the breeding season the vents of males become characteristically swollen. Males and females of the North American salamandrid genus *Taricha* have rough skin when they are not breeding, but smooth skin during the breeding season. Males of the aquatic stage of the North American species *Notophthalmus viridescens* develop larger tail fins and dorsal fins during the breeding season. Males of North American plethodontids and ambystomatids develop various courtship glands. These glands may appear on the head, on the neck, under the jaws, or on the base of the tail. Salamanders of the family Plethodontidae show sexual dimorphism in the size, structure, and number of their teeth. In some males the teeth are long and even protrude from the lips. Others have single-cusped teeth during the breeding season; the cusps are used to deliver secretions from mental glands (chin glands) that stimulate the female.

Courtship in salamanders functions in two ways: (1) to persuade the female to take part in the process; and (2) to effect transfer of sperm from the male to the female. Salthe (1969) outlined the five generalized stages of courtship in salamanders that practice internal fertilization.

1. The male locates and approaches a potential mate and frequently nudges or rubs her with his snout.
2. After making sure that the potential mate is a female, he either captures her or blocks her path and continues his rubbing or tail movements.
3. The male moves away and the female follows, but this is not present in all groups.
4. The male deposits a spermatophore.
5. The male leaves the spermatophore, and the female follows to find the spermatophore.

Eggs and Larvae

Salamander eggs are surrounded by a vitelline membrane and one or more jelly layers below that; in some species, other associated membranes are present as well. The membranes and jelly layers keep the eggs from drying out and protect the developing embryo against various small predators. It is probable that these membranes also protect the embryos from a variety of pathogens.

Salamanders that have internal fertilization may place their eggs together in a clutch and attend them or distribute them to various locations without attendance; the latter method is more prevalent. Salthe (1969) recognized three main sites of egg deposition in salamanders. (1) Lentic (still-water) sites are frequented by most species of *Ambystoma* that attach the eggs to twigs and stems. The attachment of such eggs in the middle of the water column keeps them from sinking to the bottom, where the oxygen supply is low. All newts, using their hind feet and lying upside down, conceal each egg in vegetation to protect it from predation by other newts. Lentic species of salamanders lay eggs that are smaller than those laid by stream-breeding (lotic) species. (2) Lotic (running-water) sites are used by such salamanders as the small plethodontid *Eurycea*, which deposits its eggs on the bottom of rocks in streams so that water moves over the eggs. The eggs have stalks and are attached one at a time by upside-down females. Species of the plethodontid *Desmognathus* conceal their eggs on the banks of streams. (3) Terrestrial sites are used by some plethodontid salamanders. Their eggs are laid in clutches, and the female usually attends the eggs.

Salamander larvae and frog larvae have structural adaptations that correlate with the habitats in which they live. At hatching time, larvae that breed in still water (pond forms) characteristically have non-functional limbs; fairly long dorsal fins, running up the back and extending almost to the head; and long, bushy gills. Moreover, many still-water salamander larvae have balancers, which are props that grow out of the side of the head. The props are lost within a week or so of hatching as the limbs and digits develop.

On the other hand, larvae of salamanders that breed in running water (stream forms) normally have reduced gills, dorsal fins that end near the rear limbs, and functional limbs at hatching. Some also have cornified (hardened) friction pads on their toes. The Mudpuppy, a salamander that is obligatorily aquatic, differs from the above in having very bushy gills. Larvae of various species of Pacific Giant Salamander (*Dicamptodon*) have highly branched gills coming from short stalks. Petranka (1998) pointed out that the presence or absence of a long dorsal fin is the single most reliable trait for distinguishing between lotic and lentic larvae.

Most frog larvae are herbivores that feed on algae and other types of aquatic plant material. Larval salamanders, however, are predators that feed mainly on small aquatic invertebrates. But larger salamander larvae may feed on salamander larvae of other or even their own species. As in many frogs, salamander larvae that live in seasonal habitats have rapid growth rates that may be as short as about 2 months in some species. Salamander species that breed in such situations usually lay large numbers of eggs. Larval salamanders in streams generally have slower growth rates and lower mortality rates than salamander larvae that utilize seasonal habitats.

Neoteny

The process of neoteny not only is of great importance in the ecological adaptations of modern salamanders but has been exceedingly important in the evolutionary history of the group. In fact, I am not aware of any other vertebrate order in which neoteny has played such an important evolutionary role. In this book we shall define the term *neoteny* mainly following Duellman and Trueb (1986); also see Gould (1977).

Here the term is defined as retardation of development to the extent that sexual maturity occurs in organisms that retain juvenile characteristics. This situation can occur at various levels. For instance, some salamanders, such as the Tiger Salamander (*Ambystoma tigrinum*), may exist as aquatic, non-metamorphosed breeding populations for generations when environmental conditions are poor. When environmental conditions improve these populations metamorphose and become land animals. In other situations, however, an evolutionary "point of no return" occurs when such animals become obligate neotenes. Mudpuppies, sirens, cryptobranchids, and amphiumids—all fully aquatic North American salamander families—became obligate neotenes millions of years ago.

As far as living obligate neotenic species in North America go, the Cryptobranchidae has one (its only species, *Cryptobranchus alleganiensis*); the Sirenidae, four; the Amphiumidae, three; the Proteidae, five; the Ambystomatidae, five; and the Plethodontidae, nine. But many other neotenic salamander species in North America have not yet reached the point of no return.

Habitats and Habits

Following metamorphosis, most salamanders move away from the breeding sites and live either on land or in various aquatic habitats. The juvenile stage is an important time for the dispersal of the species. Juveniles of most pond-breeding salamanders live underground while they mature and are rarely seen by humans. On the other hand young efts of the family Salamandridae are often encountered on the ground during the daylight. When pond breeding salamanders reach sexual maturity they begin to make seasonal migrations to the breeding ponds, some of them returning year after year to the original ponds. On the other hand, salamanders are usually able to quickly utilize new ponds.

Terrestrial salamanders pursue their main activities on the surface of the ground at night. At this time they forage for invertebrate food, defend their territories, and seek mates. They are also subject to predation during this activity cycle. Most species of North American salamanders belong to the family Plethodontidae. Aside from a few odd neotenic, subterranean forms that have gills, adult plethodontids lack both gills and lungs. Thus, plethodontid skin must be kept moist to facilitate gas exchange. In other words, with regard to breathing, the skin of these species must act as a thin, "external lung," with no supplementary help from conventional gills or lungs.

Many species of plethodontids live on the woodland floor where they must seek out moist places. These forms tend to restrict their foraging to the ground surface at night when the relative humidity is high. If the woods dry out, these salamanders must react by going underground or by moving under masses of leaf litter or into decaying logs. Mole Salamanders (Family Ambystomatidae) solve the dehydration problem by burrowing into the soil or entering the burrows of rodents and other small animals.

Many plethodontid salamanders exhibit territoriality. Adults tend to establish small home ranges and to defend them from other members of the same species. Studies of the Red-backed Salamander (*Plethodon cinereus*) demonstrate that adults of this species mark their territories with body secretions and fecal droppings. These territories are defended against adults of the same sex. Territorial battles may simply be confined to defensive posturing. On the other hand, this can change into chasing and biting behavior that may result in the loss of toes or even portions of tail.

The Salamander Skeleton

From the standpoint of the vertebrate paleontologist, the skeleton is the most important part of the units that make up the salamander body, as it contains the parts that are most likely to be preserved as fossils. Here we present an overview of the salamander skeleton, with its major elements discussed and labeled.

Of the three orders of modern amphibians (lissamphibians), the salamander has the most generalized skeleton (Fig. 8). The skeletal system of salamanders presented here is divided into three units: (1) the skull, including the hyobranchial apparatus; (2) the vertebral skeleton; and (3) the appendicular skeleton.

The skull is relatively open in salamanders, but it is not as widely open as in typical frogs. The skull is extremely closed and compact in caecilians. The vertebral column is moderately long in most salamanders, but it is very long in sirenids and amphiumids. It is very short in frogs and very long in caecilians. A long tail is present in salamanders; the tail is very short or absent in the caecilians and is absent in

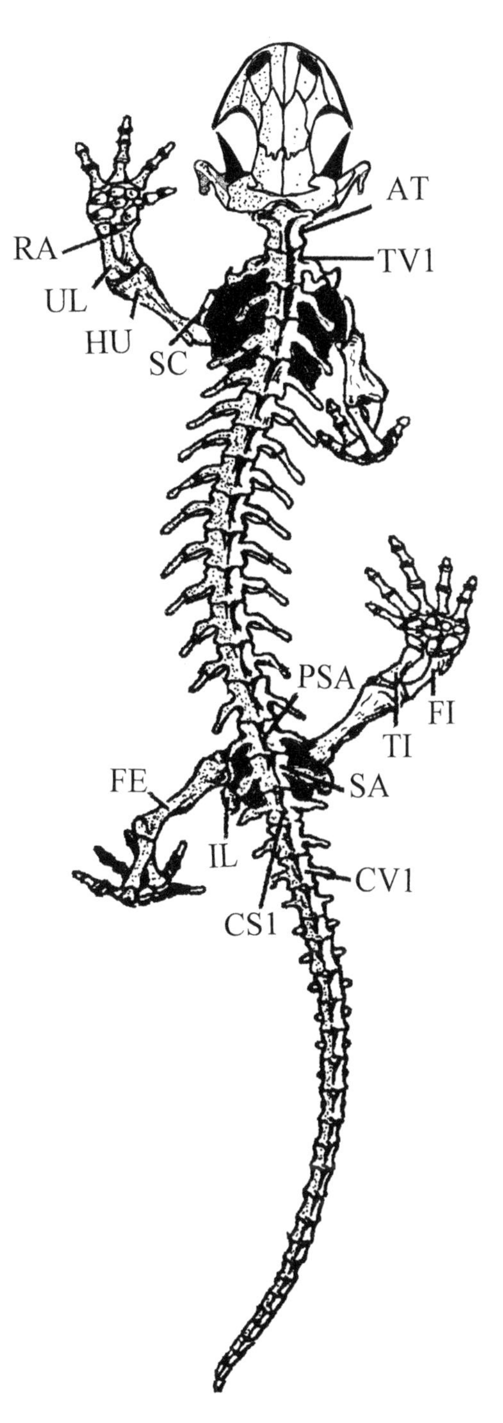

FIGURE 8. Skeleton of a generalized salamander (assembled from various sources). Abbreviations: AT, atlas (sometimes called cervical vertebra); CS1, first caudosacral vertebra; CV1, first caudal vertebra; FE, femur; FI, fibula; HU, humerus; IL, ilium; PSA, presacral vertebra; RA, radius; SA, sacrum (sacral vertebra); SC, scapula; TI, tibia; TV1, first trunk vertebra; UL, ulna.

the frogs. The limbs and limb girdles are well developed and of the "normal vertebrate variety" in most salamanders, but they are obsolete and non-functional in sirens and amphiumids. The limb girdles are highly modified for a hopping gait in the anurans and are absent in the caecilians. The next sections have a moderately detailed account of the three units of the vertebrate skeleton and are followed by a more detailed account of the salamander bones that are often used in the identification of fossils.

Skull and Hyobranchium

The skull and associated hyobranchium are complex and diverse in the salamanders. The skull houses the central nervous system, as well as the organs of sight, smell, hearing, and equilibrium. The hyobranchium (Fig. 9) is a complex of bony and cartilaginous elements in the throat that forms the structural base of the tongue, as well as the source of attachment of muscles responsible for the intake of air and the catching, holding, and swallowing of food. In most adult salamanders the hyobranchium is formed only of parts of the hyoid and the first two branchial arches.

The structure of the hyobranchium is quite variable in the salamanders (see Duellman and Trueb, 1986, fig. 13-7, p. 302), but the basic structures are as follows (Fig. 9). There is usually one long element, the basibranchial I (BI), in the middle of the unit. This unit is often called the copula. Often there is a second, reduced, basibranchial (BII) (see Duellman and Trueb, 1986, fig. 13-7B). Basibranchial I is a major element to which other parts of this complex are attached. Anterolaterally, basibranchial I supports a pair of horns consisting of anterior and posterior radials (AR and PR), elements that are imbedded in the tongue musculature of the salamander in life. The ceratohyals (CRH) are very large elements (except in plethodontids) that are ventral to the hypobranchials and bear a ligamentous connection to the suspensorium (muscles associated with opening and closing the jaws). The hypobranchials (HYI and HYII) act as rod-like connections between basibranchial I medially and the ceratobranchials distally.

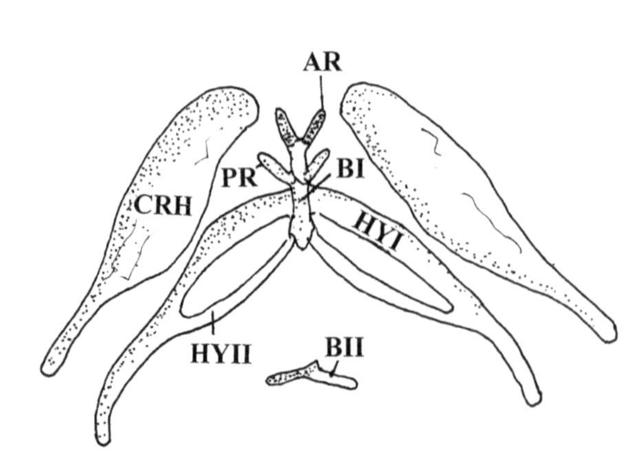

FIGURE 9. Hyobranchial apparatus of *Salamandra* sp. in dorsal view. Abbreviations: AR, anterior radius; BI, basibranchial I; BII, basibranchial II; CRH, ceratohyal (the ceratohyals are laterally displaced in this figure; normally they overlap); HYI, hypobranchial I; HYII, hypobranchial II; PR, posterior radius. Modified from Francis (1934).

The skull is composed of two types of bone, dermal bones and endochondral bones. Developmentally, dermal bones originate directly in the dermis (deep layer) of the skin, without any preformation in cartilage; and endochondral bone is preformed in cartilage in the internal part of the body.

Endochondral bones form the bony structure that protects the brain (neurocranium), the auditory (hearing) capsules, the middle ear (sound-conducting) bones, the bony portions of the hyobranchium, and the elements that take part in the articulation of the jaw with the skull. The dermal components of the salamander skull, in general, are external to the endochondral ones. These dermal bones function to protect the endochondral bones that surround the important sense organs. The primary bony parts of both the upper and the lower jaws are composed of dermal bone. Several dermal bones at once brace and suspend the upper jaw against the braincase. Dermal bones of the salamander skull are much more commonly found as fossils than are endochondral bones.

Samples of salamander skulls have been described since the late 1800s, many in a typological way (one specimen typifying a group of organisms). But series of skulls need to be examined for individual variations at all taxonomic levels. This would immeasurably help to establish a true phylogeny of the Caudata.

Dermal roofing bones of the salamander skull are relatively small and few. The temporal region of the skull is open, the orbits are large, and the nasal area is poorly roofed. Moreover, the maxillary area (arcade) is incomplete. The neurocranium is poorly developed. The architecture of the skull in salamanders is very diverse, which in a large sense is related to the variety of habitats, both terrestrial and aquatic, that these animals occupy. For instance, cryptobranchids, salamandrids, and ambystomatids tend to have short skulls that are broadly rounded anteriorly. Plethodontids tend have longer skulls than the former groups, but they are only slightly less broadly rounded anteriorly. Amphiumids have long, narrow skulls that are much more narrowly rounded anteriorly than those in the above groups. Sirenids have narrow skulls that taper anteriorly to a point. Troglodytic (cave) salamanders in North America (mostly plethodontids) may have highly modified skulls that retain larval characteristics.

Figure 10A depicts the skull of Jordan's Salamander (*Plethodon jordani*) in dorsal view, and Fig. 10B shows it in ventral view. Jordan's salamander is a member of the advanced family Plethodontidae and is fairly typical of this family, which con-

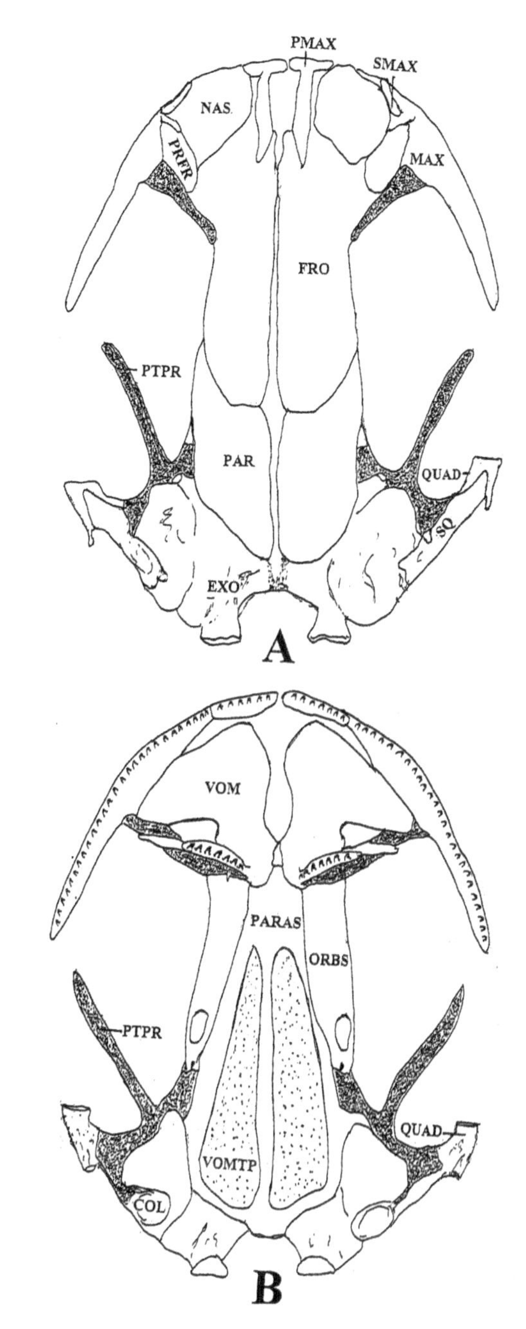

FIGURE 10. Skull of *Plethodon jordani* (Plethodontidae) in (A) dorsal and (B) ventral views. Abbreviations: COL, columella; EXO, exoccipital; FRO, frontal; MAX, maxilla; NAS, nasal; ORBS, orbitosphenoid; PAR, parietal; PARAS, parasphenoid; PMAX, premaxilla; PRFR, prefrontal; PTPR, pterygoid process of pterygoid bone; QUAD, quadrate; SMAX, septomaxilla; SQ, squamosal; VOM, vomer; VOMTP, vomerine tooth patch areas (lightly stippled). Deeply shaded areas are cartilaginous elements.

tains most of the species of North American salamanders. The major bones of the skull are featured. The skull of the Plethodontidae tends to be open in the palatal and orbital (eye) region. The elements composed of cartilage are shaded in Fig. 10. The anterior (toward the head) frontals and posterior (toward the tail) parietals (Fig. 10A) form most of the dermal skull roof in salamanders and are the only paired roofing bones found in all salamanders (Duellman and Trueb, 1986). In *P. jordani* and many other salamanders the frontals are significantly longer than the parietals. The premaxillary bones are slender, T-shaped, and separate in *P. jordani,* and they overlap the frontals posteriorly. The separation of the premaxillary bones is a rather typical condition in salamanders, but fusion of these elements can occur in some forms, even within the Plethodontidae. The Red Hills Salamander (*Phaeognathus hubrichti*), for instance, has the premaxillary bones fused, relatively thick, and not T-shaped. Nasal bones are found in all salamanders, but they are subject to much variation in shape. The nasals are roughly round in *P. jordani,* and their main articulations are with the septomaxilla and maxilla laterally (toward the side) and with the prefrontal and frontal posteriorly (toward the tail). This is relatively typical in plethodontids. The nasals always lie anterior (toward the head) to the prefrontals in salamanders, if prefrontals are present. The prefrontals are somewhat teardrop-shaped bones that are surrounded by the nasals, maxillaries and frontals, and they form part of the anterior wall of the orbit. Prefrontals are present in all salamanders except sirenids, proteids, and some plethodontids. The septomaxillary is a small, elongate bone in *P. jordani* and is lateral to the nasals. Notice that a lacrimal bone is not present in the skull roof of *P. jordani.* This small bone is found only in the hynobiids (Old World group) and the dicamptodontids and rhyacotritonids of western North America. The lacrimal, when present in salamanders, forms part of the anterior margin of the orbit. The orbit is partially open in *P. jordani.* The backwardly directed prongs of the maxillary extend posteriorly in this species to form the anterior margin of the orbit. But the cartilaginous pterygoid process of the pterygoid does not extend far enough to meet the maxilla. The pterygoid process of other salamanders (e.g., *Stereochilus marginatus*) extends far enough to close the orbital region. The squamosal bone of *P. jordani* braces the quadrate to the posterior part of the braincase, and the quadrate attaches to the lower jaw, as in other amphibians and reptiles. The exoccipitals connect the salamander's head to the vertebral column by means of two somewhat stalked occipital condyles, paired processes at the very posterior end of the skull.

We now turn to the skull of *Plethodon jordani* in ventral view (Fig. 10B). Like all other salamanders, this species has three large dermal bones—paired vomers and a parasphenoid—that make up most of the ventral part of the skull. The vomers in *P. jordani* are specialized, as they are in some other plethodontids. In *P. jordani,* they are anteriorly wide, somewhat bell-shaped bones, with short horizontal rows of teeth. But the vomers also have long, club-shaped processes that emerge posteriorly to override the parasphenoid as anteriorly tapered patches of vomerine teeth (VOMTP in Fig. 10). As the reader will notice, this situation is difficult to reproduce on a two-dimensional drawing. Of interest is that in the

Mudpuppy (*Necturus*) there are no maxillary bones, and the vomer forms a functional part of the upper jaw. The parasphenoid protects the braincase ventrally and occurs in all salamanders. This bone is usually wider anteriorly and narrows posteriorly in the region of the orbitosphenoids. The orbitosphenoids form the lateral walls of the braincase. In the adult salamander the columella is bony and fuses with the wall of the otic (ear) region.

We now turn to a discussion of the skull of the California Giant Salamander (*Dicamptodon ensatus*). This splendid salamander was at one time included in the family Ambystomatidae (Mole Salamanders), but it has now been relegated to its own family, the Dicamptodontidae, although it is still considered to be related to the Ambystomatidae. The skull of *D. ensatus* (Fig. 11) contrasts sharply with that of Jordan's Salamander (*Plethodon jordani*) (see Fig. 10). Tihen (1958, p. 22) succinctly stated that the "most distinctive feature of the skeleton of *Dicamptodon* is the solidity and rigidity of the skull." This character state has to do not only with the larger, more massive bones in *Dicamptodon* but also with the firm and extensive suture connections that occur in the genus.

Notice that in the preorbital region of *Dicamptodon*, in dorsal view, the premaxillae are large, massive, and solidly fused with the other dorsal roofing bones. Moreover, in ventral view, the prevomer and the vomer are large and joined together at the midline, forming a solid palatal shelf. In *Plethodon jordani* the premaxillae are small, T-shaped, unfused bones, mainly separated from other dorsal roofing bones in the area, as are the vomers in the palatal area. In *Dicamptodon* the orbit is fully closed in most specimens (Tihen, 1958), as the fully ossified pterygoid process makes contact with the end of the maxilla. In *P. jordani* the orbit is open, as the cartilaginous pterygoid process does not make contact with the maxilla. In *Dicamptodon* the base of the pterygoid bone itself is mas-

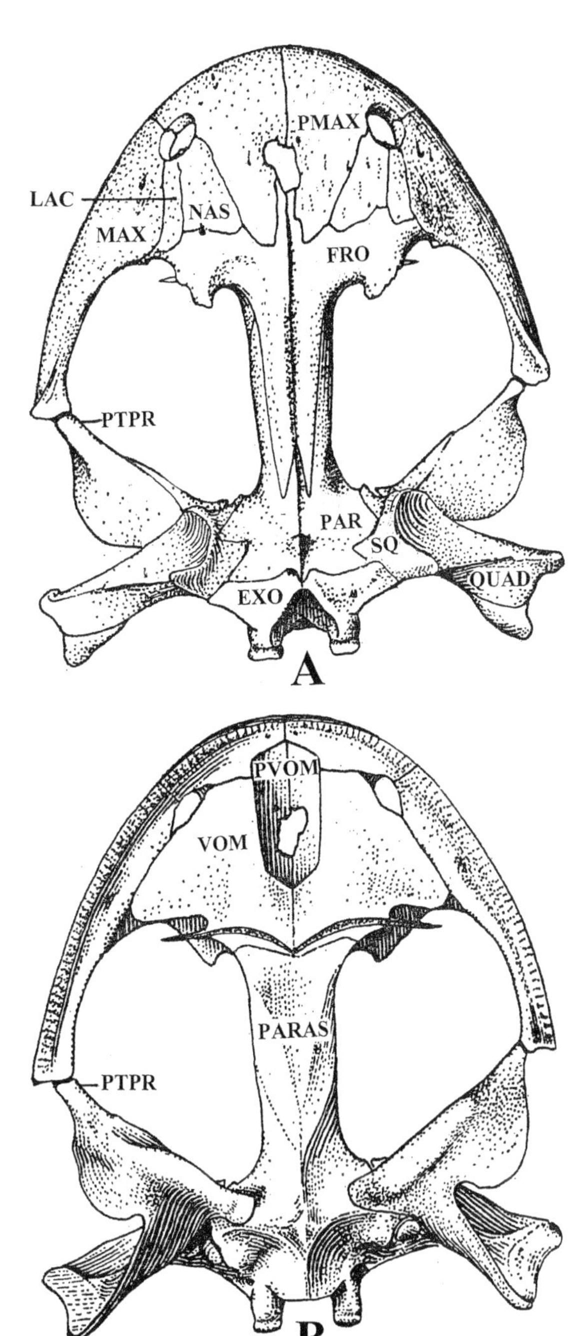

FIGURE 11. Skull of *Dicamptodon ensatus* (Dicamptodontidae) in (A) dorsal and (B) ventral views. Abbreviations: EXO, exoccipital; FRO, frontal; LAC, lacrimal; MAX, maxilla; NAS, nasal; PAR, parietal; PARAS, parasphenoid; PMAX, premaxilla; PTPR, pterygoid process of pterygoid bone; PVOM, prevomer; QUAD, quadrate; SQ, squamosal; VOM, vomer.

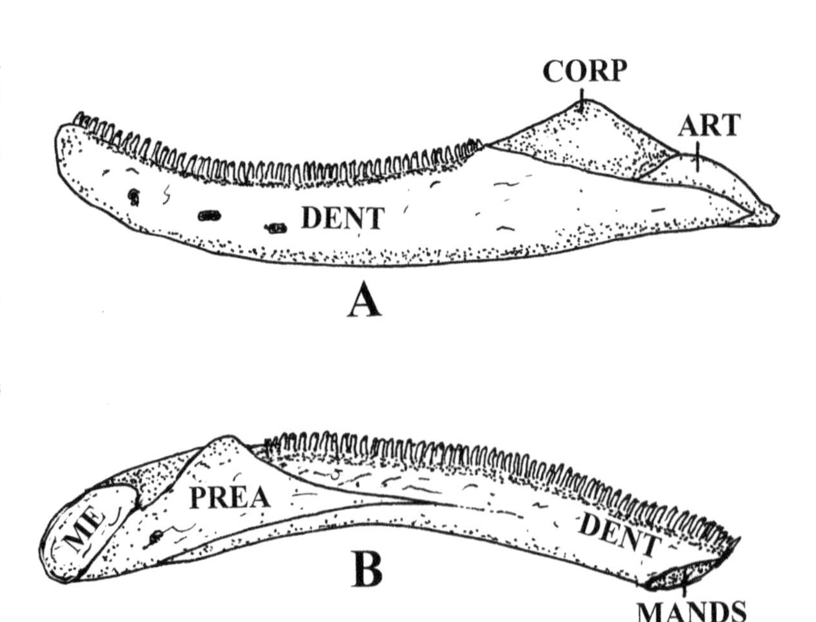

FIGURE 12. Left mandible of a generalized salamander of the family Salamandridae in (A) labial and (B) lingual views. Abbreviations: ART, articular; CORP, coronoid process of the prearticular bone; DENT, dentary; MANDS, mandibular symphysis area; ME, Meckel's cartilage area; PREA, prearticular.

sive and fully ossified, whereas in *P. jordani* the base of the pterygoid is cartilaginous.

Other differences in the two skulls include the fact that in *Dicamptodon* a lacrimal bone is present in the dorsal part of the skull (LAC in Fig. 11) and forms a very small part of the anterior wall of the orbit. This bone does not occur in *Plethodon jordani.* In fact, the lacrimal occurs only in the Hynobiidae (an Old World family), the Dicamptodontidae, and the Rhyacotritonidae. The fact that the teeth are compressed and blade-like also separates *Dicamptodon* from *P. jordani.*

The mandible (lower jaw) of the salamander (Fig. 12) is composed of two identical bony units that fuse anteriorly at the junction of an area called the mandibular synthesis (MANDS, Fig. 12). Mainly, the mandible is composed of two dermal bones that surround Meckel's cartilage, which is a cartilaginous rod that ossifies anteriorly in the area of the mandibular synthesis. The tooth-bearing bone of the lower jaw in most salamanders consists of one element, the dentary; the proteid *Necturus,* in contrast, bears teeth on the coronoid. On the medial side of each mandibular unit, the prearticular occurs posteriorly and supports musculature of the mastication process. In the posterior part of the articular unit, a portion of Meckel's cartilage may ossify to form the articular.

TEETH

The teeth of most salamander species are small and very numerous, as well as being short and bicuspid. Nevertheless, the teeth of the maxillary bones of some male plethodontids are long, have single cusps, and are directed straight out, rather than downward. Replacement teeth, both on the jaws and on other elements that bear teeth, move upward or downward to replace older teeth as they are lost.

Bones for the Paleontologist

Most of us who study the paleontology of small amphibians and reptiles dream of finding entire, articulated remains of these animals. I have been doing such studies for 50 years and have never come across such a fossil. Usually you find mixed remains of the small broken bones and teeth of salamanders, frogs, snakes, and lizards or pieces of turtle shell after hours of digging and sifting through fossiliferous sediments. Sometimes, however, you are lucky enough to come across unbroken bones or teeth in these animals.

In the following pages we shall discuss individual bones that have been used in the study and identification of salamander fossils. Each of these bones will be figured separately; names of parts that are useful in the orientation and description of the individual bones accompany each figure. We start with bones of the head and lower jaw, move on to the vertebral column, and finally end up with the important bones of the limbs and limb girdles. The hyobranchial skeleton and permanent cartilaginous units are not discussed here.

Complete salamander skull bones are rarely found, as they tend to be thin and subject to breakage during the fossilization process. Figure 13 depicts some of the skull bones that are occasionally recovered during fossil excavations in North America. The bones in Fig. 13 are drawn somewhat diagrammatically and are based on the skull of the primitive hynobiid species *Salamandrella keyserlingii*. The maxilla is a tooth-bearing bone of the upper jaw and together with the premaxilla makes up the preorbital margin of the skull. It is usually found in bits in paleontological excavations, but if this bone is fairly complete, it is useful in the identification of genera or species of salamanders.

The premaxilla is a small bone that is sometimes found in fossil deposits in North America, and it is often a diagnostic one (also see Tihen, 1958, fig. 2, p. 8). The posterior process of the premaxilla in *Salamandrella* and the Hynobiidae is very short compared with that of other salamander families. These posterior processes may be either fused or separate in salamanders. Notice that they are massive and fused in *Dicamptodon* (Dicamptodontidae) (see Fig. 11A) and slender and separate in *Plethodon* (Plethodontidae) (see Fig. 10A).

The frontal is the anterior of the two major roofing bones in the salamander skull. Relatively large pieces or even unbroken salamander frontals may sometimes be recovered in the field. If complete, they may help in the identification of salamanders to the family and even the generic level. Frontal bones tend to make up a more significant part of the skull roof in advanced families, such as the plethodontids (see Fig. 10A; also see Duellman and Trueb, 1986, figs. 13-3 and 13-4, pp. 296–297). Frontal bones may be either fused or separate in various salamander groups.

The parietal is the more posterior of the two major roofing bones of the salamander skull. Relatively large portions of parietal bones may also occasionally found in the field. If complete parietal bones are found, they also may be somewhat useful in the identification of salamanders to the familial or generic level. Parietals may have a rather complicated structure, as in *Salamandrella* (Fig. 13A); or they may be rather squarish, as

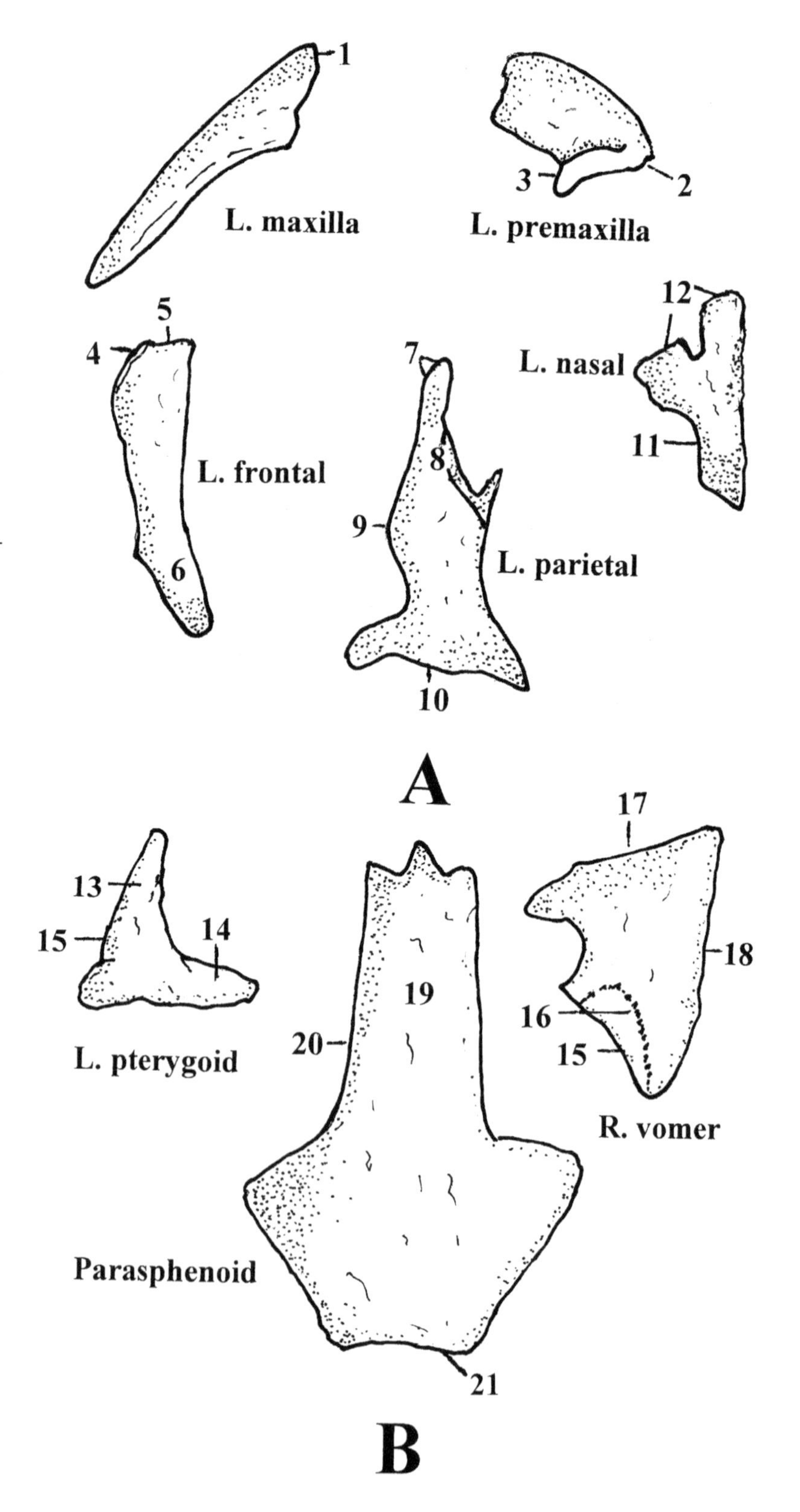

FIGURE 13. Selected skull bones of *Salamandrella* (Hynobiidae) in (A) dorsal and (B) ventral views. Legend: 1, articular surface for left premaxilla; 2, articular surface for right premaxilla; 3, posterior process; 4, articular surface for prefrontal; 5, articular surface for nasal; 6, parietal process; 7, articular surface for orbitosphenoid; 8, articular surface for frontal; 9, lateral border; 10, posterior border; 11, anterolateral border; 12, articular surfaces for premaxilla; 13, pterygoid process of pterygoid; 14, squamosal process; 15, anterolateral surface; 16, vomerine tooth row; 17, anteromedial surface; 18, medial surface; 19, anterior process; 20, lateral surface; 21, basal plate.

in *Taricha*, a North American salamandrid (see Duellman and Trueb, 1986, fig. 13-3e, p. 296). Parietals may be either fused, as in *Taricha*, or separate, as in *Plethodon jordani* (see Fig. 10A).

The nasal bone is a small roofing bone that forms part of the roof of the preorbital area of the skull. It is often excluded from the rim of the orbit by the small prefrontal (see Fig. 10A). It appears to be one of those small salamander bones that "gets lost" in fossil digs. On the other hand, if an entire skull roof is available, the placement and shape of the nasal bone may be a fairly important character.

We now turn to the bones of *Salamandrella* that make up the floor of the skull (Fig. 13B). In salamanders the pterygoid is an element that may be composed of either bone or cartilage. Notice that it is bony in *Salamandrella* (Fig. 13B) and *Dicamptodon* (see Figs. 11A, 11B) and cartilaginous in the plethodontid *Plethodon jordani* (see Figs. 10A, 10B). The pterygoid has a slender extension called the pterygoid process that forms the posterior rim of the orbit. This process may extend to the maxilla to close the orbital area, as in *Dicamptodon* (see Figs. 11A, 11B); or the pterygoid process may not reach the maxilla, resulting in an open orbital area, as in *P. jordani* (see Figs. 10A, 10B). When the salamander pterygoid bone is found as a fossil—and this seldom occurs—it usually has a complicated enough structure to be useful in the identification of salamander fossils.

Vomers in salamanders are paired elements that form a part of the floor of the braincase. They are important salamander fossil elements and are found in fairly good condition from time to time. Vomers may be relatively short, as in *Dicamptodon* (see Fig. 11B), or long, as in *Plethodon jordani* (see Fig. 10B). They nearly always bear teeth, some of which occur in rows and others of which occur in large patches. Sometimes large patches of teeth extend almost to the posterior end of the skull, as in *P. jordani*. If fairly complete vomers are found, the conformation and location of the circles and patches of vomerine teeth may be very helpful in the identification of salamander taxonomic groups.

The undivided parasphenoid bone forms the posterior floor of the salamander braincase. It is not infrequently found as a fossil and may be useful for separating salamander groups (see Tihen, 1958, fig. 6, p. 13). It is longer than the vomer complex in some salamanders, such as *Dicamptodon* (see Fig. 11B), but in other salamanders, such as most of the Plethodontidae, the vomer complex extends above the parasphenoid and then under it to near the posterior end of the floor of the braincase (see Fig. 10B).

The mandible (see Fig. 12) is composed mainly of two fused bones, the rather small prearticular and the large tooth-bearing dentary. Other than the atlas and individual trunk vertebrae, the dentary is probably the element most widely used for the identification of fossil salamanders in North America. This is primarily because the mandible is probably the most robust bone of the salamander skeleton. Thus the salamander mandible not only is more likely to fossilize than other bones but also is more likely to remain complete during the fossilization process.

Individual mandibular teeth of salamanders (Fig. 12) are seldom found as fossils. They are almost microscopic and likely to be missed by the vertebrate paleontologist in the field or to slip through the sieve into the spoil pile in the lab. Most salamander tooth crowns (see Fig. 6) separate from the pedicels that bear them during the fossilization process. But if tooth crowns are recovered, they may sometimes be identified to family or genus or to larval stages within genera (see Tihen, 1958, fig. 1, p. 7).

The vertebral column of salamanders (see Fig. 8) bears an atlas (Fig. 14) that directly joins the skull to the vertebral column. Differing from all other amphibians, the salamander has an atlas bearing four points of articulation with the back of the skull. The salamander atlas bears two anteriorly directed, concave cotyles (6, Fig. 14B) that articulate with two posteriorly directed, knoblike, convex condyles (see Fig. 11, not labeled) of the skull. Between the cotyles of the atlas is a structure called the odontoid process (also interglenoid tubercle) that projects into the foramen magnum (opening into the skull) and bears articular facets (9, Fig. 14B) that articulate with the lateral walls of the foramen magnum. All of these articulations facilitate upward and downward movement of the salamander head, but rotation and lateral movements of the head are restricted.

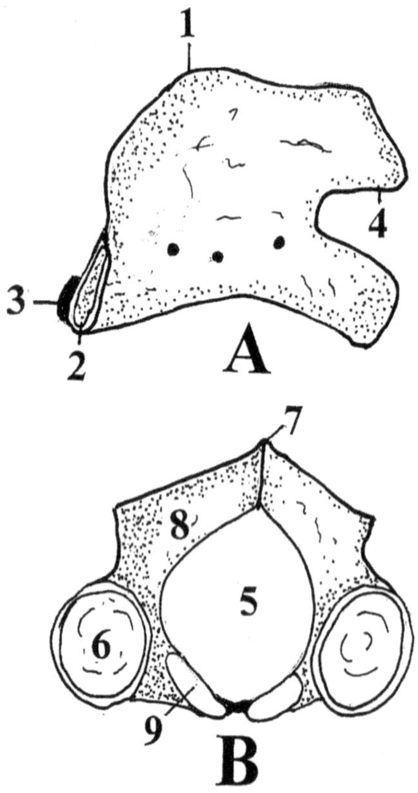

Posteriorly, the atlas joins the first trunk vertebra. The trunk vertebrae extend posteriorly to a vertebra called the presacrum, which lies just anterior to the single vertebra that forms the sacrum. The sacrum supports the attachment of the pelvic girdle to the body. Behind the sacral vertebra lies the first and second caudosacral vertebrae, the remainder of the vertebrate being a long string of caudal vertebrae that form the skeleton of the tail.

The vertebral column is sturdy, and the individual vertebrae are strong and dense enough to become fossilized in such a manner that the original structure is often well preserved. When the atlas is found, it is useful in the identification of fossil salamanders, but because trunk vertebrae are more numerous, they have been more widely used.

Figure 15 shows essential structures of a trunk vertebra of the *Ambystoma maculatum* group (Ambystomatidae). The vertebra is longer than wide. Anteriorly, two prezygapophyseal articular facets (2, Fig. 15A) face upward to articulate posteriorly with the down-turned postzygapophyseal articular facets (3, Fig. 15B) of the next trunk vertebra. A set of double bicapitate (two-headed) rib-bearers (5, Figs. 15C,

FIGURE 14. Atlas of *Salamandra salamandra* (Salamandridae) in (A) lateral and (B) anterior views. Legend: 1, neural spine; 2, left cotyle; 3, left portion of odontoid process; 4, left postzygapophysis; 5, neural canal; 6, right cotyle; 7, neural spine; 8, right side of neural arch; 9, right portion of odontoid process. Modified from Francis (1934).

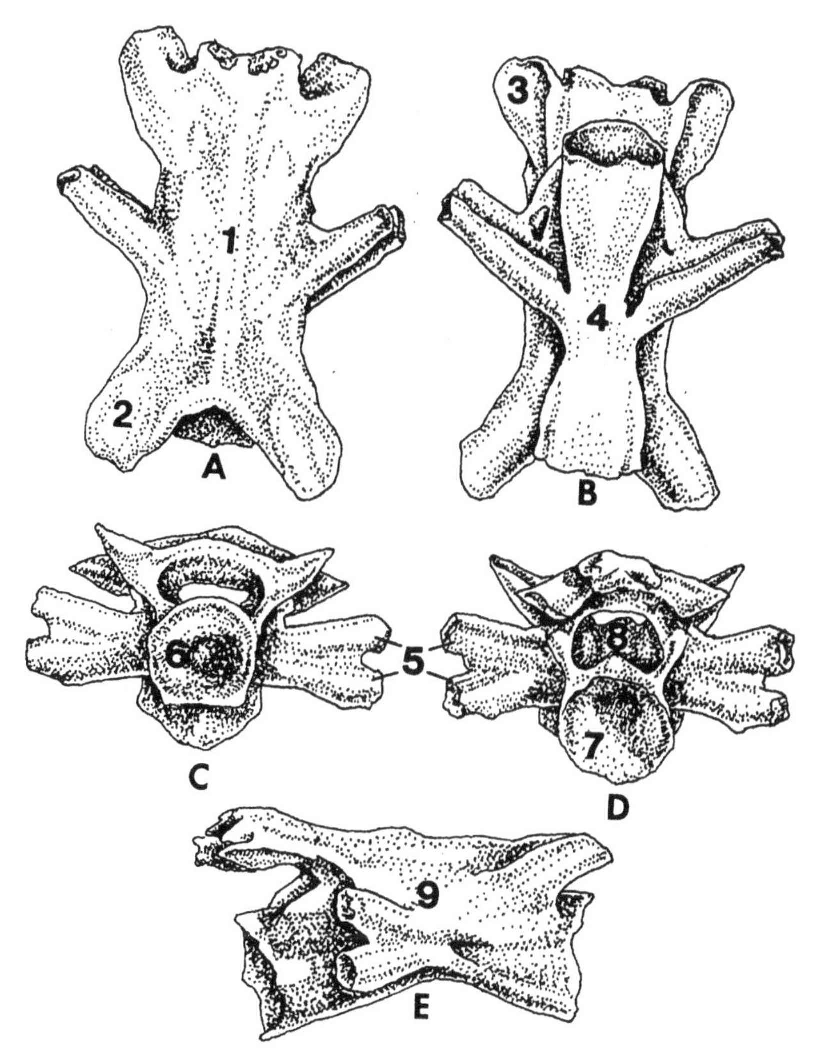

FIGURE 15. Salamander trunk vertebra of the *Ambystoma maculatum* group in (A) dorsal, (B) ventral, (C) anterior, (D) posterior, and (E) lateral views. Legend: 1, neural spine; 2, prezygapophyseal articular facet; 3, postzygapophyseal articular facet; 4, ventral view of centrum; 5, bicapitate rib-bearers; 6, anterior cotyle; 7, posterior cotyle; 8, neural canal; 9, lateral wall of neural arch.

15D) occur on each side of the vertebra in all views. Notice that the rib-bearers are deflected posteriorly. Two-headed ribs (see Fig. 7A) articulate with these rib-bearers in salamanders. The neural arch (9, Fig. 15E) rises up to form the side walls and roof of the vertebra. The top of the vertebra has a thin neural spine at its middle that runs the length of the vertebra (1, Fig. 15A). Some authors use the term neural crest for this structure if it rises high from the vertebra (e.g., in the Amphiumidae). The centrum of the vertebra (4, Fig. 15B) is a rather tube-like structure constricted in the middle. A cotyle (6, Fig. 15C; 7, Fig. 15D) occurs at each end of the vertebra. This is called the amphicoelous condition. The neural canal through the vertebra (8, Fig. 15D) contains the spinal cord.

Some of the features on salamander trunk vertebrae that have been helpful in the identification of caudate fossils include (1) the shape of the vertebrae (e.g., long or short; Tihen, 1958); (2) the shape of the post- or prezygapophyseal articular facets (oval, round, etc.); (3) the shape of the opening of the neural canal, both anteriorly and posteriorly; (4) the

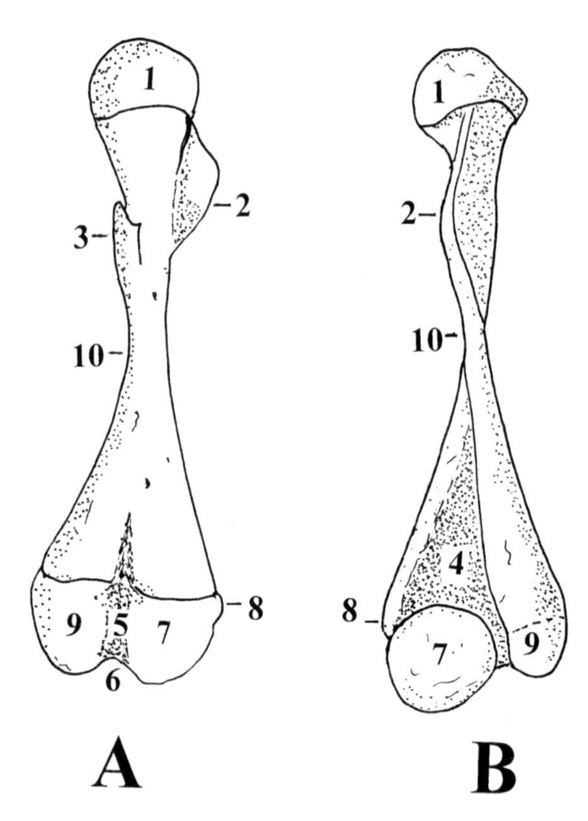

FIGURE 16. Right humerus of *Salamandra salamandra* (Salamandridae) in (A) flexor and (B) extensor views. Legend: 1, head; 2, ventral crest; 3, dorsal crest; 4, cubital fossa; 5, olecranon fossa; 6, trochlear groove; 7, radial condyle; 8, lateral epicondyle; 9, ulnar condyle; 10, midpoint of humeral shaft. Modified from Francis (1934).

height and shape of the neural spine (neural crest); (5) the relative length of the rib-bearers and the extent of their fusion with one another; (6) the presence of a notch on the posterior end of the neural arch; (7) the distribution of muscles as determined by various scars and depressions; (8) the location and number of various foramina; and (9) the presence of accessory structures, such as alar and aliform processes.

Some of the skeletal elements of the pectoral girdle are seldom used in the identification and study of North American salamander fossils. Some of these elements are cartilaginous and are unlikely to fossilize. Moreover, the carpals, metacarpals, and phalanges of the forelimb are very small and seldom turn up as fossils; and if they do turn up, not only are they difficult to orient on the skeleton, but once oriented they also have very few definitive characters.

Bones of the pectoral girdle that are large enough and compact enough to be found somewhat regularly as fossils include the humerus, the ulna, and the radius. Of these three bones the humerus is the most widely used in fossil salamander studies. Two views of the right humerus of the Fire Salamander (*Salamandra salamandra*), an Old World species, are presented in Fig. 16. The head of the humerus (1, Fig. 16) articulates with the shoulder girdle proximally; and the radial condyle (7, Fig. 16) and lateral condyle (8, Fig. 16) of the humerus articulates with the radius and ulna distally. The head and the radial and lateral condyles are poorly ossified until rather late in the lives of salamanders, when they tend to become more osseous. But even then these structures are often not completely bony. I have found many fossil salamander humeri with their proximal and distal ends missing.

Notice the marked flexure of the shaft of the humerus (10, Fig. 16B) and its ventral crest (2, Figs. 16A, 16B). The extent of this flexure appears to be variable. Notice the hook-like dorsal crest (3, Fig. 16A) as well. The shape of this hook may be variable even within the Salamandridae.

Please observe how the shape of several structures of the humerus changed as the figured bone was rotated for drawing. For instance, the head of the humerus has a different shape in Figs. 16A and 16B. This illustrates the point that fossil salamander bones embedded in rock, preserved in one dimension in shales, or preserved as mere impressions cannot convey the true shape of structures on the bones. This also illustrates how trying to identify fossils on the basis of flat pictures in books can lead to misidentifications! Many vertebrate paleontologists are now publishing photographs in stereo pairs, which at least gives more depth to the fossils. As was the case with the salamander vertebra discussed above, various pits and depressions occur on the long bones of both the pectoral and the pelvic girdles. Note the cubital fossa of the humerus (4, Fig. 16B); in another view of the same bone, the olecranon fossa (5, Fig. 16A) appears.

The ulna and radius are depicted in Figs. 17A and 17B, respectively; like the humerus, each bone is shown in a different view. The ulna and the radius articulate with the distal end of the humerus proximally and with the bones of the manus (hand) distally (see Fig. 8). Both of these bones are fairly sturdy, but again, both have poorly ossified areas at each end. It is very difficult to identify salamanders, even to the generic or familial level, on the basis of either one of these bones, especially if it is incomplete.

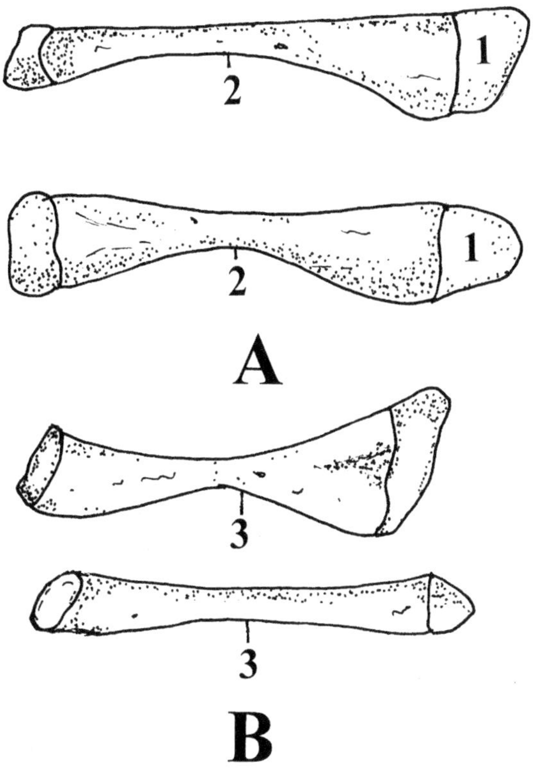

FIGURE 17. (A) Right ulna and (B) right radius of *Salamandra salamandra* (Salamandridae). Upper figures in both (A) and (B) in extensor view. Lower figures in both (A) and (B) in preaxial view. Legend: 1, olecranon process; 2, ulnar shaft; 3, radial shaft. Modified from Francis (1934).

Now we turn to the bones of the pelvic girdle and hind limbs. As in the pectoral and forelimb skeleton, many elements of the posterior girdle and limb skeleton are not often used in the identification of fossil salamanders. These include cartilaginous elements, as well as the tarsals, metatarsals, and phalanges. Bones of the pelvic girdle that are large and compact enough to be found as fossils on a regular basis include the ilium, femur, tibia, and fibula. Of these the femur is used the most often in fossil salamander studies.

As was the case with the humerus, two views of the right femur of *Salamandra salamandra* are presented. The head of the femur (3–5, Fig. 18B) articulates with the ilium and pubis of the pelvic girdle proximally, and the fibular and radial condyles (8, 9, Fig. 18B) articulate with the fibula and tibia distally. Both the head of the femur on the proximal end of the bone and the two condyles on the distal end are poorly ossified in salamanders until fairly late in life. Thus, many fossil salamander femora are found with the head and condyles missing. Still, useful characters remain on what is left of this bone. Notice in Fig. 18B the marked proximal flexure of the shaft of the femur and recall that a similar situation existed in the humerus. Moreover, the shape and position of the hook-like trochanter (6, Fig. 18B) may be a good character. Also notice how the shape of the structures on the femur change when the bone is rotated.

Figure 18A illustrates a salamander ilium in dorsal view. The ilium is the bone that is most frequently used for identification of frogs in North

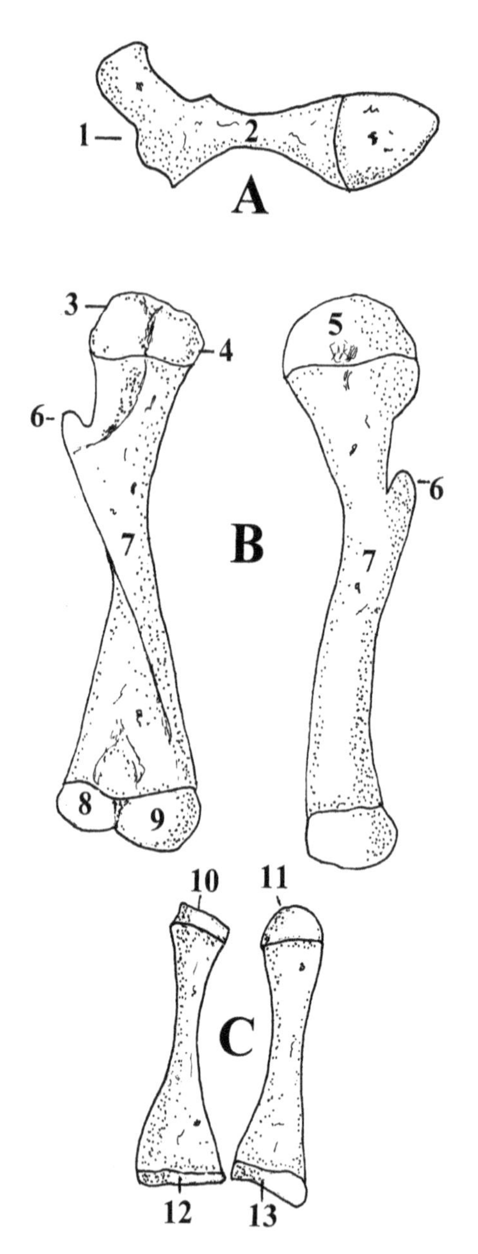

FIGURE 18. Pelvic appendages of *Salamandra salamandra* (Salamandridae). (A) Right ilium in dorsal view. (B) Right femur in flexor (left) and preaxial (right) views. (C) Tibia (left) and fibula (right), both in extensor view. Legend: 1, proximal articular surface for attachment to ischium and pubis; 2, ilial shaft; 3, ligamental attachment to pubis; 4, ligamental attachment to ilium; 5, head of femur; 6, trochanter; 7, midpoint femoral shaft; 8, fibular condyle; 9, radial condyle; 10, tibial area of attachment to pes (foot); 11, fibular area of attachment to pes; 12, tibial articular attachment to radial condyle of femur; 13, fibular articular attachment to fibular condyle of femur. Modified from Francis (1934).

America (Holman, 2003), but this is hardly the case in salamanders. In frogs the ilium is long and expanded proximally and is the site of important muscle attachments associated with the unique hopping and leaping gait of anurans. In salamanders the ilium is much less complicated. Moreover, the distal process for articulation with the femur is poorly ossified and often not present in salamander fossils. Figure 18C illustrates the tibia and fibula of *Salamandra salamandra*. These elements occur from time to time in the fossil salamander record but are not particularly useful in the identification process.

EARLY EVOLUTION OF THE SALAMANDERS

Earlier we discussed characters that separate the lissamphibians (frogs, salamanders, and caecilians) from the other amphibians and how salamanders differ from frogs and caecilians. However, the origin and interrelationships of the lissamphibians are still major unresolved problems in the study of the various vertebrate groups (e.g., Schoch and Carroll, 2003). Three hypotheses on the origin of lissamphibians have been debated (Laurin, 2002; Schoch and Carroll, 2003). (1) Many researchers (e.g., Bolt, 1991; Trueb and Cloutier, 1991; Milner, 1993) believe that lissamphibians form an assemblage that diverged from a single group of Paleozoic amphibians, the temnospondyls (see a variety of these ancient amphibians in Figs. 2–4). (2) Contrasting distinctly with this view, Laurin and Reisz (1997) and Laurin (1998) suggested a common ancestry from a distinct lineage, the lysorophid lepospondyls (Early Permian elongate little amphibians with reduced limbs and skull bones). Finally, others (see Schoch and Carroll, 2003, for references) have suggested that caecilians are allied with the lepospondyls; frogs, with the temnospondyls; and salamanders, with either the lepospondyls or the temnospondyls. This debate continues to go forward.

The early evolution of salamanders

themselves is more poorly known than that of frogs, but Schoch and Carroll (2003) recently suggested putative ancestors of salamanders from strata at the Permian–Pennsylvanian boundary in Germany. Here more than 600 fossils of the neotenic genus *Apateon* demonstrated how this form changed from hatching to metamorphosis. The fossils of *Apateon*, a temnospondyl of the family Branchiosauridae (this family is not figured in Figs. 2–4), showed a sequence of ossification of individual bones and a changing structure of the skull that closely parallel those observed in the development of primitive living salamanders.

These fossils indicated how features of the salamander skull might have evolved relative to their feeding specializations. Of great interest is that fossil larvae of *Apateon* share many unique, derived (advanced) characters with the modern salamander families Hynobiidae (Old World group), Salamandridae, and Ambystomatidae that have not been documented in any other group of Paleozoic amphibians. In short, this suggests a temnospondyl, perhaps even a branchiosaurid temnospondyl ancestry for salamanders. There is even the possibility that *Apateon* itself is the ancestral genus.

Salamander fossils first appear in the Early Jurassic, but a poorly preserved amphibian from the Triassic, *Triassurus*, may indeed be a salamander. Thus we can say that the origin of salamanders must have been at least as early as the Early Jurassic and possibly earlier, if *Triassurus* is a true salamander. Most of the Jurassic salamander fossils seem to lack one or more of the characters of the so-called crown-clade caudates, the group that includes all of the living salamander families.

Turning to details of these Triassic and Jurassic fossils, we note that *Triassurus sixtelae* (Fig. 19), from the Late Triassic of Uzbekistan, was described as a salamander by Ivachnenko (1978). Estes (1981) suggested it was a larval temnospondyl. Milner (1994, 2000) pointed out that *Triassurus* appeared to have the open cheek structure, short ribs, and long humeri and femora of the salamanders, the latter two elements being unknown in Middle and Late Triassic temnospondyls. It is possible that *Triassurus* is an early salamander, but the vertebrae of this genus are not completely ossified as in all "true" salamanders. Moreover, the skeleton of this animal is very incomplete, and the figure of the specimen by Ivachnenko appears to be somewhat simplified. I and many others would not be willing to state that this fossil unequivocally represents a true salamander.

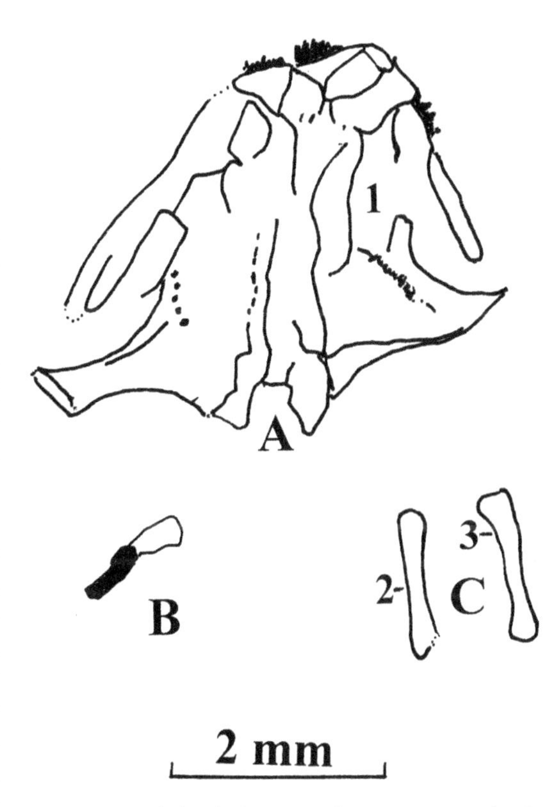

FIGURE 19. Skeletal elements of *Triassurus sixtelae* from the Late Triassic of Uzbekistan. (A) Skull showing (1) open cheek region. (B) Relatively short rib from pectoral region. (C) Relatively long humerus (2, left; 3, right). Scale bar applies to all figures. Modified and re-drawn from Ivachnenko (1978).

Curtis and Padian (1999) studied an Early Jurassic microvertebrate fauna (vertebrate fauna consisting of bones of very small vertebrates) from the Late Jurassic Kayenta Formation of northeastern Arizona. The same formation previously yielded the oldest frogs and caecilians in North America. Here two salamander atlantes (atlases) were unearthed from the Harvard Gold Spring Quarry, which appears to be of Sinemurian and (or) Pliensbachian (both Early Jurassic) age (see Fig. 23). This material represents the earliest verified North American fossil salamanders.

The family Karauridae is recognized on the basis two genera, *Karaurus* and *Kokartus*, both of which are better preserved than *Triassurus*. *Karaurus sharovi* (Fig. 20) is based on a single specimen from the Late Jurassic of Kazakhstan (Ivachnenko, 1978). *Kokartus honorarius* (Fig. 21) is based on material from the Middle Jurassic of Kyrgyzstan (former Kirghiz SSR) reported by Nessov (1988) and then further detailed by Nessov et al. (1996). Both were somewhat robust animals that are said to resemble modern terrestrial salamanders (Milner, 2000) and were about 200 mm long. This puts them in about the size class of the large Tiger Salamanders (*Ambystoma tigrinum*) and Giant Salamanders (*Dicamptodon*) of the United States (Petranka, 1998). Both *Karaurus* and *Kokartus*, however, had heavily sculptured skull roofs, and I am not aware that any modern salamanders have such heavy coossified thickening. Many fossil and modern frogs, however, have such skulls (see Duellman and Trueb, 1986; Roček and Lamaud, 1995; Sanchiz, 1998; Holman and Harrison, 2002, 2003), and these skulls are used for specific purposes, such as blocking the entrance of burrows that certain frogs inhabit during dry spells.

Karaurus is probably more primitive than *Kokartus*. In *Karaurus*, a

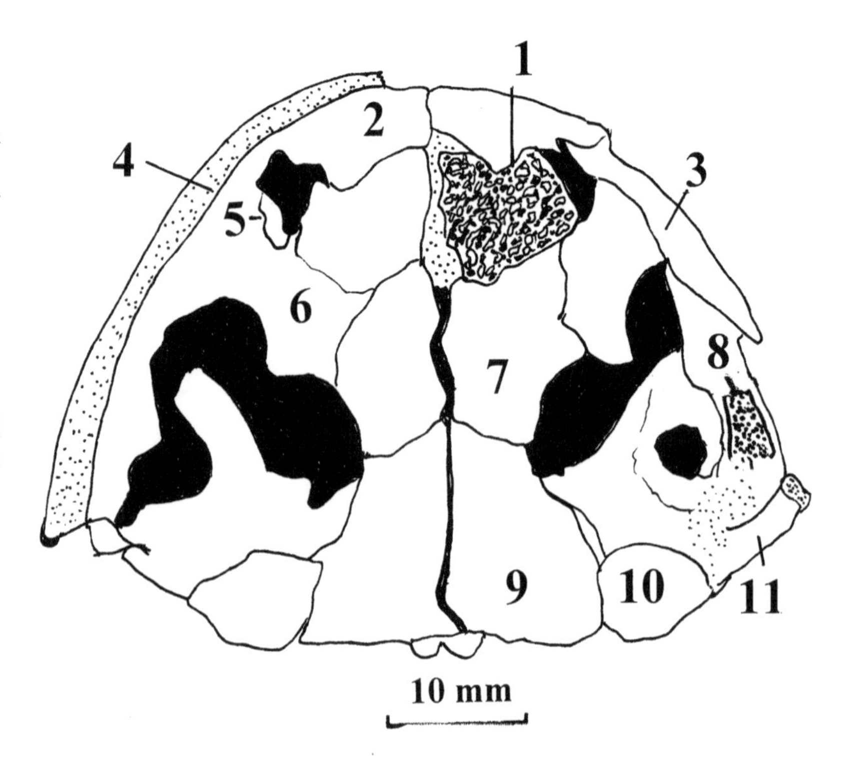

FIGURE 20. Semidiagrammatic dorsal view of the skull of *Karaurus sharovi* from the Late Jurassic of Kazakhstan. The original skull was crushed, so several bones are out of place. All the bones in the original skull except 4 and 11 had dermal sculpturing as in 1. Legend: 1, nasal bone; 2, premaxillary; 3, maxillary; 4, dentary; 5, lacrimal; 6, prefrontal; 7, frontal; 8, angular (exposed from mandible); 9, parietal; 10, squamosal; 11, quadrate.

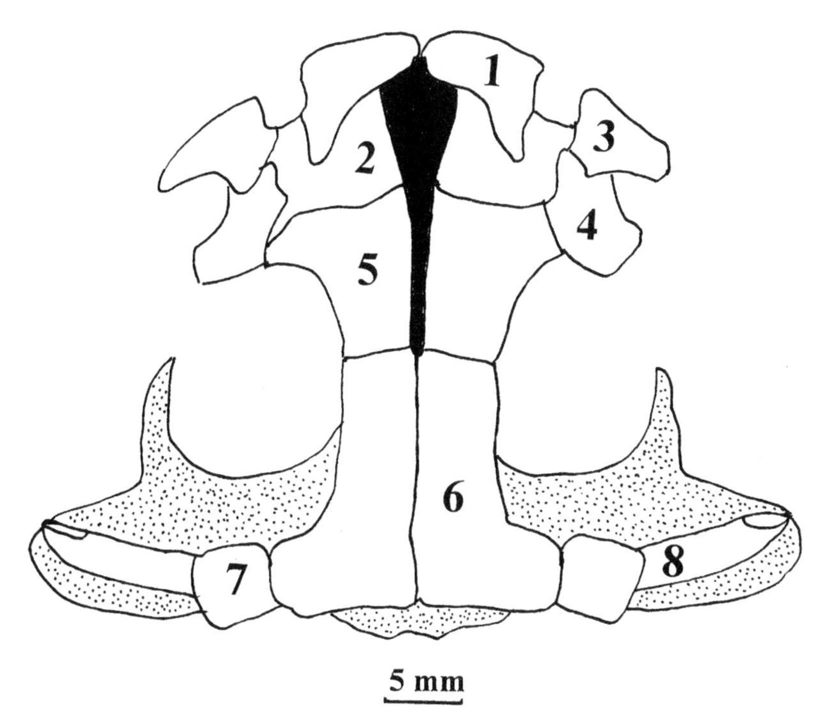

FIGURE 21. Semidiagrammatic dorsal view of *Kokartus honorarius* from the Middle Jurassic of Kyrgyzstan). Legend: 1, premaxilla; 2, nasal; 3, maxilla; 4, prefrontal; 5, frontal; 6, parietal; 7, squamosal; 8, quadrate.

continuous area of dermal sculpture passes over both the parietal and the squamosal bones. This covering would have blocked the posterior passage of a specific muscle (adductor mandibulae internus) to either the occipital region of the skull or to the atlas, as in all living salamanders. *Karaurus* retains a lacrimal bone (present in some salamandrids) and a separate angular bone in the mandible (also in hynobiids and cryptobranchids). *Kokartus* appears to have been neotenic in retaining a tooth-bearing coronoid bone in the lower jaw and a shortened maxillary bone.

Four true salamanders have been unearthed from the Middle Jurassic Bathonian age (169–164 million years before the present [Ma BP]) strata of Kirtlington in Oxfordshire, England (Evans and Milner, 1994). Other material has emerged from the Bathonian of Skye, Scotland (Evans and Waldman, 1996). The fact that four taxa of salamanders exist in these faunas is demonstrated by four distinct atlas vertebrae, but figuring out what other elements are associated with the atlantes in these assemblages has been a difficult problem.

A salamander fossil, referred to as Salamander A, is a large animal (Evans and Milner, 1994, 1995) and the most common amphibian from the Kirtlington assemblage in England and is represented by many scattered elements. Salamander A has been found in the Bathonian locality in Skye, Scotland, as well, where it is also represented by scattered elements (Evans and Waldman, 1996). This salamander is of special interest because the vertebrae appear to be at least partly composed of endochondral bone, as in modern salamanders. The vertebrae of Salamander A also resemble those of modern salamanders in having bone that appears to have been preformed in cartilage. Some of the elements of Salamander A appear to be similar to those of *Kokartus*, a neotenic taxon that is a

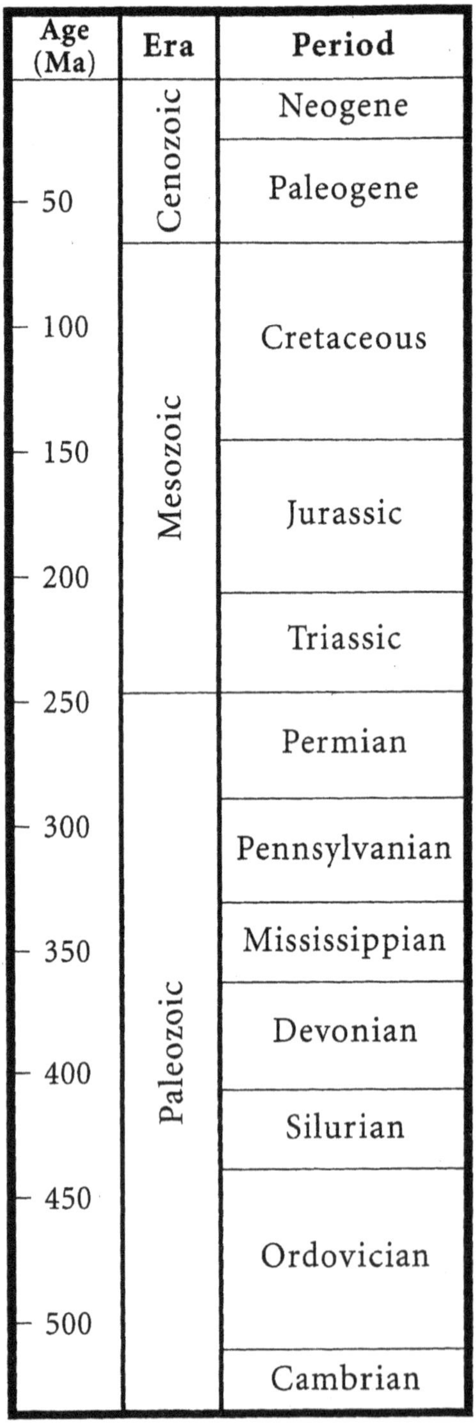

Era	Period
Cenozoic	Neogene
	Paleogene
Mesozoic	Cretaceous
	Jurassic
	Triassic
Paleozoic	Permian
	Pennsylvanian
	Mississippian
	Devonian
	Silurian
	Ordovician
	Cambrian

Age (Ma): 50, 100, 150, 200, 250, 300, 350, 400, 450, 500

FIGURE 22. Eras and periods (other than the Quaternary: see Fig. 24) of the geologic time scale. 1 Ma = 1 million years.

member of the Karauridae. A fossil salamander designated as Salamander B from the Kirtlington locality in England is possibly a member of the enigmatic family Batrachosauroididae. This family will be discussed later in the book.

Two salamanders from the Kirtlington Middle Jurassic in England have been given the names *Marmorerpeton kermacki* and *Marmorerpeton freemani* by Evans et al. (1988). Unlike the vertebrae of Salamander A, the bones of *Marmorerpeton* are composed of dermal (membrane) bone. A partially articulated specimen of *Marmorerpeton* has been reported from the Middle Jurassic locality of Skye in Scotland, as well as from other Middle Jurassic fossil assemblages. A possibility is that *M. freemani* is distinct enough to be designated as a new genus. It is also possible that *M. kermacki* will turn out to be similar to *Karaurus* (Evans and Waldman, 1996).

An articulated salamander skeleton from the Late Jurassic Morrison Formation in Dinosaur National Monument of North America was reported by Evans et al. (2005). This fossil is said to show a combination of primitive and derived (advanced) characters that led to its being named *Iridotriton hechti* (a new genus and species). Some of its characters, including the presence of spinal nerve openings in the tail, suggest it was a salamandriform (see Milner, 2000). All salamandriforms have the center of their diversity in either North America or Europe and include such living (crown clade) families as the Proteidae, Plethodontidae, and Ambystomatidae.

Chronological Terms Used in the Book

The system of chronology used in this book follows the globally standardized era, period, epoch, and age (for the Mesozoic era) units of the *1999 Geological Time Scale* issued by the Geological Society of America (Palmer and Geissman, 1999). The three post-Precambrian eras (oldest to

Epoch/Age	Dates (Ma BP)	Epoch/Age	Dates (Ma BP)
Late Cretaceous		Middle Jurassic	
Maastrichtian	71.3–65.0	Callovian	164–159
Campanian	83.5–71.3	Bathonian	169–164
Santonian	85.8–83.5	Bajocian	176–169
Coniacian	89.0–85.8	Aalenian	180–176
Turonian	93.5–89.0	Early Jurassic	
Cenomanian	99.0–93.5	Toarcian	190–180
Early Cretaceous		Pliensbachian	195–190
Albian	112–99.0	Sinemurian	202–195
Aptian	121–112	Hettangian	206–202
Barremian	127–121	Late Triassic	
Hauterivian	132–127	Rhaetian	210–206
Valanginian	137–132	Norian	221–210
Berriasian	144–137	Carnian	227–221
Late Jurassic		Middle Triassic	
Tithonian	151–144	Ladinian	234–227
Kimmeridgian	154–151	Anisian	242–234
Oxfordian	159–154	Early Triassic	
		Olenekian	245–242
		Induan	248–245

FIGURE 23. Internationally recognized ages of the Mesozoic in millions of years before the present (Ma BP). After *Geologic Time Scale* (Palmer and Geissman, 1999).

youngest) are Paleozoic, Mesozoic, and Cenozoic. Only the Mesozoic and Cenozoic (Figs. 22–24) will be discussed here, as salamanders are restricted to these time units. Periods are divided into epochs, which may be divided into two subdivisions, Early and Late, or three subdivisions, Early, Middle, and Late. Unlike the daily time units that we use, which can be broken down into subdivisions of equal duration, classical geologic subdivisions are of unequal duration because they are defined on the basis of the relative sequence of rocks and fossils that make up geological time, rather than on the basis of actual time.

Age systems, especially those of the Cenozoic, may differ locally. Here I shall use the internationally recognized age units for the Mesozoic (Fig. 23), but I shall use the North American land-mammal ages (NALMAs) for the Cenozoic (Fig. 24), as the internationally recognized age units are mainly based on marine sequences. The NALMAs were originally based on the concept that large land mammals were not only wide ranging, but often restricted to narrow units of time (Wood et al., 1941).

THE NORTH AMERICAN PLEISTOCENE

There are about as many fossil records of salamanders in the North American Pleistocene as in all the other North American geologic units put together. This record has been understated in recent comprehensive fossil amphibian studies. To rectify this, I am giving salamanders of this epoch due attention in this book.

Two systems of dividing the North American Pleistocene into temporal units exist. The older system, the Pleistocene glacial- and interglacial-age system, relies mainly on the glacial and interglacial sedimentary record. The second system, the NALMA system, briefly discussed above, relies on the biochronology (dating of biological events using biostratigraphic or other objective paleontological data) of land mammals.

With the older system, we find that before the 1840s, North American scientists were prone to attributing what was actually glacial deposition to the effects of flooding. But later, Louis Agassiz's studies showed that an-

FIGURE 24. Periods and epochs of the Cenozoic and North American land-mammal ages (after Hulbert, 2001). Please note the very short Rancholabrean age of the Pleistocene.

Millions of Years Ago	Period	Epoch		Age
	Quaternary	Pleistocene		Rancholabrean
	Quaternary	Pleistocene		Irvingtonian
	Neogene	Pliocene	E | L	Blancan
5	Neogene	Miocene	Late	Hemphillian
10	Neogene	Miocene	Late	Clarendonian
15	Neogene	Miocene	Middle	Barstovian
	Neogene	Miocene	Middle / Early	Hemingfordian
20, 25	Neogene / Paleogene	Miocene / Oligocene	Early / Late	Arikareean
30	Paleogene	Oligocene	Early	Whitneyan
	Paleogene	Oligocene	Early	Orellan
35	Paleogene	Eocene	Late	Chadronian
	Paleogene	Eocene	Middle	Duchesnean
40	Paleogene	Eocene	Middle	Uintan
45	Paleogene	Eocene	Middle / Early	Bridgerian
50	Paleogene	Eocene	Early	Wasatchian
55	Paleogene	Paleocene	Late	Clarkforkian
60	Paleogene	Paleocene	Late	Tiffanian
	Paleogene	Paleocene	Early	Torrejonian
	Paleogene	Paleocene	Early	Puercan

cient glacial landforms and deposits existed far south of existing glaciers and that this indicated former cold climates. Stratigraphic studies then determined that layers with weathered zones of organic soils and containing plant remains existed between layers of glacial sediments, such as sands and gravels. These early scientists postulated that the organic layers formed during unglaciated intervals and that the glaciers must have advanced and retreated several times.

A little after the turn of the 19th century, four major glacial drift sheets were identified, each of them separated from the other layers on the basis of organic layers or fossils that indicated an interglacial environment. The classical North American glacial and interglacial sequence that evolved from these studies is presented below—the oldest at the bottom.

Wisconsinan glacial age
Sangamonian interglacial age
Illinoian glacial age
Yarmouthian interglacial age
Kansan glacial age
Aftonian interglacial age
Nebraskan glacial age

Today, the classical glacial and interglacial ages that occurred before the late part of the Illinoian are considered poorly defined and highly questionable. At best, strata of the Nebraskan and Kansan glacial ages are difficult to define because they have been exposed to long periods of weathering and erosion, as well as the scouring effects of later glacial activity, especially that of the Wisconsinan. Since the Wisconsinan is the youngest glacial age and has not been followed by striking glacial events in the North American mid-latitudes, its strata contain the most detailed records of Pleistocene events.

It has been shown that several warmer intervals (interstadials) within the generally cold Wisconsinan led to the temporary withdrawal of the ice. These intermittent warm spells have been given names, especially in areas where Pleistocene glaciers were active. In the Toronto, Ontario, area, for instance, two interstadials are recognized in the Wisconsinan: the Port Talbot interstadial, which has yielded a biological assemblage dating somewhat earlier than 54–45 thousand years (ka) before the present (BP); and the Plum Point interstadial, which has yielded an assemblage that lived about 34–24 ka BP.

Turning to the Pleistocene NALMAs, we find that two are currently recognized (see Fig. 24): the Irvingtonian, which is the older and much longer of the two; and the short Rancholabrean. The Irvingtonian land-mammal age was originally defined on the basis of a mammalian fauna from a gravel pit southeast of Irvington, in Alameda County, California. Recent work by Bell and Mead (2000) on the dispersal of microtine rodents from Eurasia into North America suggests that the Irvingtonian began about 1.9 Ma BP and lasted to about 150 ka BP.

Three Irvingtonian subunits, I, II, and III, are also based on rodent-dispersal studies. Irvingtonian I is considered to have begun about 1.9 Ma BP (a little before the 1.8 Ma BP date recognized by the international geological community for the beginning of the Pleistocene) and to have lasted until about 850 ka BP. Irvingtonian II is thought to have begun about 850 ka BP and to have lasted to about 400 ka BP. Irvingtonian III is positioned between about 400 and 150 ka BP.

The Rancholabrean land-mammal age was originally defined on the basis of the famous Rancho La Brea faunal assemblage in Los Angeles. Bell and Mead's (2000) rodent-dispersal study suggests that the Rancholabrean began about 150 ka BP and lasted until about 10 ka BP, when the Pleistocene is considered to have ended. The period from about 10 ka BP until the present is called the Holocene by most of the international geological community, but some workers express doubt that the Holocene is truly a discrete unit of geologic time, believing we are still in the Pleistocene.

Comments on the Identification of Fossil Salamanders

The identification of fossil salamanders is a daunting task. Here I will make some simple suggestions for anyone interested in studying and identifying fossils salamanders. First, obtain a mounted, articulated salamander skeleton from one of the biological supply houses. Type in "biological supply house" on an Internet search engine and go from there. You need to become familiar with all the bones in the skeleton. You can find laboratory manuals on the subject, and biological supply house specimens often come with a drawing of the skeleton that identifies the individual bones. The next part may be difficult, depending upon where you live in North America.

Try to procure the dead body of a salamander, the larger the better. In some areas Tiger Salamanders (*Ambystoma tigrinum*) make annual nightly migrations in the early spring or fall, so you might find usable roadkill. Please avoid busy roads and highways in these endeavors. You can often find the dead bodies of Mudpuppies (*Necturus maculosus*) washed up on the beaches of slow-moving rivers or in clean lakes. Some biological supply houses may sell fresh bodies of large salamanders.

Put the dead salamander in a large jar filled with water and screw the lid on tight. Let the jar sit in some warm, protected place at room temperature for a few weeks. Bacterial decay of the soft tissue of the animal will then occur, unless some inhibiting mold or fungus is accidentally introduced into the jar. Wearing rubber gloves, unscrew the top of the jar and pour, with extreme care, the thickened liquid through a large tea strainer or other type of sieve into a safe holding receptacle. Do this under a laboratory hood or in the outdoors, far away from people. A disarticulated salamander skeleton will appear in the strainer. Wash these bones thoroughly in the sieve and put these remains back in the jar for about another two weeks at room temperature. At the end of the two weeks, repeat the pouring off procedure. Be extremely careful not to spill any of the liquid on yourself or your clothes, or you will not be welcome anywhere.

After the second pouring off, the bones are usually ready to be placed in a tray to dry. You may wish to soak the bones in a weak solution of ammonia for a few days before they are dried, to cut down on any lingering odor on the salamander skeletal remains. You will find that, unlike many of the disarticulated skeletons produced by enzymes or chemicals, these bacterially macerated specimens will last indefinitely if stored in a reasonably cool, dry place.

Put the clean, dry, disarticulated specimen in a sturdy box, and attach labels inside and outside that identify the species. The inside label should also give, at least, either the snout-to-vent length (SVL) or the total length (TL) of the original specimen, the sex, the locality where the animal was collected, the date, and the name of the collector, if this is known. At this point you are ready to compare the individual bones of your disarticulated skeleton with those of the articulated one. The large, individual bones of the body are relatively easy to identify, but learning to identify the individual bones of the skull is much more difficult.

When you can identify the main individual body elements and most of the bones of the skull of your first disarticulated salamander, it is time to try to find other roadkill of the same species and prepare a few more skeletons. When the new skeletons are ready, lay out a series of the same bones from different individuals for comparison. First, determine which bones have a relatively simple structure and which have a complicated one. It will probably be obvious that bones such as the parietal or parasphenoid of the skull are rather simple, but the humeri and femora are more complex. In some cases you will find out that the complex bones are more useful for the identification of genera and species than the simpler ones.

Next, study the series of bones to determine the amount of intraspecific (within a species) variation there is. You may note that a cavity or a prominence on the humerus of one individual may be round, while in another individual of the same species the same structure is oval. Or, in the case of a jawbone, there may be four foramina (small openings) in a specific individual in one area and one or two in the same area in another individual of the same species. Usually, you will be very surprised at how much individual variation there is.

After the above exercise, try to obtain a disarticulated skeleton of another genus of salamander to compare with the first one. In beginning studies, it is useful to compare salamander genera that are morphologically distinct from one another; for instance, comparing the disarticulated bones of a Tiger Salamander (*Ambystoma tigrinum*) with those of a Mudpuppy (*Necturus maculosus*) would be a good exercise. Soon you will realize that some of the individual bones of these two animals are quite different from one another. On the other hand, if you compare the individual bones of species within the genus *Ambystoma* or the genus (*Necturus*) you will see that bones appear to be similar or even identical. Intensive studies are usually warranted to differentiate such elements. This can be frustrating, and you may come to the conclusion that bones in salamander species within the same genus cannot be differentiated. You may be right or wrong.

When you are engaged in the identification of fossil salamanders, it is of utmost importance to have a modern caudate skeletal collection at hand. You should not try to identify fossil salamanders solely on the basis of written descriptions and figures in this book, or for that matter, any other book. Interspecific, intracolumnar (e.g., within the vertebral column in salamanders), ontogenetic (changes in development with age), and pathological conditions occur in salamander bones, both fossil and modern. Without a series of comparative skeletons, you might misinterpret these characters as being useful in the description of new species or higher taxonomic levels. Such mistakes have often been made in the past and are still being made.

The first step in the identification of a fossil salamander is to determine what each skeletal element or elements each fossil represents. The next step is to determine the caudate family, genus, and species, in that order. If you are stymied at one level or the other, it is necessary to search the literature to find papers on the subject that will guide you. For instance, if you determine that a vertebra is in the genus *Ambystoma* but you have no idea where it belongs within the genus, consult Tihen (1958), who has written an excellent paper on the osteology of ambystomatid salamanders that includes comparison of vertebrae and other elements of taxa within the group.

The identification of a single salamander bone, or even a fairly complete caudate skeleton, to the specific level is a daunting task that will take hours, days, and even many weeks. A rule of thumb that most professionals use is to look for good characters first and then substantiate them by measurements and ratios later. It is a mistake to come to an early conclusion that there are no differences between the bones of two species and resort to the interpretation of "blind" measurements alone. On the other hand, it is a mistake to assume that an osteological difference between a specimen or two of different species is valid unless several more specimens of each species are studied.

Once you decide to use measurements and ratios to distinguish salamander bones from one another, you will soon discover that such elements have curves and other irregularities that make them difficult to measure accurately. For this reason you need to establish landmarks that can be reference points for comparative measurements. Esteban et al. (1995) indicated how they used some simple landmarks on frog bones, and Sanchiz et al. (1993) indicated how they used some simple landmarks and statistics to establish the taxonomic relationships of an Oligocene frog. Actually, few morphometric (related to the characterization of the form of organisms for quantitative analysis) or even simple statistical studies have been reported in papers dealing with the identification of fossil salamanders. An apparent weakness derives from the fact that fossil salamanders often occur one at a time in fossil deposits; thus variation cannot be evaluated. I hope that more fossil populations of individual salamander species will be excavated in sites of the future.

CHAPTER

2 Systematic Accounts

This chapter details the occurrence of fossil salamanders in North America in systematic order. In the headings and lists that follow, the symbol # before a taxon indicates that the family or genus is extinct, and the symbol * indicates an extinct species of an extant genus. English names of North American salamander taxa are now capitalized. Capitalization of English names has been done in ornithology for years and has recently been adopted by the Committee on Standard English and Scientific Names, whose members represent the three most prominent herpetological societies in North America (the Society for the Study of Amphibians and Reptiles, the American Society of Ichthyologists and Herpetologists, and the Herpetologists League) (Crother, 2000). This capitalization rule has also been adopted by the Center for American Herpetology, a group that has published a separate list of the common and scientific names of North American amphibians and reptiles (Collins and Taggart, 2002). I have chosen to follow Crother (2000) and subsequent changes for the common and scientific names of North American living salamanders that are represented by fossils. Genera and species within families appear in alphabetical order when possible.

Names and dates that appear after the scientific names designate the describer of the taxon involved and the date of the publication in which that taxon was described. A date in parentheses indicates that the present genus was given another generic name after the one proposed by the original author. Only fossil salamanders from North America, exclusive of Mexico, are detailed in the following pages.

Order Caudata Oppel, 1811
Salamanders
Family Sirenidae Gray, 1825
Sirens

In many ways the sirens are the most interesting family of caudates. In fact, they are so different from the other salamanders that not so long ago they were thought to belong to a separate order of amphibians, the Trachystomata (Goin and Goin, 1962). Just the fact that adult sirens do not have pedicellate teeth—possibly the most significant single character that separates the lissamphibians (frogs, salamanders, and caecilians) from other amphibians—is something to be reckoned with. Many things are still to be learned about this family, both in the living species and in the fossil record.

Living species of the Sirenidae occur in the eastern United States as far west as south-central Texas, and they occur down into Mexico in extreme northern Tamaulipas. Fossil sirenids in North America are first known in the Late Cretaceous in Wyoming, and then they spread across the continent to Florida.

There is a good possibility that sirenids occurred in Gondwana in the Cretaceous. Isolated elements of large eel-like salamanders include *Noterpeton* from the Maastrichtian (ca. 71–65 Ma BP) Cretaceous of Bolivia, *Kababaisha* from the Cenomanian (ca. 99–93.5 Ma BP) Cretaceous of the Sudan, and a specifically unnamed *Kababaisha* from the Cretaceous of Niger. Rage et al. (1993) put *Noterpeton* in a new family, Noterpetontidae, but Evans et al. (1996) suggested that these Gondwana forms represent a lineage of sirenid salamanders. If this is true, these Bolivian and African salamanders represent the only known Gondwanan range extension of Mesozoic Laurasian salamanders that subsequently became extinct.

Some osteological characters used to define the living members of the group (Duellman and Trueb, 1986) are as follows. The pelvic girdle and the hind limbs are absent. The short dorsal processes of the premaxillae do not separate the nasals. Septomaxillae, coracoids, and the second ceratobranchials are present. The maxillary teeth are absent in *Pseudobranchus*, but they are free and tooth-bearing in *Siren*. The pterygoids are reduced, and the angular is fused with the prearticular. The exoccipital, prootic, and opisthotic are not fused. The columella is free from the operculum. The teeth are not pedicellate. Premaxillary teeth are absent and are replaced by horny beaks. The vertebrae are amphicoelous (concave at both ends), and all but the first two spinal nerves exit intervertebrally. Edwards (1976) and Estes (1981) considered the following characters of the Sirenidae to be derived (not from an ancestral character state): (1) The front limbs are small, and the hind limbs are absent. (2) The angular, the septomaxilla, and the lacrimal are absent. (3) The maxilla is present or absent. (4) The teeth are not pedicellate. (5) The spinal nerves all exit intervertebrally, except for the second spinal nerve. (6) The ypsiloid cartilage is absent. (7) One or three larval gill slits may be present. It is interesting to note that living sirens may estivate (have a dry season or summer dormancy) for long periods during conditions of drought. Sirens produce a mucoid cocoon from their skin secretions for estivation that is similar to that produced by African lungfishes (Estes, 1981).

As far as I am aware, the latest diagnosis of the Sirenidae has been provided by Gardner (2003a), modified as follows. The following combination of derived characters provisionally supports the monophyly (origin from a single ancestor) of the Sirenidae. The teeth are non-pedicellate and have single cusps. When present, the palatine teeth are arranged in multiple, parallel rows. The dentary has its ventral margin behind the symphysis, and the area for the attachment of the postdentary bones is sharply deflected ventrally. The spinal nerve foramen pierces the posterior half of the neural arch wall in the vertebrae posterior to the atlas. The trunk and anterior caudal vertebrae laterally bear three alar processes (dorsal and ventral laminae of Estes, 1981) that are associated with the transverse processes. These vertebrae also have a pair of interzygapophyseal ridges dorsolaterally. The neural spine (neural crest of Gardner, 2003a) splits posteriorly to form aliform processes, a V-shaped structure in dorsal view. The transverse processes are bicapitate (double headed) on the anteriormost trunk vertebrae and unicapitate (single headed) on the rest of the vertebrae. Ribs are restricted to the anterior part of the trunk series and articulate with the double-headed vertebrae in this region. Plesiomorphic (primitive) characters are (1) amphicoelous vertebrae posterior to the atlas; and (2) paired anterior basapophyses on the trunk vertebrae.

The history that relates to the placing of sirens in their own order—namely, the Trachystomata—and then classifying them with the Caudata again later is of interest. While describing new fossil sirenid remains in Florida in the 1950s, Coleman Goin and Walter Auffenberg (see Goin and Auffenberg, 1955) were surprised at how similar the vertebrae of sirenids were to those of an elongate group of ancient Paleozoic amphibians called aistopods (order Aistopoda). For a while, Goin and Auffenberg speculated that sirens themselves might actually be aistopods. This generated much excitement among us graduate students at the University of Florida, and the possibility of having 330 million year old living fossils swimming around in our local lakes made us ecstatic. In fact, if the aistopod hypothesis were true, there would have been a group of living vertebrate fossils to put the coelacanths to shame! Goin and Goin (1962, pp. 64–65), in the first college textbook on herpetology ever published in North America, put the sirenids in a separate amphibian order, the Trachystomata, stressing that sirenids showed "marked resemblance to the aistopod–nectridian stock." By the third edition of this text (Goin et al., 1978), the authors had dropped the idea of an aistopod origin for the sirenids, but they hung on to the ordinal separation of the group. Other authors (e.g., Estes, 1965a, 1981; Wake, 1966; Edwards, 1976; Naylor, 1978b) together put the Trachystomata matter to rest, at least for the time being, and the consensus now is that the sirenids are indeed salamanders. Nevertheless, they are the most distantly related to other living salamanders (see cladograms in Fig. 1).

Zug (1993, p. 345) stated that "in spite of their locally high abundance and widespread distribution, their biology is poorly known. Females lack spermatotheca for storage of sperm, and males have no obvious structures for producing spermatophores, thus external fertilization is assumed the reproductive mode. Courtship behavior has not been observed; eggs are laid singly or in small clusters attached to vegetation." Eggs and nests of sirenids have been rather commonly found (see Petranka, 1998), and bite marks on both males and females have also been found (Raymond, 1991), possibly made during courtship. But we still do not know much about how these animals reproduce, although it is assumed that external fertilization takes place.

In Field and Lab. If you are looking for modern sirenid specimens to skeletonize, don't waste your time looking for roadkill, as all Sirenidae are obligatorily aquatic and only occasionally wriggle overland. You might get lucky and hook one on a worm if you like fishing in swamps, but the best way to catch one is to use a square hand-dredge with a screen bottom. Get a dedicated herpetologist friend to push mats of vegetation and swamp-bottom muck into the dredge from above, and you'll need a strong person to pull the dredge and its contents onto the bank. Next, sort through the vegetation with a hand net ready, as the slimy sirenids can effortlessly squirm through your fingers. As you prepare large *Siren* for the maceration jar, do not even think about frying removed muscles; big sirens look like eels, but they taste like mud.

Genus #*Habrosaurus* Gilmore, 1928

Genotype. Habrosaurus dilatus Gilmore, 1928.

Revised Diagnosis. This revised diagnosis is modified from Gardner (2003a). *Habrosaurus* is a genus of the Sirenidae that differs from the other two genera of the family, *Pseudobranchus* and *Siren*, on the basis of five derived character states, as follows. (1) The crowns on the marginal and palatal teeth are expanded and bear a low median crest that extends across their occlusal (aligned) surface. (2) Wear facets are variably developed on the crowns of marginal and palatal teeth. (3) A posteriorly open notch occurs in the dentary immediately behind and below the occlusal margin. (4) A deep groove occurs in the labial surface of the dentary below and parallel to the posterior part of the occlusal margin. This groove is anteriorly elongate, and in some large individuals the posterior part of this groove is roofed labially by bone, forming a canal. (5) The articular surface of the atlas with the skull continues across paired anterior cotyles and the lateral and anterior surfaces of the odontoid process. Primitive character states and one uncertain character state are also included in the diagnosis of Gardner (2003a).

#*Habrosaurus dilatus* Gilmore, 1928
(*Adelphesiren olivae* Goin and Auffenberg, 1958)

Figs. 25–28

Holotype. Partial left dentary with four complete teeth (USNM 10749, United States National Museum).

Type Locality. Lance Formation, Niobrara County, Wyoming (Gilmore, 1928).

Horizon. Late Cretaceous.

Other Material. A very extensive list of University of California Museum of Paleontology (UCMP) *Habrosaurus dilatus* material from various localities in the Late Cretaceous Lance Formation of Wyoming was provided by Estes (1964). Other museum sources for *H. dilatus* material ranging from the Late Cretaceous to the Late Paleocene of Montana and Wyoming were given by Estes (1981). These museums were the American Museum of Natural History (AMNH); University of Minnesota (UM);

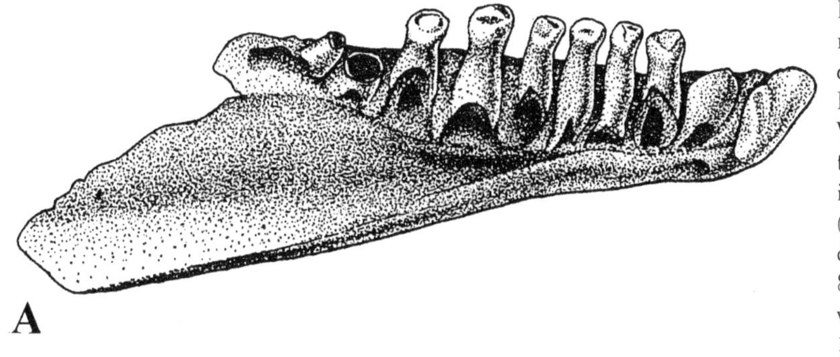

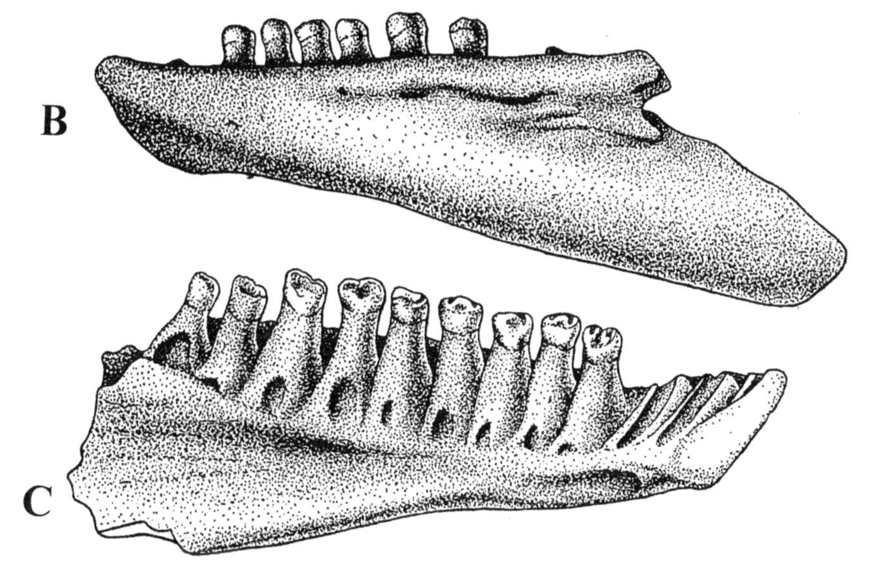

FIGURE 25. Dentaries of #*Habrosaurus dilatus* from the Late Cretaceous of Wyoming. Left dentary (AMNH 8106) in (A) lingual and (B) labial views. Left dentary (AMNH 8108) in (C) lingual view. All figures, × 3.

Carnegie Museum of Natural History (CM); Princeton University (PU); University of Alberta Laboratory for Vertebrate Palaeontology (UALVP); Museum of Natural History, University of Kansas (KU); and Museum of Comparative Zoology, Harvard University (MCZ).

Gardner (2003a) did recognize Late Paleocene records for *Habrosaurus dilatus* and did not include referred jaws, palatal bones, or vertebrae currently housed in the AMNH and UCMP from localities in the type area of the Lance Formation (see Estes, 1964).

Gardner (2003a) provided a list of additional specimens of *Habrosaurus dilatus*, mainly from the UALVP. The present known distribution of this species is quoted directly from Gardner (2003a, p. 1101): "Late Maastrichtian–middle Palaeocene, North American Western Interior: Lance Formation, Wyoming and Frenchman Formation, Saskatchewan (both late Maastrichtian or Lancian in age); Bug Creek Anthills (late Maastrichtian or Lancian and early Palaeocene or Puercan in age), Hell Creek Formation, Montana; and Lebo Formation (middle Palaeocene or Torrejonian in age), Montana."

Revised Diagnosis. This revised "differential" diagnosis is directly from Gardner (2003a, pp. 1101–1102): "Species of *Habrosaurus* with marginal teeth differing from those on comparable-sized jaws of the new

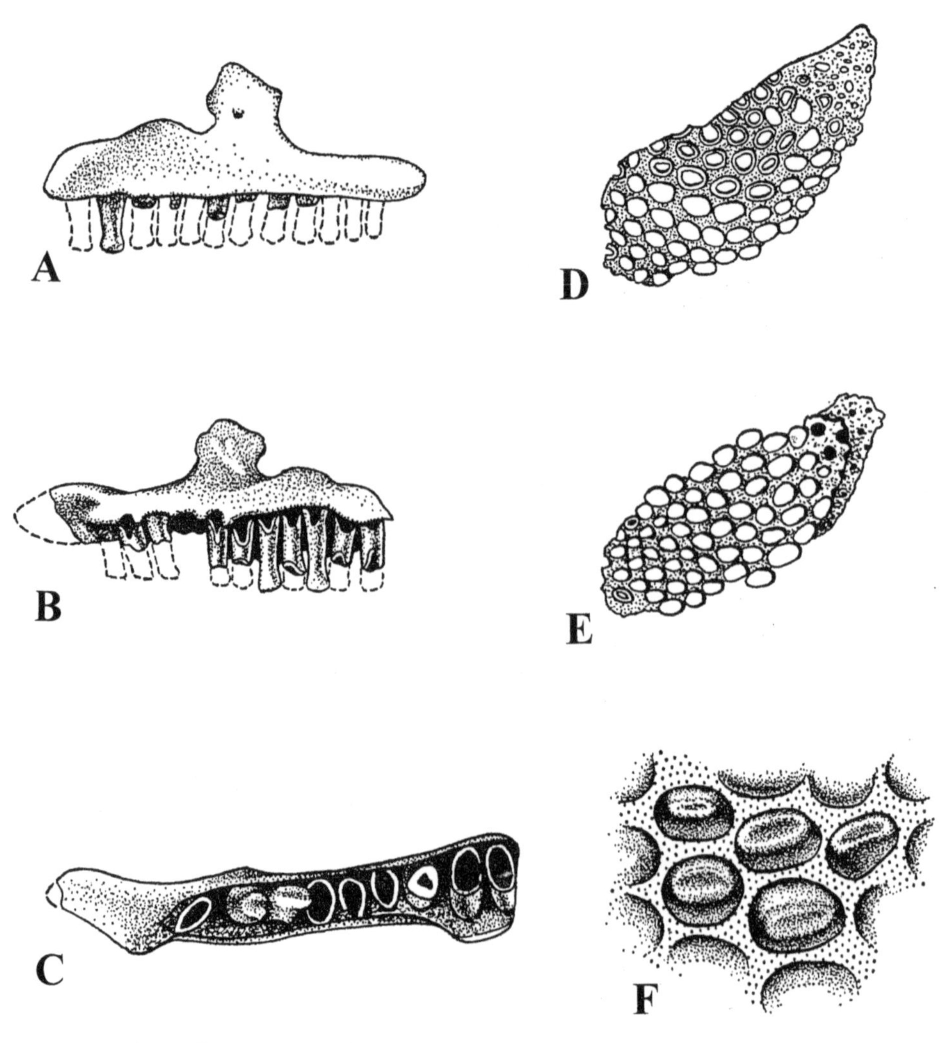

FIGURE 26. Maxillae and teeth of #*Habrosaurus dilatus* from the Late Cretaceous of Wyoming. (A) Left maxilla (UCMP 46043) in labial view. (B) Left maxilla (UCMP 55717) in lingual view. (C) Right maxilla (UCMP 55714) in occlusal view. (D) Right palatal tooth plate (AMNH 8120) in ventral view. (E) Right palatal tooth plate (AMNH 8119) in ventral view. (F) Magnified view of palatal tooth plate of AMNH 8119, showing several teeth. (A)–(C) and (F), ×9; (D) and (E), ×3.

Judithian species described below, as follows: teeth relatively stouter and about 90 per cent as long; neck between shaft and crown more constricted; crowns expanded labiolingually and bulbous; crowns on adjacent teeth more closely spaced, occasionally contacting one another or slightly overlapping; and teeth on premaxilla and anterior part of dentary develop prominent wear facets on occlusolingual surface of crown and, in larger individuals, crowns of these teeth may be ground flat."

Description. The following description is paraphrased from Estes (1964, 1981). This description is mainly from his material from the Late Cretaceous Lance Formation, Wyoming. The dentary has 10–16 pleurodont (arising from the inside of the lower jaw) teeth. These teeth are

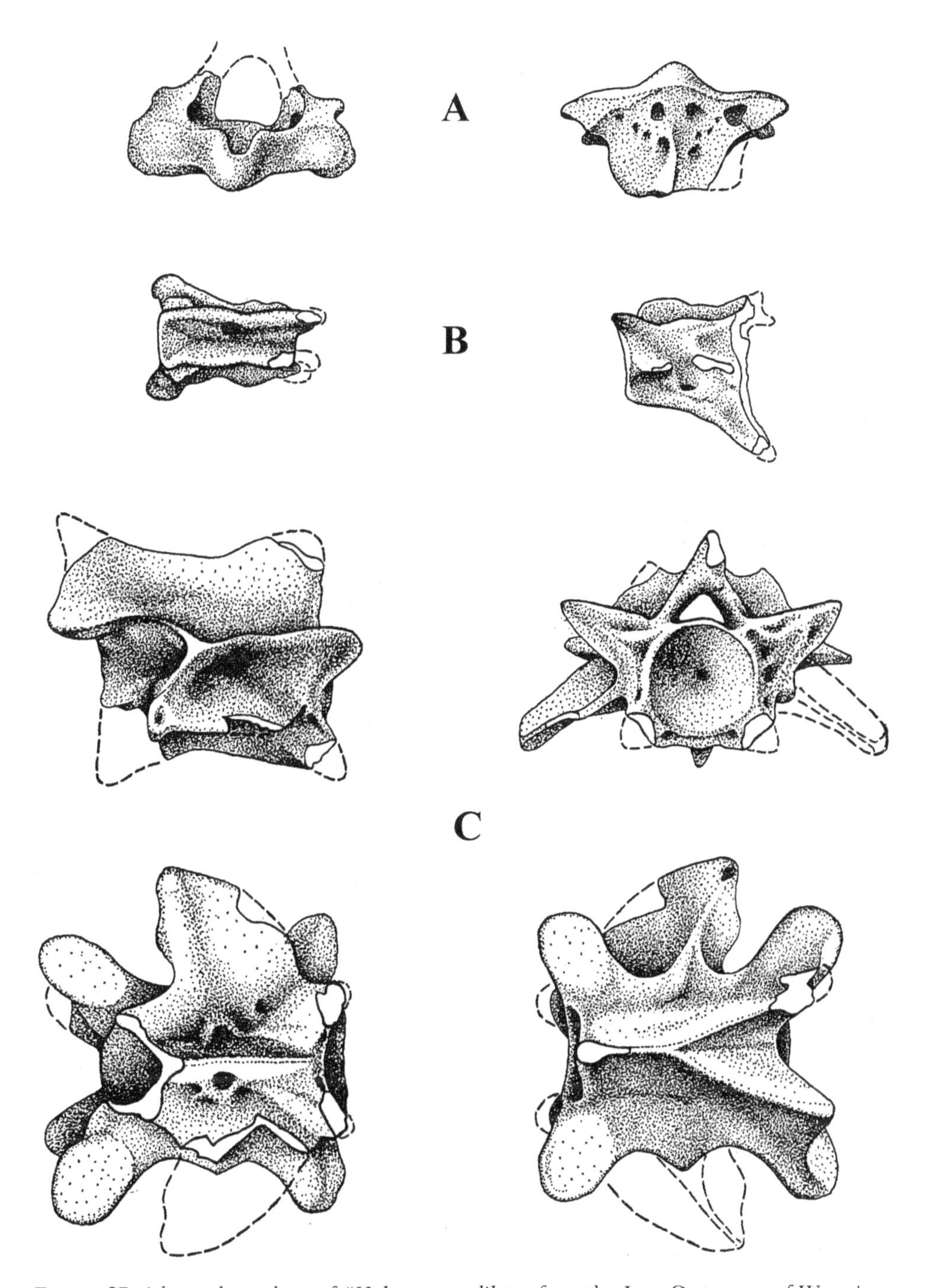

FIGURE 27. Atlas and vertebrae of #*Habrosaurus dilatus* from the Late Cretaceous of Wyoming. (A) Atlas (UCMP 49519) in anterior (left) and ventral (right) views. (B) Caudal vertebra (AMNH 8115) in ventral (left) and lateral (right) views. (C) Trunk vertebra (UCMP 46627) in lateral (upper left), anterior (upper right), ventral (lower left), and dorsal (lower right) views. All figures, ×3.

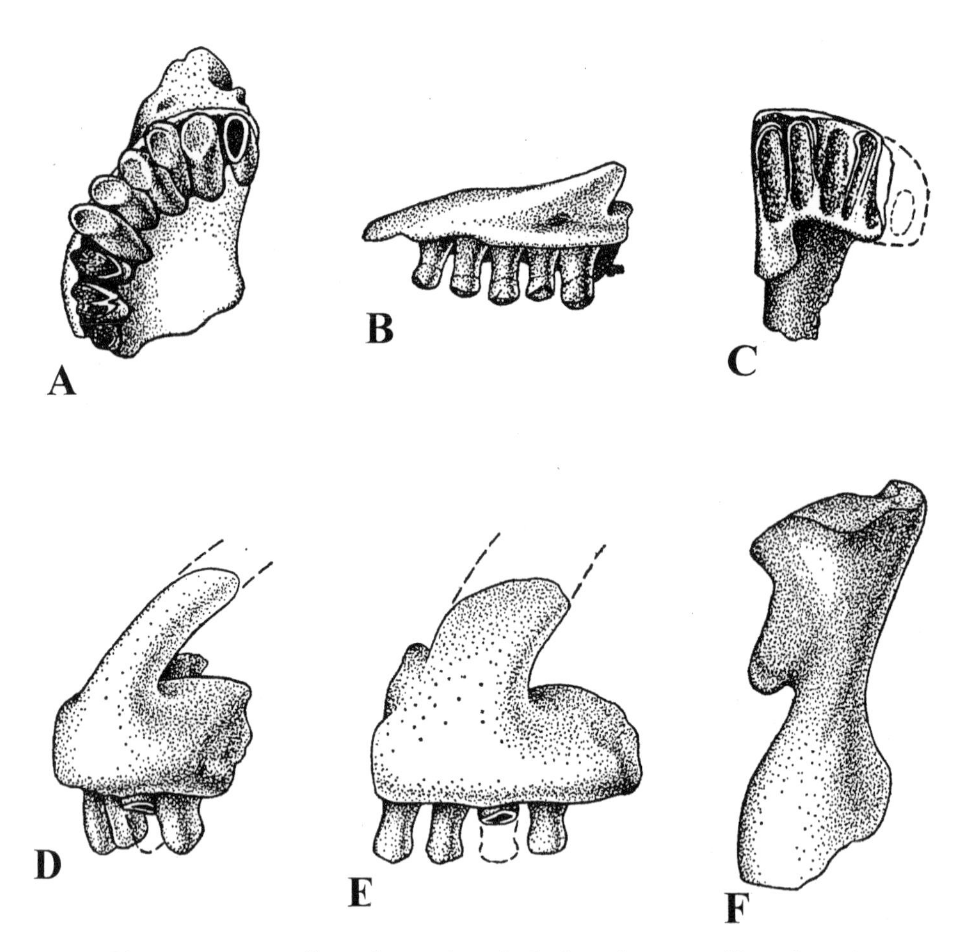

FIGURE 28. Vomer, premaxilla, and ceratobranchial of #*Habrosaurus dilatus* from the Late Cretaceous of Wyoming. Right vomer (AMNH 8116) in (A) ventral and (B) labial views. Left premaxillary (AMNH 8118) in (C) ventral view. Left premaxillary (UCMP 46073) in (D) lateral and (E) anterior views. Ceratobranchial (AMNH 8117) in (F) ventral view. All figures, ×9.

shovel shaped, with expanded, bulbous crowns. The crowns are separated from the tooth shaft by a prominent constriction. The maxilla is of reduced size, compared with the other tooth-bearing bones in *Habrosaurus*, and its 8–12 teeth are smaller as well. The maxilla has a blunt anterior end, which is flattened and roughened for its attachment to the premaxilla. The premaxilla is robust and bears only three to five teeth, which are identical to those of the dentary. A prominent, posterior tapering nasal spine is present. It was suggested that the premaxilla has a ligamentous attachment with the maxilla. A large posterior projection on the premaxilla articulates with a corresponding notch on the vomer.

The vomer is a short bone that bears a curved row of nine teeth. These teeth are attached to an external ridge that disappears posteriorly. The vomers meet on the midline in a suture that is about the entire length of the bone and are separated only by the posterior processes of the premaxillae. The palatopterygoid tooth plate is oval and bears teeth in a pattern of broad, sinuous curves. The crown structure of these teeth is similar to that of the jaw teeth. But these palatopterygoid teeth differ from the mandibular teeth in that they have a constriction that separates the crown from the tooth shaft. The teeth on these tooth plates become smaller and have recurved, pointed crowns posteriorly. No indications of tooth replacement are noted on these tooth plates.

Moving to the vertebral column, we see that the anterior trunk vertebrae have a narrower angle between the aliform processes. Moreover, their cotyles are more teardrop shaped, and the neural arches are higher. In addition, the zygapophyses are placed higher in relation to the dorsal edge of the centrum, and the centrum and the neural arches are more laterally compressed and are sometimes flattened ventrally.

The centra of the more posterior trunk vertebrae range from about 5 to 19 mm in length and are about twice as long as they are high. The anterior cotyles are generally round. The basapophyses originate just anterior to the subcentral foramina and extend anterolaterally at about a 45-degree angle, increasing in size, projecting anterior to the centrum, and ending as blunt points. A subcentral keel is present and has its edge, in lateral view, either straight or slightly concave. Smaller vertebrae have lower and blunter keels than larger vertebrae. The centra of these vertebrae are broadly excavated ventrally. Prominent subcentral foramina, as well as prominent transverse processes, are present. Ventral laminae originate anteriorly just below the base of the anterior zygapophyses on the border of the anterior cotyle; then these laminae curve posterolaterally, as a thin sheet, to the top of the transverse process. A small foramen for the spinal nerve is present at the posteroventral region of the neural arch. In most specimens, a zygapophyseal ridge drops abruptly down, meeting the dorsal lamina of the transverse process at a point slightly more than halfway back between the pre- and postzygapophyses. The neural spine of these vertebrae is prominent. The aliform processes have an angle between them of 45–55 degrees. The floor between these processes is marked with a low midline ridge.

The anterior caudal vertebrae resemble the trunk vertebrae, except for the presence of a hemal arch. The posterior caudal vertebrae have small zygapophyses and more sharply rising aliform processes. They have simple hemal arches formed of two unfused, posteroventrally directed processes.

General Comments. Habrosaurus dilatus was originally believed to be a lizard by its describer, C. W. Gilmore (1928), whose only specimens were toothed dentaries. Sirenid vertebrae were later described at the same locality by Goin and Auffenberg (1958) and assigned to *Adelphesiren olivae*. Estes (1964) had abundant Lance Formation material and was able to demonstrate that *A. olivae* was a synonym of *H. dilatus*. Estes also pointed out at the time that *H. dilatus* differed from modern sirens mainly in the structure of the skull. *Habrosaurus* has a number of characters that resemble those of *Pseudobranchus* rather than those of *Siren*, although the ancient sirenid (estimated by Estes, 1964, to be in excess of 1600 mm in the largest specimen) is much larger than *Pseudobranchus*, as is the modern species *Siren lacertina*.

Estes (1964, p. 95) stated that the "relationships between the Lance

Formation fauna and that of the present southeastern United States suggests that *Habrosaurus* is probably close to, if not on the line leading to modern sirens, which differ from *Habrosaurus* primarily in features related to a greater degree of neoteny."

About the feeding habits of *Habrosaurus dilatus*, Gardner (2003a) pointed out that its marginal, vomerine, and palatal teeth are robust and have expanded crowns that often have a prominent wear facet. Moreover, the teeth on the jaws and vomer are packed in a single row, and the teeth on the palatine are packed in parallel, tight rows, forming a broad-toothed pavement. On the basis of these structures, Gardner considered Estes's (1964) suggestion that the diet of this salamander may have consisted largely of mollusks and hard-bodied arthropods reasonable. Extant sirenids prey on gastropods and bivalves, but they have a different set of adaptations for this method of feeding.

#*Habrosaurus prodilatus* Gardner, 2003a

Holotype. A partial right premaxilla preserving an intact tooth row with three complete teeth, one nearly complete tooth, and the bases of three teeth (UALVP 43906, University of Alberta Laboratory for Vertebrate Palaeontology).

Type Locality. Dinosaur Park Formation, Irvine microvertebrate locality, Alberta, Canada.

Horizon. Late Cretaceous (middle Campanian).

Other Material. Referred specimens are all from the holotype locality and consist of premaxillae (UALVP 43902–43905), dentaries (UALVP 43907–43909), and a centrum from an atlas (UALVP 40051).

Diagnosis. The diagnosis is directly from the differential diagnosis given by Gardner (2003a, p. 1113): "Species of *Habrosaurus* with marginal teeth differing from those on comparable-sized jaws of *H. dilatus*, as follows: teeth more gracile and about 10 per cent longer; neck between shaft and crown less constricted; crowns compressed labiolingually and more chisel-like; crowns on adjacent teeth more widely spaced and do not contact one another or overlap; and wear facets, where developed, less extensive and restricted to occlusalmost tip of crown."

Description. The description is modified from Gardner (2003a). The five partial premaxillae from the type locality collectively show most of the premaxillary structures, with the exception of the distal part of the dorsal portion (pars dorsalis) and the vomerine process. The two most complete specimens (UALVP 43906 [holotype], 43904) resemble *Habrosaurus dilatus* in several characters, some of which are rather trivial. None of the *Habrosaurus prodilatus* premaxillary specimens, however, have the elongate groove on the inner face of the pars dorsalis of the premaxilla as seen in some premaxillae of *H. dilatus*.

Collectively, the three dentary specimens of *Habrosaurus prodilatus* show that the dentary resembles that of *Habrosaurus dilatus* in being moderately robust in build and broadly curved labially; in having the occlusal edge straight in lateral view and the subdental shelf weakly developed anterior to the Meckelian groove; and in having the dentary symphysis developed as a thin, vertical plate.

The marginal teeth of *Habrosaurus prodilatus* are known only for the premaxilla and the dentary. These teeth resemble the teeth of *Habrosaurus dilatus* in being moderately pleurodont and non-pedicellate; in having a lingual resorption pit at the base of most teeth; in having the tooth shaft broadest at its base and tapered toward the crown; in having a crown with a single cusp that is mesiodistally expanded, with a low medial crest extending mesiodistally across its apex.

In other characters, however, the marginal teeth of *Habrosaurus prodilatus* are distinctly different from those of *Habrosaurus dilatus*. In lingual view the tooth shaft is narrower and less swollen, with the mesial and distal sides nearly straight to slightly convex and tapering to a weakly constricted neck. The crown resembles a chisel and is labiolingually compressed, with the labial and lingual faces declining steeply away from the medial crest. The crowns of adjacent teeth are well separated from one another. Roughly one-half of the intact premaxillary teeth have a weak wear facet in the form of a shallowly beveled surface that extends mesiodistally across the tip of the crown. This wear facet does not occur on the remaining intact premaxillary teeth or on the eight complete dentary teeth. Two of the premaxillae (UALVP 43904, 43906) have an intact tooth row, with six and seven loci, respectively.

An atlas is represented by a dorsoventrally crushed centrum about 2–4 mm in midline length and 2.8 mm in intercotylar width. Each cotyle is broader than high in anterior view, with the articular surface slightly convex. The odontoid process is short and wide anteriorly, broadly rounded distally in dorsal or ventral view, and deep in anterior view. The posterior cotyle is subcircular in posterior view and is deeply excavated, with a notochordal pit opening in its posterior half. The base of the neural arch is present on the right side of the atlas and encloses a spinal foramen.

General Comments. Gardner (2003a) pointed out that the rarity of *Habrosaurus prodilatus* in the richly fossiliferous Irvine microvertebrate locality and other productive sites in the Dinosaur Park Formation of southwestern Alberta implies that this species may have been a rare component of these faunas in Late Cretaceous (middle Campanian) times. The middle Campanian Late Cretaceous age of *H. prodilatus* makes it the earliest known sirenid (see Fig. 23).

Genus *Pseudobranchus* Gray, 1825

Dwarf Sirens

Genotype. Pseudobranchus striatus (LeConte, 1824).

Dwarf Sirens occur today in the extreme southern United States, mainly in Florida, but with some populations in southern South Carolina and southern Georgia. Only two living species are currently recognized, *Pseudobranchus axanthus* and *Pseudobranchus striatus*, neither of which has been recorded in the fossil record. Characters given by Duellman and Trueb (1986) that separate *Pseudobranchus* from the other living sirenid genus, *Siren*, are as follows. Maxillae are absent in *Pseudobranchus*, and maxillae are present, free, and tooth bearing in *Siren*. One gill slit is present in *Pseudobranchus*, and three are present in *Siren*. *Pseudobranchus* has only three toes on each forefoot; *Siren* has four. *Pseudobranchus* reaches a length of only 250 mm, whereas *Siren* reaches a length of 950 mm.

Fossil *Pseudobranchus* and *Siren* species have been described mainly on the basis of individual trunk vertebrae. Goin and Auffenberg (1955, p. 498) gave vertebral characters that distinguish the two genera: "*Siren* differs from *Pseudobranchus* in that the lower margin of the centrum is nearly straight as seen from the side while in *Pseudobranchus* it is quite concave. Furthermore, in *Siren* the zygapophyseal ridge is gently curved or nearly straight and continues forward to meet the transverse process near the base of the prezygapophysis while in *Pseudobranchus* the ridge curves downward to meet and fuse with the transverse process in a shallow V posterior to the base of the prezygapophysis. In *Pseudobranchus* the zygapophyseal ridge has more of a tendency to flare where it fuses with the transverse process than it does in *Siren*."

**PSEUDOBRANCHUS ROBUSTUS* GOIN AND AUFFENBERG, 1955

FIG. 29A

Holotype. A middle thoracic (trunk) vertebra (MCZ 2279, Museum of Comparative Zoology, Harvard University).

Type Locality. Haile Pit VIIA, a little south of the village of Haile, Alachua County, Florida.

Horizon. Late Pleistocene (Rancholabrean NALMA).

Other Material. Material referred by Goin and Auffenberg (1955) includes five vertebrae (MCZ 2280) from the type locality and one vertebra (MCZ 2281) from the Pleistocene (Rancholabrean NALMA) at Kanapaha, Alachua County, Florida. Holman (1962) collected additional

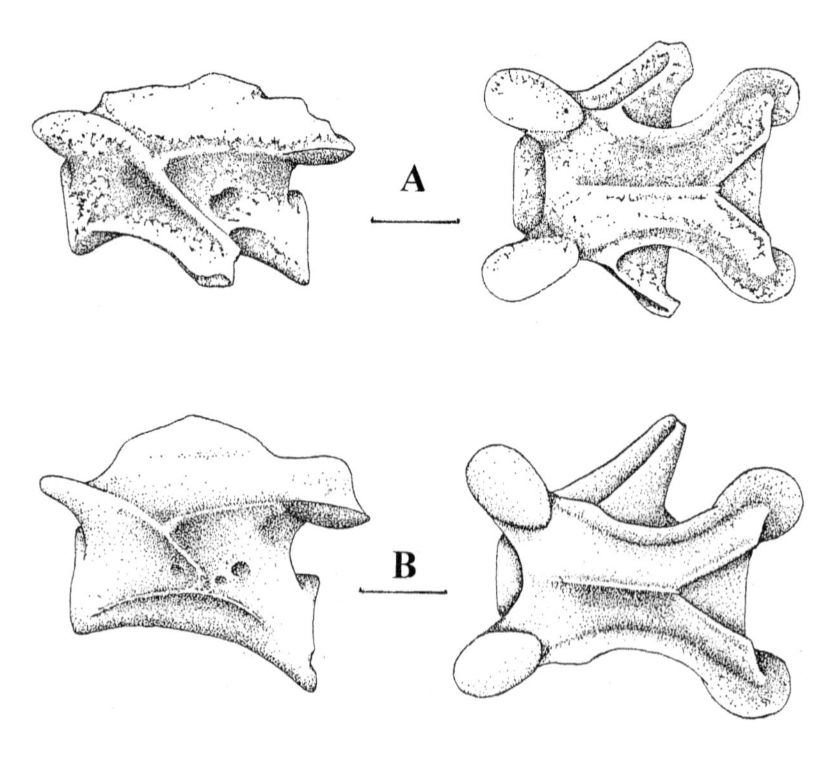

FIGURE 29. Holotype trunk vertebrae of extinct species of *Pseudobranchus.* (A) **Pseudobranchus robustus* from the Late Pleistocene of Florida in lateral view (left) and dorsal view (right). (B) **Pseudobranchus vetustus* from the Late Miocene of Florida in lateral view (left) and dorsal view (right). Scale bars = 1 mm for both groups.

material of *Pseudobranchus robustus* from the Reddick I (Pleistocene: Rancholabrean NALMA) locality 1 mile (about 1.6 km) southeast of Reddick, Marion County, Florida. These three unnumbered vertebrae are presumably now at the Florida Museum of Natural History (UF) after having been transferred there from the Florida Geological Survey (FGS). Lynch (1965) referred eight vertebrae (all assigned to UF 9101), all from the Pleistocene (Rancholabrean NALMA) Arredondo Pit II deposit, Alachua County, Florida, to *Pseudobranchus* cf. *Pseudobranchus robustus*.

Diagnosis. This diagnosis is directly from Goin and Auffenberg (1955, pp. 505–506): "A *Pseudobranchus* with large, massive articulating facets on the zygapophyses and with the margins of the zygapophyseal ridges pronouncedly concave as seen from above. It differs from the modern species in the wider angle between the aliform processes and in the more concave sides of the zygapophyseal ridges as seen from above. From the Pliocene species described below it differs in the stronger concavity of the zygapophyseal ridges and in the more widely flaring, relatively shorter aliform processes."

Description of the Holotype. The following description is modified from Goin and Auffenberg (1955). The centrum is longer than high. The cotyles are oval and wider than they are high. The centrum has an elevated, ridgelike, medial ventral keel, on either side of which is a moderately large subcentral foramen. The margin of the ventral keel is strongly concave. The total length of the neural arch is greater than that of the centrum, and its width at the narrowest part of the zygapophyseal ridges is slightly greater than the width of the centrum. The neural canal is in the shape of an inverted crescent anteriorly, is nearly rounded posteriorly, and has a very low median epipophyseal ridge on its floor.

The articulating surfaces of the prezygapophyses are oval, longer than wide, and directed more anteriorly than laterally. The articulating surfaces of the postzygapophyses are ovoid. The zygapophyseal ridges are well developed and are markedly concave in dorsal view. In lateral view, the zygapophyseal ridge forms a very shallow V with the apex at the point where the dorsal portion of the transverse process meets the zygapophyseal ridge. The aliform processes are well developed, vertical in their position, and somewhat rectangular in lateral view. In dorsal view the aliform processes form an anteriorly pointing V. The floor between the aliform processes is present and has a nearly straight posterior margin. The neural spine is well developed, but its dorsal margin is broken.

The transverse processes are well developed and are formed of two plate-like portions, of which the ventral portion is larger than the dorsal. The ventral portion is a wing-like piece that extends from near the anterior margin of the side of the centrum for about four-fifths of the length of the centrum. The posterior margin of the transverse process is approximately perpendicular to the long axis of the centrum. A lateral foramen is present in the angle between the dorsal and ventral parts of the transverse process, and another lies somewhat ventral and posterior to the angle between the dorsal portion of the transverse process and the zygapophyseal ridge.

Other Material. The six vertebrae of *Pseudobranchus robustus* (other than the holotype) reported by Goin and Auffenberg (1955) were described as being remarkably constant in specific characters. In fact, these authors reported that most of the variation seemed to be due to either erosion or fragmentation in the fossils. All six vertebrae have widely flaring aliform processes, and the zygapophyseal ridges are very concave in dorsal view. These ridges all have the same massive structure that is so apparent in the type specimen. The most pronounced variation noticed in the six other vertebrae was the shape of the margin of the median subventral keel.

The material assigned to *Pseudobranchus robustus* from the Reddick I (Pleistocene: Rancholabrean NALMA) locality near Gainesville, Florida, by Holman (1962) consisted of two middle trunk vertebrae, as well as a caudal vertebra previously unreported from this region of the *P. robustus* vertebral column. Holman reported that this caudal vertebra differed markedly from that in modern *Pseudobranchus striatus* and further strengthened the status of the extinct form. He pointed out that the zygapophyseal ridges in the fossil have their margins much more concave than in modern *P. striatus*, and the angle between the aliform processes is greater.

Measurements of the caudal vertebra of *Pseudobranchus robustus* are as follows: length of centrum along mid-ventral line, 1.9 mm; width of vertebra at narrowest point of zygapophyseal ridges, 1.0 mm; height of vertebra from lower margin of centrum to a line drawn between facets of postzygapophyses, 1.1 mm; distance from tips of prezygapophyses to tips of postzygapophyses, 2.7 mm; angle between aliform processes, 69 degrees; width of anterior glenoid cavity, 0.9 mm; height of anterior glenoid cavity, 0.7 mm.

Lynch (1965) measured five of the eight vertebrae he collected from the Pleistocene (Rancholabrean NALMA) site at Arredondo Pit II, Florida, and compared these measurements with those of Goin and Auffenberg (1955) and Holman (1962). Lynch pointed out that the range in measurements of his specimens overlapped those of *Pseudobranchus robustus* and suggested that further studies on the status and relationships of this species are necessary. Estes (1981) continued to cautiously recognize *P. robustus*. Milner (2000) did not comment on the specific status of fossil *Pseudobranchus*.

General Comments. I currently consider *Pseudobranchus robustus* a somewhat questionable species and believe there is much more comparative work to be done with this taxon. Most of the modern skeletons that *P. robustus* has been compared with have been *Pseudobranchus striatus* specimens from the vicinity of Gainesville in north-central peninsular Florida. Three subspecies of living *P. striatus* and two subspecies of living *Pseudobranchus axanthus* inhabit Florida (Conant and Collins, 1998; Crother, 2000). At the least, skeletal specimens of *P. axanthus* should be compared with *P. robustus*.

**Pseudobranchus vetustus* Goin and Auffenberg, 1955

Fig. 29B

Holotype. A thoracic (trunk) vertebra (MCZ 2282, Museum of Comparative Zoology, Harvard University).

Type Locality. Haile Pit VI, a little south of the village of Haile, Alachua County, Florida.

Horizon. Late Miocene (late Hemphillian NALMA).

Other Material. Referred material by Goin and Auffenberg (1955) includes six vertebrae (MCZ 2283) from the type locality.

Diagnosis. This diagnosis is directly from Goin and Auffenberg (1955, p. 509): "A *Pseudobranchus* in which the neural arch stands high on the centrum. It differs from *robustus* in the less concave zygapophyseal ridges as seen from above and in the reduced angle between the aliform processes. From the Recent species it differs in having a higher neural arch."

Description of the Holotype. The following description is modified from Goin and Auffenberg (1955). The centrum is longer than it is high. The cotyles are round. The centrum bears an elevated, ridgelike, medial ventral keel on either side of which is a relatively large subcentral foramen. The margin of the ventral keel is strongly concave. The total length of the neural arch is greater than the length of the centrum. The neural canal is pentagonal and is wider than it is high. It is roughly rounded posteriorly and has a very low epipophyseal ridge on its floor.

The prezygapophyseal articular facets are oval and longer than they are wide. They are directed more anteriorly than laterally. The postzygapophyseal articular facets are ovoid. Zygapophyseal ridges are well developed and are markedly concave in dorsal view. In lateral view, the zygapophyseal ridge forms a very shallow V with the apex at the point where the dorsal portion of the transverse process meets the zygapophyseal ridge.

Aliform processes are well developed and vertical in their position. In lateral view they appear somewhat rectangular. In dorsal view they form an anteriorly pointing V. The floor between the aliform processes is present and has a concave posterior margin as seen from above.

Transverse processes are well developed and composed of two platelike portions, of which the ventral portion is larger than the dorsal one. The ventral portion forms a wing-like structure that extends from the anterior margin of the side of the centrum for about two-thirds of the length of the centrum. The dorsal portion of these processes is a flat plate that extends from a point on the zygapophyseal ridge a little posterior to the posterior margin of the prezygapophyses ventrally, and then it extends posteriorly to the posterior margin of the ventral portion to which it is fused. The posterior margin of the transverse process is not quite perpendicular to the long axis of the centrum. Laterally, a foramen is present in the angle between the dorsal and ventral portions of the transverse process, and another foramen lies somewhat posterior and ventral to the angle between the dorsal portion of the transverse process and the zygapophyseal ridge.

Other Material. The six vertebrae referred to *Pseudobranchus vetustus* by Goin and Auffenberg (1955) appeared to them to be consistent with the specific characters of the holotype. The flare of the aliform processes is similar to that of the holotype, and the zygapophyseal ridges exhibit about the same degree of concavity as seen in the type specimen in dorsal view, except in two vertebrae that are apparently from the anterior part of the body. In those two vertebrae the sides are straighter than in the holotype. The prezygapophyseal articular facets are similar to those of the type specimen. In three of the five vertebrae that have the postzygapophyseal articular facets intact, they are somewhat narrower and longer than in the holotype. The point of fusion between the dorsal portion of the transverse process and the zygapophyseal ridge is similar to that of the holotype in five of the six referred vertebrae. In the other specimen, a vertebra from a more anterior part of the column, the transverse process joins the zygapophyseal ridge near the base of the prezygapophysis.

The degree of concavity of the medial subventral keel is similar to that of the holotype in three vertebrae, but in two it is more nearly straight. In two of the six vertebrae, the aliform processes are not as high as in the holotype and the other four referred specimens. In one referred vertebra with a complete neural spine, that spine is highest anteriorly, and the anterior margin of this spine is almost vertical.

General Comments. Some North American Hemphillian herpetofaunas, especially ones from the late Hemphillian NALMA, contain fossils of species that are currently living (e.g., Holman, 2000a, 2003). As in the case of *Pseudobranchus robustus*, I would suggest that *Pseudobranchus vetustus* at least be compared with both species of living *Pseudobranchus*.

Genus *Siren* Linnaeus, 1766

Sirens

Genotype. Siren lacertina Linnaeus, 1766.

Modern salamanders of the genus *Siren* belong to one of two species. The Lesser Siren, *Siren intermedia* Barnes, 1826, occurs today mainly in the southeastern states and Gulf Coast region and then, in the central part of its range, north to Illinois, Indiana, and extreme southwestern Michigan. In the western part of its range it extends to eastern and south-central Texas. Two subspecies, an eastern form and a western one, are recognized. *Siren intermedia* ranges to about 686 mm in length.

The Greater Siren, *Siren lacertina*, occurs only in the southeastern coastal plain of the United States, from Maryland to Alabama. Subspecies of the Greater Siren are not currently recognized, but two distinct allopatric (separated) forms of *S. lacertina* are known: one from south Texas and adjacent Mexico and another from peninsular Florida (Crother, 2000). As far as I am aware, only the Greater Siren is known as a fossil. Characters separating both living and fossil *Siren* from *Pseudobranchus* were given in the preceding *Pseudobranchus* account.

**Siren dunni* Goin and Auffenberg, 1957

Figs. 30A–30D

Holotype. A dorsal (trunk) vertebra (YPM 3873, Yale Peabody Museum, Yale University), collected by O. C. Marsh, August 13–14, 1873.

Type Locality. Cottonwood White Layer, Henry's Fork, Bridger Basin, Sweetwater County, Wyoming.

Horizon. Middle Eocene (Bridgerian NALMA) (*Orohippus* Faunal Zone).

Other Material. In addition to the holotype, two other vertebrae

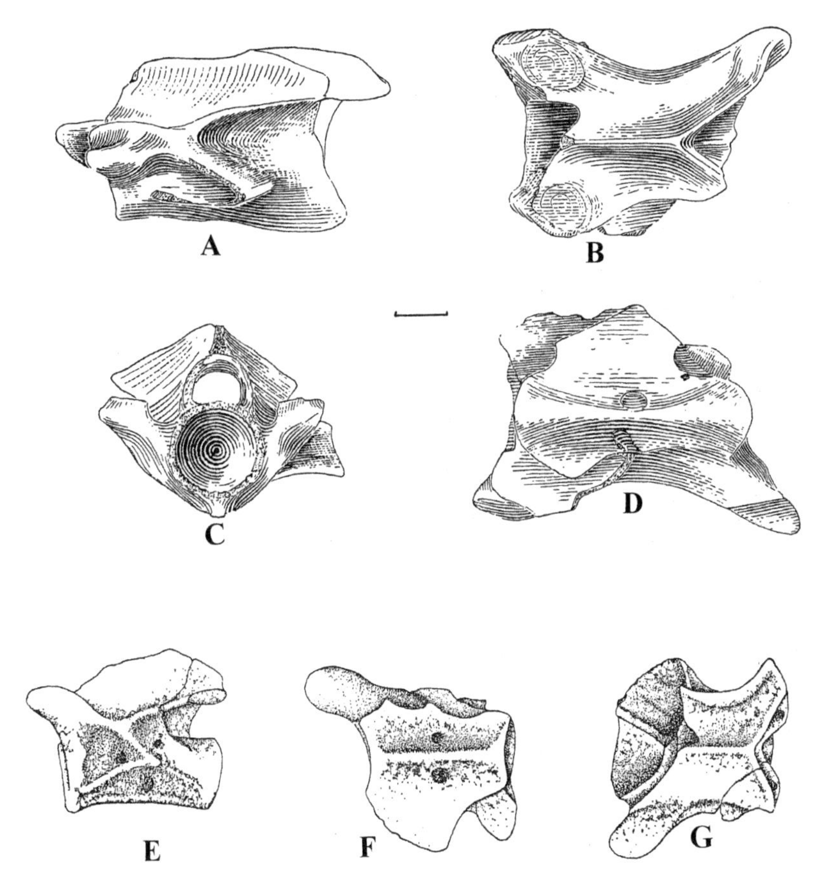

FIGURE 30. Holotype trunk vertebrae of extinct species of *Siren*. **Siren dunni* from the Middle Eocene of Wyoming in (A) lateral, (B) dorsal, (C) anterior, and (D) ventral views. **Siren hesterna* from the Early Miocene of Florida in (E) lateral, (F) ventral, and (G) dorsal views. Scale bars = 1 mm for both groups.

(YPM 3874, 3875) are referred to this species. The first (YPM 3874) is from Dry Creek, Bridger Basin, Sweetwater County, Wyoming, and was collected by "Lamonthe and Chew"; the other (YPM 3875) is from the type locality and was collected by O. C. Marsh on September 15, 1871.

Diagnosis. The diagnosis is directly from Goin and Auffenberg (1957, pp. 83–84): "A small *Siren* with the neural arch standing high above the centrum and with the zygapophysial [zygapophyseal] ridge nearly straight as seen from the side. From the Recent species it differs in having the zygapophysial [zygapophyseal] ridges more concave as seen from above and in the nearly straight horizontal position of the zygapophysial [zygapophyseal] ridges as seen from the side. From *S. hesterna* of the Miocene of Florida it differs in the reduced angle between the aliform processes and in having a better developed floor between the aliform processes. From *S. simpsoni* of the Pliocene [= Late Miocene] of Florida it differs in that the dorsal wing of the transverse process originates near the posterior margin of the centrum and swings up gradually to meet the zygapophysial [zygapophyseal] ridge at an angle of about 40 degrees whereas in *simpsoni* the dorsal wing originates well forward of the posterior margin of the centrum and rises abruptly, meeting the zygapophyseal ridge at an angle of about 60 degrees."

Description of the Holotype. The following description is modified from Goin and Auffenberg (1957). The centrum is longer than it is high. The anterior cotyle is somewhat eroded but appears to be oval and wider than it is high. The posterior cotyle is also oval and wider than it is high. The centrum has an elevated, ridgelike median ventral keel. On either side of this structure is a relatively large subcentral foramen. The margin of the ventral keel in lateral view is nearly straight.

The total length of the neural arch is greater than the length of the centrum, and its width at the narrowest part of the zygapophyseal ridges is slightly greater than the width of the centrum. The neural canal is stirrup shaped, both anteriorly and posteriorly. There is no medial ridge on the floor of the neural canal. The prezygapophyseal articular facets are broken on the left side and damaged on the right side. These facets are ovoid, longer than wide, and directed more anteriorly than laterally. The postzygapophyseal articular facets are broken off on the left side and are ovoid on the right side. The zygapophyseal ridges are well developed and markedly concave in dorsal view. In lateral view, the zygapophyseal ridge is nearly straight, but it is slightly depressed where it joins the dorsal portion of the transverse process.

The aliform processes are well developed and vertical in position. They appear somewhat rectangular in lateral view. In dorsal view the aliform processes form an anteriorly pointing V. The floor between the aliform processes is eroded, so its full extent is not visible, but it extends at least two-thirds of the distance to the posterior margins of these processes. The neural spine is well developed, but its dorsal margin is eroded.

The transverse processes are well developed and are composed of two plate-like portions, of which the ventral portion is larger than the dorsal portion. The ventral portion is a wing-like structure that extends from close to the anterior margin of the side of the centrum for about four-fifths of the length of the centrum. The tip of this ventral portion is broken to the extent that its transverse extent is not determinable. The dorsal portion of the transverse processes is a flat plate that extends from the zygapophyseal ridge, somewhat behind the posterior margin of the prezygapophyses, ventrally and posteriorly to the posterior margin of the ventral portion to which it is fused. The remnant of the posterior margin of the transverse process forms a concave curve. Laterally, a foramen occurs in the angle between the dorsal and ventral portions of the transverse process, and another foramen lies somewhat ventral and posterior to the angle between the dorsal portion of the transverse process and zygapophyseal ridge.

Other Material. The two other vertebrae referred to *Siren dunni* are somewhat fragmentary, but they do yield some information about the vertebral variation in this species. In both specimens the angle between the aliform processes is similar to that of the holotype, but in one of the referred vertebrae the aliform processes are higher, although the authors (Goin and Auffenberg, 1957) pointed out that this may be because of erosion in the specimen. In both of these vertebrae the floor between the aliform processes is more extensive than in the holotype, but this also may be due to erosion in these specimens. In one specimen the zygapophyseal ridges in lateral view are slightly sinuous, rather than nearly straight as in the holotype. The other referred vertebra is similar to the holotype in this respect. In this same specimen, the lower plate of the transverse process is longer than in the holotype and extends for nearly the entire length of the centrum. The authors stated that in other characters that can be determined the referred vertebrae are essentially the same as the type specimen.

General Remarks. Estes (1981) commented that *Siren dunni* differs from the extinct sirenid genus *Habrosaurus dilatus* from the Cretaceous and Paleocene and resembles Miocene and later sirenids in lacking discrete basapophyses. Estes also pointed out that this is a range extension for *Siren*, which currently occurs in eastern and coastal regions of the United States, south into northeastern Mexico, and in the Mississippi River drainage north to Lake Michigan.

**Siren hesterna* Goin and Auffenberg, 1955

Figs. 30E–30G

Holotype. A posterior thoracic (trunk) vertebrae (MCZ 2278, Museum of Comparative Zoology, Harvard University).

Type Locality. Thomas Farm Local Fauna, Gilchrist County, Florida.

Horizon. Early Miocene (Hemingfordian NALMA).

Other Material. No other material of this species is known.

Diagnosis. The diagnosis is directly from Goin and Auffenberg (1955, p. 504): "A small *Siren* with strongly diverging zygapophyses, with a high neural arch, and with a very wide angle (123 degrees) between the aliform processes. The forward sweeping margin of the transverse process and the widely diverging aliform processes serve to distinguish it from *lacertina, intermedia* and *simpsoni.*"

Description of the Holotype. This description is modified from Goin and Auffenberg (1955). The centrum is longer than it is high. The posterior cotyle is ovoid. The centrum has a median ventral keel, on either side of which is found a large subcentral foramen. The margin of the ventral keel is nearly straight. The total length of the neural arch is greater than the length of the centrum, and its width at the narrowest portion of the zygapophyseal ridges is slightly greater than the width of the centrum. The neural canal is somewhat rounded posteriorly, and it lacks a well-developed median epipophyseal ridge on its floor.

The prezygapophyseal articular facets are oval. They are longer than they are wide, and they are directed more anteriorly than laterally. The postzygapophyseal articular facets are broken. Zygapophyseal ridges are well developed and are moderately concave in dorsal view. In lateral view the zygapophyseal ridge is nearly straight, but it slants downward anteriorly, meeting the transverse process near the base of the prezygapophysis. The aliform processes are well developed. They are vertical in position, and in lateral view they are somewhat rectangular. In dorsal view they form an anteriorly pointed V. A floor is present between the aliform process, but the posterior margin of this floor is eroded. The neural spine is well developed, but its margin is eroded to the extent that its form cannot be determined.

The transverse processes are well developed and composed of two plate-like portions, of which the ventral portion appears to be larger than the dorsal one. The ventral portion is a wing-like structure that extends from the anterior margin of the side of the centrum for about three-fourths the length of the centrum. The dorsal portion is a flat plate that extends from the posterior margin of the prezygapophysis downward and backward to the ventral portion that it fuses with. As well as can be determined, the transverse process slants upward from the horizontal axis. A laterally placed foramen is present in the angle between the dorsal and ventral portions of the transverse process, and another foramen lies between the dorsal portion of the transverse process and the zygapophyseal ridge.

General Comments. Estes (1981) pointed out that *Siren hesterna* strikingly differs from other sirens in the very wide angle between the aliform processes and in the anterolateral orientation of the posterior border of the transverse processes. He mentioned that other sirens from the Miocene of Texas differ from *S. hesterna* and that by comparison with other fossil sirenids, *S. hesterna* probably deserves generic recognition. On the other hand, this species is based on a single vertebra whose condition is possibly pathological. As far as I can ascertain, no other material of *S. hesterna* is available.

Siren lacertina Linnaeus, 1826

Greater Siren

Figs. 31A–31D

Fossil Localities. **Pleistocene (Rancholabrean NALMA):** Arredondo site, Alachua County, Florida—Goin and Auffenberg (1955), Lynch (1965), Holman (1995a). Haile (Rancholabrean complex) site, Alachua County, Florida—Goin and Auffenberg (1955). Kanapaha I site, Alachua County, Florida—Goin and Auffenberg (1955). Reddick I site, Marion County, Florida—Gut and Ray (1963), Holman (1995a). Vero Beach, strata 2 and 3, Indian River County, Florida—Hay (1917), Weigel (1962), Holman (1995a). Wall Pit, Alachua County, Florida—Goin and Auffenberg (1955).

Siren lacertina is the largest living member of the family Sirenidae and appears so eel-like that people have mistakenly eaten them for eels—or at least have taken one bite of them before turning green. If these folks had looked at them more closely they would have found two small, four-toed front legs; so close to the head and gill region that they might not be noticed. There are no hind legs in sirens. Greater Sirens normally range from 510 to 760 mm long, with the record being 978 mm (about 38.5 inches) long (Conant and Collins, 1998). The general background color of these animals is olive to grayish, with the back being darker than the sides. Sometimes black spots occur on top of the head or on the back and sides.

These animals live in a variety of mainly still or slow moving shallow bodies of water and often lurk in thick mats of vegetation. They may make a yelping sound when grabbed by a human or other predator. Greater Sirens are true filter feeders and gulp down a variety of freshwater animals, such as crayfish, worms, mollusks, and minnows; so their digestive tracts may be rather full of aquatic vegetation.

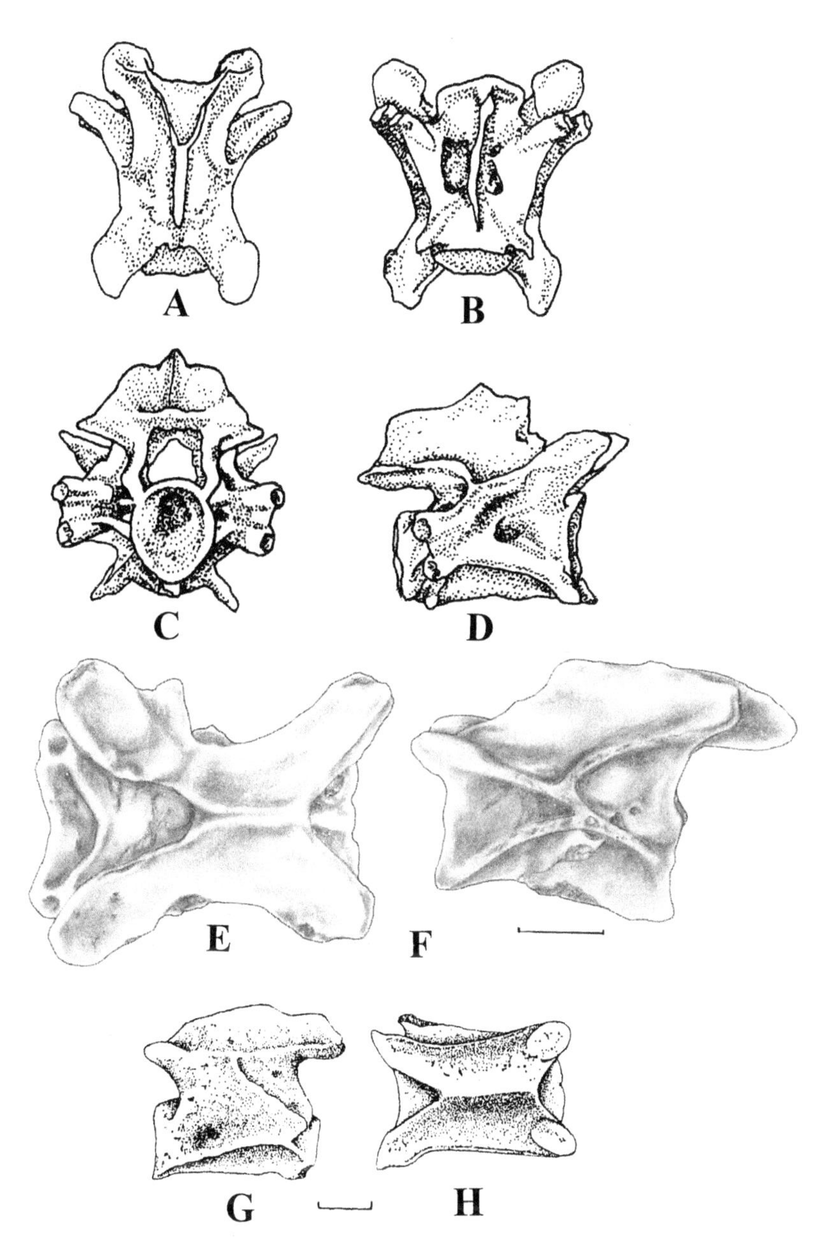

FIGURE 31. Trunk vertebra of modern *Siren lacertina* in (A) dorsal, (B) ventral, (C) posterior, and (D) lateral views. Trunk vertebra of **Siren miotexana* from the Middle Miocene of Texas in (E) dorsal and (F) lateral views. Trunk vertebra of **Siren simpsoni* from the Late Miocene of Florida in (G) lateral and (H) dorsal views. Scale bars = 1 mm and apply to (E)–(H). Scale bar not available for (A)–(D).

Identification of Fossils. Very large (more than 10 mm in greatest length) vertebrae with aliform processes and a mainly straight, rather than a deeply concave, lower border of the centrum probably represent *Siren lacertina.*

General Remarks. People that have kept *Siren lacertina* in aquaria have found them to be very interesting pets. One woman wrote me several letters about a large Greater Siren that she had kept for several years. She once sent a picture of her pet pushing a ball around the bottom of her aquarium and reported that the animal often engaged in that behavior.

**Siren miotexana* Holman, 1977a

Figs. 31E, 31F

Holotype. A trunk vertebra (SMPSMU 63673, Shuler Museum of Paleontology, Southern Methodist University).

Type Locality. Moscow Local Fauna, Fleming Formation, near the town of Moscow, northern Polk County, Texas.

Horizon. Early Middle Miocene (early Barstovian NALMA).

Other Material. Referred material consisted of a trunk vertebra (SMPSMU 63674) from the Town Bluff site in east-central Tyler County, Texas (Miocene: early Barstovian NALMA); and four fragmentary vertebrae (SMPSMU 63675) from the Trinity River site near Coldspring in western San Jacinto County, Texas (Miocene: early Barstovian NALMA). SMPSMU 61869 from the Trinity River site near Coldspring (Holman, 1966a) was originally designated as *Siren* sp. but is here recognized as *Siren miotexana.* The "four fragmentary trunk vertebrae" of *S. miotexana* listed as SMPSMU 67374 by Holman (1996b, p. 6) are actually the same specimens as SMPSMU 63675 above.

Diagnosis. The diagnosis is taken directly from Holman (1977a, p. 392): "A small *Siren* that differs from *S. hesterna* Goin and Auffenberg from the Middle Miocene of Florida in having the angle between the aliform processes 70 degrees (123 degrees in *S. hesterna*). It differs from *S. simpsoni* Goin and Auffenberg from the middle Pliocene of Florida in having (1) the angle between the aliform processes wider (63 degrees in *S. simpsoni*), (2) a curved zygosphenal ridge [actually zygapophyseal ridge] (straight in *S. simpsoni*), and (3) a more depressed neural arch. It differs from Holocene *S. lacertina* Linnaeus in being (1) much smaller, and (2) in having a curved zygosphenal ridge [actually zygapophyseal ridge] (straight in *S. lacertina*). It differs from Holocene *S. intermedia* LeConte in having (1) the centrum straight in lateral view (curved in *S. intermedia*), and (2) with a wider angle between the aliform processes (45 degrees in *S. intermedia*)."

Description of the Holotype. This description is slightly modified from Holman (1977a). In anterior view, the cotyle is round, and the neural canal is subrounded. The neural arch has its anterior portion broken off but is moderately vaulted. The prezygapophyses are moderately tilted upward. The transverse processes are broken. In dorsal view, the vertebra is longer than it is wide. The prezygapophyseal articular facets are ovoid. The aliform processes are low, but they are distinct; they form an angle with one another of 70 degrees. The vertebra is constricted medially to about the width of the centrum.

In lateral view, the zygapophyseal ridge is curved downward, joining the transverse process at the base of the prezygapophyses. The posterior part of the neural arch is upswept. The neural spine is broken. The centrum is straight. In posterior view, the neural arch is moderately vaulted. The cotyle is round and slightly larger than the subrounded neural canal. The left postzygapophyses is broken. In ventral view, the bottom of the centrum is keeled. There is a large foramen on either side of the keel. The left prezygapophysis is broken, but the right one has an ovoid articular face.

Other Material. There is surprisingly little variation in the referred material. The angle between the aliform processes in the vertebra from the Town Bluff site (SMPSMU 63674) is 70 degrees, as in the holotype. This angle can be measured in only one of the four fragmentary vertebrae from the Trinity River site, but in that vertebra the angle is 72 degrees.

General Remarks. The fossil vertebrae were referred to the genus *Siren* rather than *Pseudobranchus* on the basis of the larger size of the vertebrae of *Siren miotexana,* as well as a character given by Goin and Auffenberg (1955), who stated that in *Siren* the lower margin of the centrum is nearly straight when seen from the side, whereas in *Pseudobranchus* it is quite concave. Goin and Auffenberg also stated that the sharp downward curve of the zygapophyseal ridge into the transverse processes in *Pseudobranchus* separates *Pseudobranchus* from *Siren,* in which this border is straight. In modern sirenid skeletons examined by the author, *Siren lacertina* appears to be rather constant in this character, whereas *Siren intermedia* and a large series of *Pseudobranchus striatus* exhibit intracolumnar variation in this character.

**SIREN SIMPSONI* GOIN AND AUFFENBERG, 1955

FIGS. 31G, 31H

Holotype. A thoracic (trunk) vertebra (MCZ 2284, Museum of Comparative Zoology, Harvard University).

Type Locality. Haile Pit VI, a little south of the village of Haile, Alachua County, Florida.

Horizon. Late Miocene (late Hemphillian NALMA).

Other Material. Referred material includes five thoracic (trunk) vertebrae and a second cervical vertebra (first post-atlantal vertebra) (MCZ 2285) from the type locality.

Diagnosis. The diagnosis is directly taken from Goin and Auffenberg (1955, p. 500): "A small *Siren* with the neural arch standing high above the centrum, a nearly straight zygapophyseal ridge as seen from the side and rather wide-flaring aliform process. It can be distinguished from the Miocene species described below by the straighter zygapophyseal ridge as seen from the side and the smaller angle of the aliform process. From *lacertina* and *intermedia* it can be distinguished by straighter zygapophyseal ridges as seen from the side and by the wide flare of the aliform processes."

Description of the Holotype. This description is modified from Goin and Auffenberg (1955). The lower half of the centrum is broken off, but the upper portion is mainly well preserved. The shape of the ventral keel cannot be determined, because of the breakage in the area. The total length of the neural arch is greater than the length of the centrum, and its width at the narrowest portion of the zygapophyseal ridges is slightly greater than the width of the centrum. The neural canal is broken anteriorly, is nearly rounded posteriorly, and has a very low median epipophyseal ridge on its floor.

The prezygapophyseal articular facets are broken off, and the postzygapophyseal articular facets are ovoid. The zygapophyseal ridges are well developed and markedly concave in dorsal view. In lateral view, the zygapophyseal ridge is nearly straight and continues forward to near the base of the damaged prezygapophyses. The aliform processes are well developed, oriented vertically, and somewhat rectangular when seen from the side. In dorsal view, they form an anteriorly pointing V. The floor between the aliform processes has a nearly straight posterior margin. The neural spine is well developed posteriorly but is broken off anteriorly.

The transverse processes are well developed and composed of two plate-like portions. The ventral portion is a wing-like structure that extends from near the anterior margin of the side of the centrum for about three-fourths the length of the centrum. The dorsal portion is a flat plate that extends from the zygapophyseal ridge somewhat behind the posterior margin of the ventral portion to which it is fused. The posterior margin of the transverse process slants posteriorly. Laterally, a foramen is present in the angle between the dorsal and ventral portions of the transverse process, and another lies somewhat ventral and posterior to the angle between the dorsal portion of the transverse process and the zygapophyseal ridge.

Other Material. The fragmentary vertebrae that were referred to *Siren simpsoni* give some idea about the variation in this species. Only two of the body vertebrae have the neural canal complete, and in both, this canal is pentagonal. In the specimen that has adequate zygapophyseal ridges, these ridges are not quite as concave in dorsal view as in the holotype. In the single specimen in which the aliform processes are discernable, it has them at about the same angle as in the type. The upper portion of the transverse process meets the zygapophyseal ridge in about the same place as in the holotype in two specimens and slightly posterior to that point in two others.

The second cervical vertebra (vertebra just posterior to the atlas) differs from those of modern *Siren* in several ways. The neural arch rises much higher above the centrum, and the total height of the vertebra is about one-fourth more than the length of the centrum, whereas in the Recent species, the height of the vertebra is about equal to the length of the centrum. This difference is reflected in the anterior part of the vertebra, which in *Siren simpsoni* is higher than it is wide and in *Siren lacertina* is wider than it is high. *Siren simpsoni* also has much shorter aliform processes than do the Recent species. In this vertebra in *S. simpsoni*, each aliform process is shorter than the neural spine, whereas in *S. lacertina* the neural spine is shorter than either of the aliform processes.

Siren sp. indet.

Two records of *Siren* unidentified to the species level are now present in the literature (recall that "*Siren* sp. indet." from the Miocene of Texas [Holman, 1966a] has been included with *Siren miotexana* above). *Siren* sp. indet. was reported from the Norden Bridge Quarry of Brown County, Nebraska (middle Miocene: medial Barstovian NALMA) by Holman and Voorhies (1985). *Siren* sp. indet. was also reported from the Leisey Shell Pit Fauna of Hillsborough County, Florida (early Pleistocene: Irvingtonian NALMA) by Hulbert and Morgan (1989). The Hulbert and Morgan record is not unexpected. On the other hand, the somewhat fragmentary specimens from north-central Nebraska nicely fill in the gap between Texas and Wyoming in the fossil record of this genus.

SUMMARY REMARKS ON SIRENIDAE

The fossil record is incomplete in the Sirenidae (as is true in all of salamanders), but it is much more complete than in several other salamander families to follow. Figure 32 indicates important differences in the skulls of *Siren* (Figs. 32A–32D), *Pseudobranchus* (Figs. 32E–32H), and #*Habrosaurus* (Figs. 32I–32L). Here, among other differences, one should note the replacement of the mass of bulbous marginal and palatal teeth in *Habrosaurus* by a horny beak with many fewer teeth in *Siren* and *Pseudobranchus*. This was interpreted by Gardner (2003a) as representing convergent strategies for achieving a crushing bite. Gardner proposed *Habrosaurus* as the most basal sirenid but argued that this extinct genus is not directly ancestral to either *Siren* or *Pseudobranchus*.

Following is a summary list of the described North American species of the Sirenidae from oldest to youngest. I have followed Gardner (2003a) for the geological age ranges in *Habrosaurus*.

#*Habrosaurus prodilatus*—Late Cretaceous (middle Campanian) of Alberta.
#*Habrosaurus dilatus*—Much later Late Cretaceous (late Maastrichtian) to the Middle Paleocene (Torrejonian) of Saskatchewan and Wyoming.
**Siren dunni*—Middle Eocene of Wyoming.
**Siren hesterna*—Early Miocene of Florida.
**Siren miotexana*—Middle Miocene of Texas.
**Siren simpsoni*—Late Miocene of Florida.
Siren intermedia—No fossil record—probably derived from one of the Miocene fossil *Siren*, all of which were small.
Siren lacertina—Late Pleistocene of Florida—probably derived from one of the Miocene fossil *Siren*.
**Pseudobranchus vetustus*—Late Miocene of Florida—possibly derived from *Siren hesterna* or *Siren miotexana*.
**Pseudobranchus robustus*—Late Pleistocene of Florida—possibly derived as a robust form of *Pseudobranchus vetustus*.
Pseudobranchus striatus—No fossil record—probably just a less robust form of *Pseudobranchus robustus*.
Pseudobranchus axanthus—No fossil record—probably just a less robust form of *Pseudobranchus robustus*.

It is a pity that so many species of *Siren* and *Pseudobranchus* had to be described on the basis of vertebrae, many of them more or less fragmental. Why the sirenids withdrew from the west during their occupation of North America is not precisely known. It may have had to do with the deterioration of the climate toward the end of the Eocene.

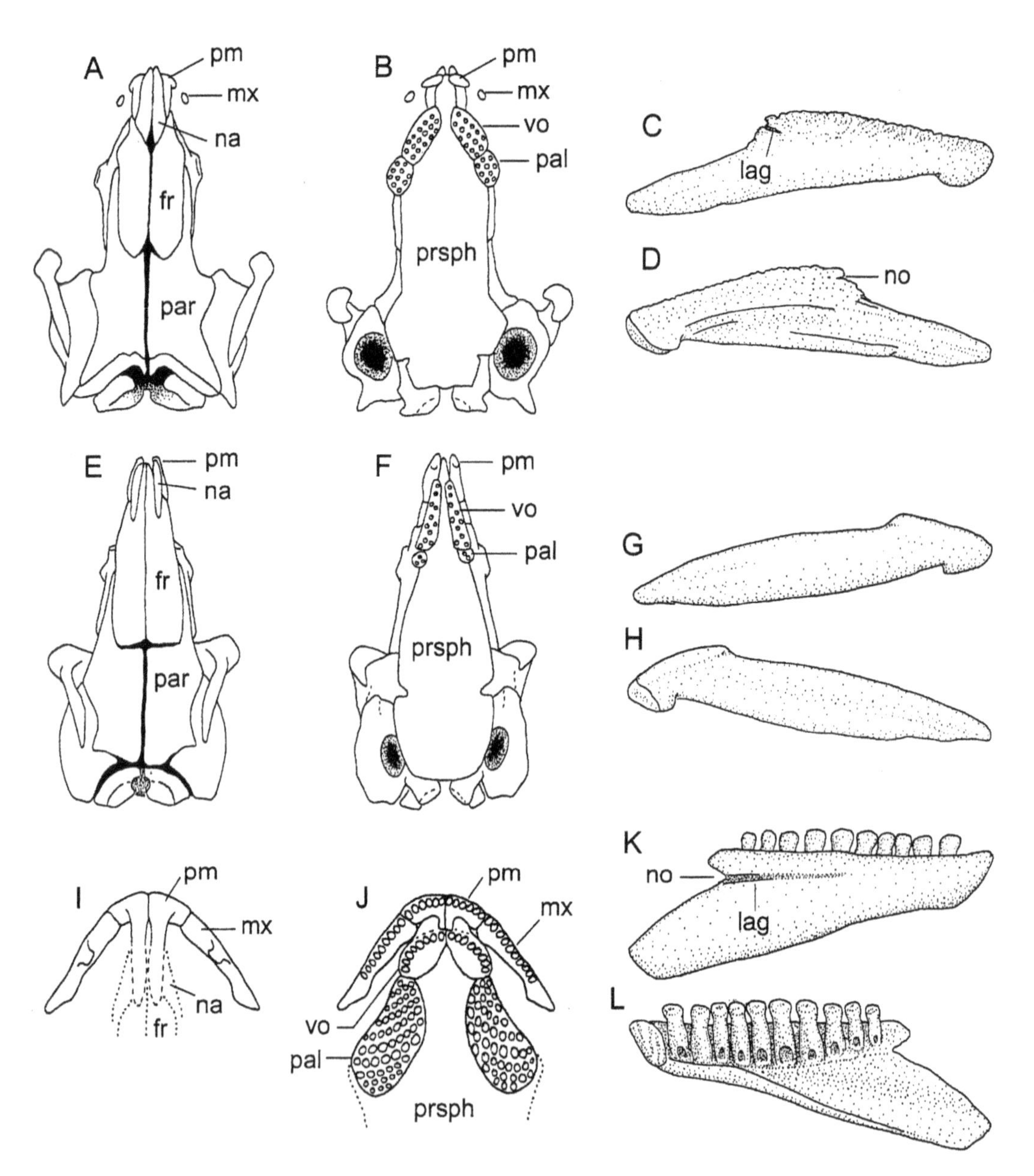

FIGURE 32. Idealized skulls and dentaries of Recent and fossil sirenid genera. *Siren:* (A) dorsal and (B) ventral views of skull; (C) labial and (D) lingual views of right dentary. *Pseudobranchus:* (E) dorsal and (F) ventral views of skull; (G) labial and (H) lingual views of right dentary. #*Habrosaurus:* (I) dorsal and (J) ventral views of anterior portion of skull; (K) labial and (L) lingual views of right dentary. Abbreviations: fr, frontal; lag, labial groove; mx, maxilla; na, nasal; no, notch; pal, palatine; par, parietal; pm, premaxilla; prsph, parasphenoid; vo, vomer.

AN EARLY EOCENE SIRENID WAITING TO BE RESTUDIED

#*PALEOAMPHIUMA TETRADACTYLUM* RIEPPEL AND GRANDE, 1998

Holotype. A somewhat complete skeleton (FMNH PR-1810, Field Museum of Natural History).

Type Locality. Locality H, Fossil Butte Member, Green River Formation, Lincoln County, Wyoming.

Horizon. Early Eocene.

Author's Comment. Rieppel and Grande (1998) assigned this fossil to the Amphiumidae, but the vertebral column of *Paleoamphiuma tetradactylum* is clearly sirenid, rather than amphiumid. The authors mistook the aliform processes of the trunk vertebrae of *P. tetradactylum* for the amphiumid posterior zygapophyseal crest (compare sirenid vertebrae figured in this book with those of Rieppel and Grande [1998, fig. 1, p. 701, and fig. 4, p. 704]). The sirenid aliform processes distinctly arise from a posterior splitting of the neural spine (neural crest of Rieppel and Grande), whereas the amphiumid posterior zygapophyseal crest arises from the dorsolateral part of the neural arch. Moreover, the characters that supposedly separate *P. tetradactylum* from "other amphiumids"—that is, (1) the posterior rib-bearers being directed posteriorly and (2) the presence of an anterior flange on the rib-bearers—are strong characters of the Sirenidae. Other characters of the hyobranchial and limb skeleton of *P. tetradactylum*, which will not be discussed here, are clearly not amphiumid. The anticipated re-study of *P. tetradactylum* will no doubt elucidate the relationships of this form with other taxa of the Sirenidae.

FAMILY CRYPTOBRANCHIDAE FITZINGER, 1826

CRYPTOBRANCHIDS

Cryptobranchid salamanders are large to giant salamanders that have a nondescript, flat, dull-colored, wrinkled body form. Hellbender is the official common name for our living North American species, *Cryptobranchus alleganiensis.* It is said the name arose from someone who once remarked that "anything that ugly is surely bent for hell." I know of no other postulated sources for that odd name.

Cryptobranchids are thought to have been derived from the Old World family Hynobiidae by developing an incomplete metamorphosis. At present, most students of salamanders recognize two genera of cryptobranchids: the Giant Salamander *Andrias,* which currently occurs in China and Japan; and the Hellbender (*Cryptobranchus*), which occurs in North America. Milner (2000) considered the two genera as forming a well-defined clade, with *Cryptobranchus* being the more derived form and *Andrias* originating in the Cretaceous from some unknown "cryptobranchoid" ancestor in East Asia. In North America, *Andrias* appeared in the Late Paleocene (probably arriving from Asia via the Bering land bridge) and lasted to the Middle Miocene. Later cryptobranchid genera have all been identified as *Cryptobranchus. Andrias* is probably the largest salamander that ever lived. Living *Andrias* reaches a total length of 1600 mm, but fossil forms of this genus go up to 2300 mm (Estes, 1981).

Some osteological characters that define the group composed of the

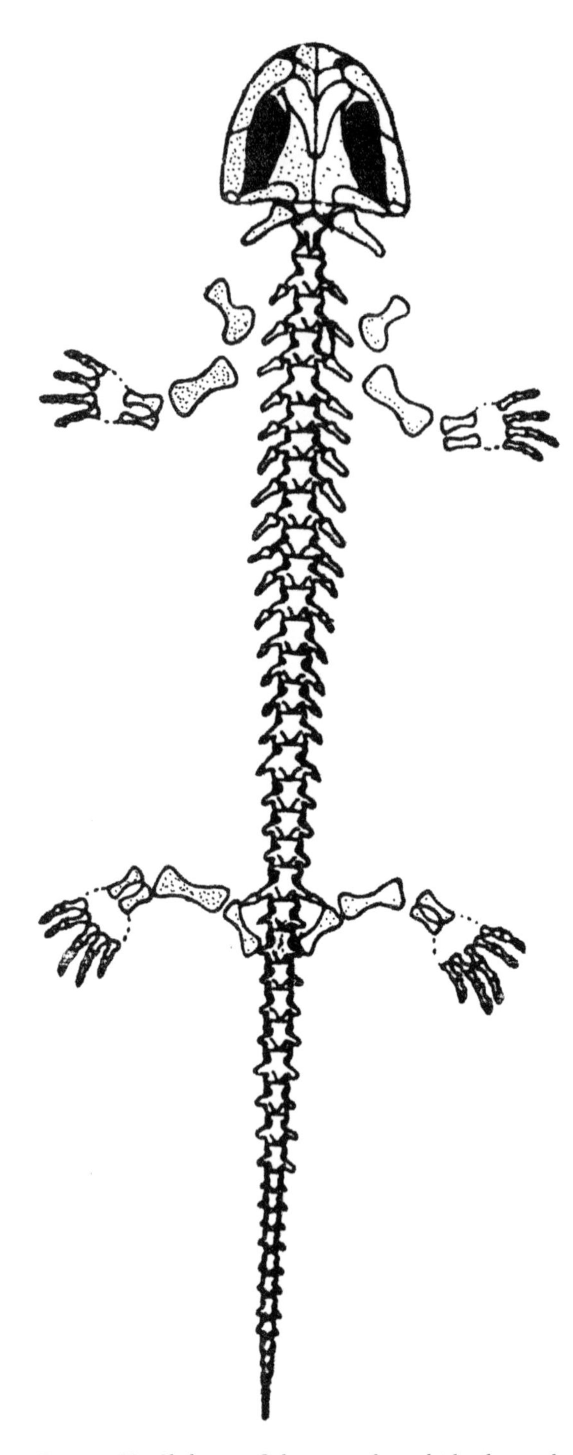

FIGURE 33. Skeleton of the cryptobranchid salamander *Andrias scheuchzeri* from the Miocene of Germany. Modified from Huene (1956). John J. Scheuchzer published on this fossil in Latin in the early 18th century and named it *Homo diluvii testis,* believing it to be a human that was a witness to the biblical flood.

living Hynobiidae and Cryptobranchidae indicate the closeness of the two families. These characters are modified from Duellman and Trueb (1986) as follows. The dorsal processes of the premaxillary bones are short and do not separate the nasals. The nasals ossify separately from two anlage (developmental centers). The angular does not fuse with the prearticular. The second ceratobranchial and the ypsiloid cartilage are both present. The vertebrae are amphicoelous. The spinal nerves of only the posterior caudal vertebrae exit intervertebrally. The ribs are single headed. A unique character in the two families is that the first hypobranchial and first ceratobranchial are fused into a single rod.

The Hynobiidae and Cryptobranchidae themselves may be separated on the basis that (1) adults have eyelids and no gill slits in the Hynobiidae, whereas adults lack eyelids and have one pair of gill slits in the Cryptobranchidae; (2) lacrimal and septomaxillary bones are present in the Hynobiidae and absent in the Cryptobranchidae; and (3) the palatal tooth pattern is transverse and does not parallel the maxillary and premaxillary teeth in the Hynobiidae, whereas the palatal teeth are in a curved row parallel to the maxillary and premaxillary teeth in the Cryptobranchidae.

GENUS *ANDRIAS* TSCHUDI, 1837
GIANT SALAMANDERS

FIG. 33

Genotype. Andrias scheuchzeri (Holl, 1831).

Diagnosis. This generic diagnosis is modified from Estes (1981). Ossification occurs only in the second and third branchial arches. The frontal is excluded from the narial opening. The prefrontal is broad and oval and does not extend farther than the posterior tip of the maxilla. The 21st vertebra usually forms the sacrum; if not, the sacrum is formed by the 20th vertebra. The angle between the neural spine and the axis of the centrum ranges from 19 to 37 degrees.

****Andrias matthewi* (Cook, 1917)**
(*Plicagnathus matthewi* Cook, 1917; *Cryptobranchus mccalli* Tihen and Chantell, 1963)

Figs. 34, 35

Holotype. A fragmentary dentary (AMNH 8303, American Museum of Natural History).

Type Locality. Lower Snake Creek Beds, Sinclair Draw, Sioux County, Nebraska.

Horizon. Middle Miocene (early Barstovian NALMA).

Other Material. Material referred to *Andrias matthewi* by Estes (1981) includes maxillae (UCMP 37165, University of California Museum of Paleontology; UNSM 61044, University of Nebraska State Museum) and a dentary (AMNH 8361), Marsland Formation (Early Miocene: late Ari-

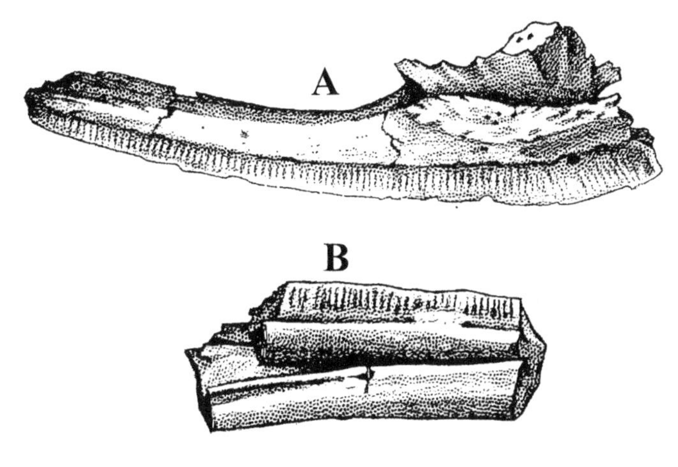

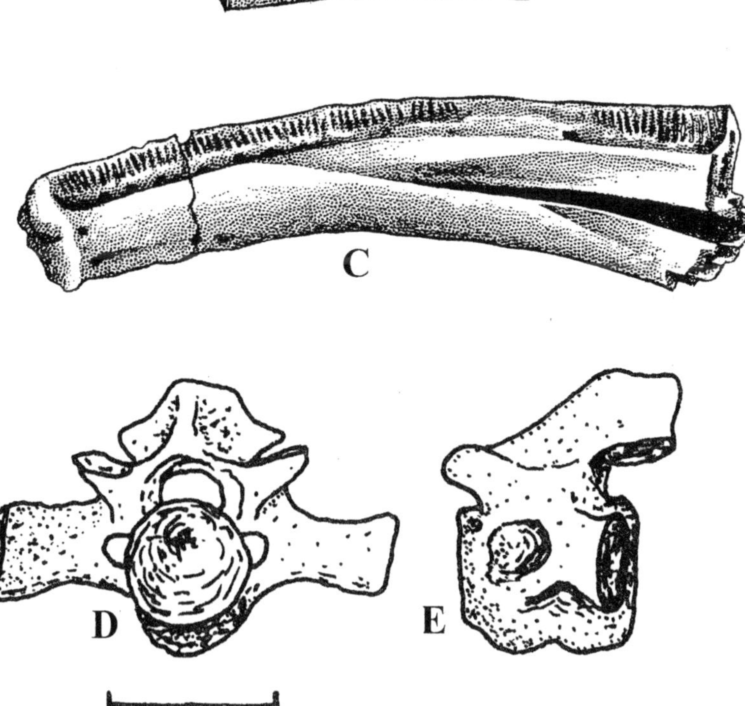

Figure 34. **Andrias matthewi* from the Miocene of North America. Lingual view of upper and lower jaw bones: (A) left maxilla, (B) left dentary, and (C) right dentary. First trunk vertebra in (D) anterior and (E) lateral views. Scale bar = 20 mm and applies to (D) and (E) only.

FIGURE 35. Artist's depiction of *Andrias matthewi* chasing a school of small fish in a river in northern Nebraska during the Middle Miocene.

kareean NALMA), Nebraska; maxillae, dentaries, vertebrae, and an exoccipital (KU 12004, Museum of Natural History, University of Kansas), Marsland Formation equivalent (Early Miocene: late Arikareean NALMA), Colorado; maxillae (UNSM 61000, 61001), Valentine Formation (Middle Miocene: medial Barstovian NALMA), Nebraska; and vertebrae (ROM 12727, Royal Ontario Museum), Wood Mountain Formation (Middle Miocene: medial Barstovian NALMA), Saskatchewan, Canada. Additional *A. matthewi* material has been recorded from the Middle Miocene of north-central Nebraska (see Voorhies, 1990).

Diagnosis. The diagnosis is directly from Estes (1981, p. 17): "Differs from other species of *Andrias* in having more slender, straight maxilla and dentary, and shallow sulcus dentalis."

Description of the Fossil Material. This description is condensed and somewhat modified from Estes (1981) and Tihen and Chantell (1963).

The dentaries, maxillae, and vertebrae represent an animal that ranged from about 430 to 1800 mm in total length (Estes, 1981). Tihen and Chantell (1963) provided a succinct description of the complete (UNSM 61000) and incomplete (UNSM 61001) maxillae from the Norden Bridge Fauna (Middle Miocene: medial Barstovian NALMA). They originally described these elements as belonging to *Cryptobranchus mccalli*, but they have subsequently been reassigned to *Andrias matthewi*. The complete maxilla has a total length of 55 mm. The height at the anterior end, excluding the teeth, is 3.6 mm. The length of the base of the ascending process is 17.3 mm. The straight-line height from the ventral border to the highest point of the ascending process is 15.5 mm. No complete teeth were present on the bone, but the bases of many of them were present, indicating that there were about 75 teeth present in life (about 13–14 teeth/cm). Tihen and Chantell reported that the maxilla was less curved than in *Cryptobranchus alleganiensis*.

The referred specimen was the anterior 8.5 mm of a left maxilla. It is thought to have represented an individual about 80% as large as the complete specimen. There were bases or spaces for 14 teeth in this specimen. In comparison, the anterior 8.5 mm of the complete specimen bore only 12 teeth.

Estes (1981) discussed the composite fossil material of *Andrias matthewi*. He reported that the dentaries, maxillae, and vertebrae from the United States indicated animals 430–1800 mm in total length. The total length of the maxillae range from 54 to 102 mm. These maxillae are not highly curved. The facial portion of the maxilla is roughly tripartite and expanded dorsally. The dental portion of the maxilla is a slender bar that is toothed along its entire surface and has a shallow dental sulcus. The posterior end of the dental portion of the maxilla bears a ridge for a ligamentary attachment to the pterygoid. The dentary bone has a shallow dental sulcus. This bone is also not highly curved. The vertebrae are amphicoelous, squarish, and relatively massive. The length of the centrum is short relative to the height of the cotyles. The rib-bearers of the vertebrae have only one head (unicapitate condition) and are dumbbell shaped. The neural spines of these vertebrae are finished in cartilage.

Naylor (1981a) reported two massive vertebrae of this species from the Middle Miocene (medial Barstovian NALMA) of Saskatchewan. These vertebrae had estimated centrum lengths of 30 and 40 mm. Naylor suggested an extrapolated total length of 2300 mm for the larger of the two vertebrae. These vertebrae may represent the largest known salamander (Estes, 1981).

General Comments. It is difficult to compare these North American disarticulated remains with the European fossils, which are more or less complete. Nevertheless, it appears that they differ from the Old World remains and from the Holocene *Andrias* in the construction of the maxilla and dentary, which indicate a somewhat different, more wedge-shaped head. Estes (1981) pointed out that whether these differences indicate a wider or a narrow skull depends on knowing the other skull bone proportions.

*Andrias saskatchewanensis (Naylor, 1981a)

Cryptobranchus saskatchewanensis Naylor, 1981a

Fig. 36

Holotype. A partial dentary (UALVP 14858, University of Alberta Laboratory for Vertebrate Palaeontology).

Type Locality. Ravenscrag Formation, southern Saskatchewan, Canada.

Horizon. Late Paleocene.

Other Material. Dentaries, maxillae, an exoccipital, and vertebrae (UALVP) are also known from the type locality. A dentary fragment from the Early Eocene Willwood Formation of Wyoming (UMMPV 71316, University of Michigan Museum of Paleontology, Vertebrate Collection) is also available.

Diagnosis. This diagnosis is taken directly from Estes (1981, p. 14): "Differs from *Cryptobranchus alleganiensis* in having a deeper flange of the dentary below the dental gutter and in being of larger size. Differs from *Andrias* spp. in having a narrower flange, and in having a shallow medial groove in the dentary that leads to or near the symphysis."

Description. This description is modified from Estes (1981) and Naylor (1981a). The symphysis of the dentary is robust but is not expanded. The tooth-bearing portion of the dentary is subequal in depth to the part of the dentary below it. The tooth bases are compressed (flattened from side to side). The dentary has a thin ventral keel that is set off from the

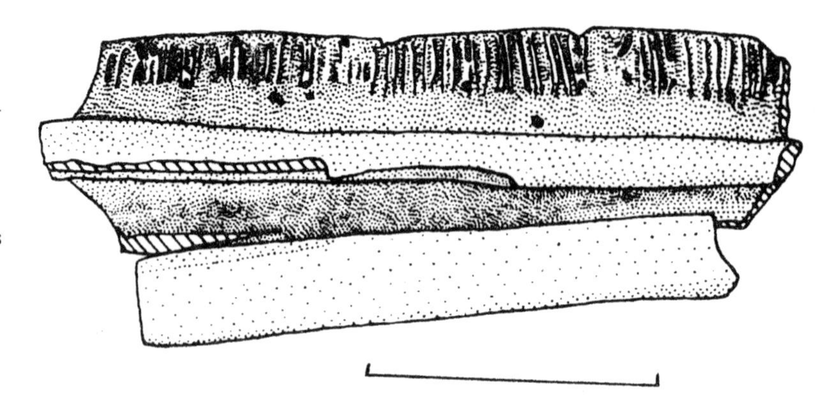

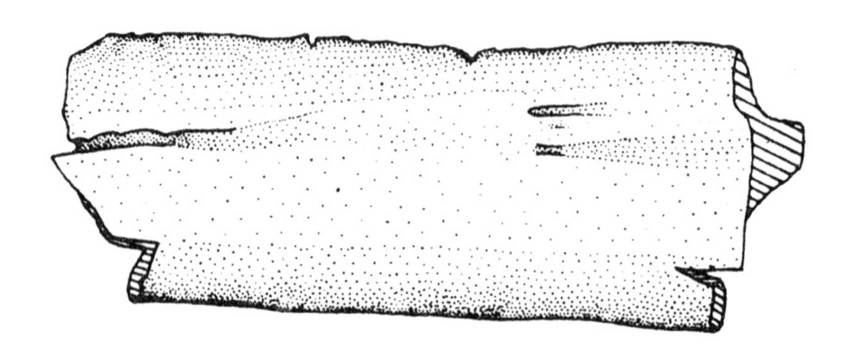

Figure 36. **Andrias saskatchewanensis* holotype dentary from the Late Paleocene of Saskatchewan. Upper, lingual view; lower, labial view. Scale bar = 10 mm and applies to both figures.

tooth-bearing portion of the bones by a thick medial ridge. The exoccipitals are typical of the family Cryptobranchidae, except that the interglenoid tubercle does not articulate with the exoccipital. The maxillae and vertebrae are poorly preserved but otherwise typical of the family.

General Remarks. This material represents the earliest cryptobranchids in North America.

Genus *Cryptobranchus* Leuckart, 1821
Hellbenders

Genotype. Cryptobranchus alleganiensis (Daudin, 1803).

The living Hellbenders are a single species, *Cryptobranchus alleganiensis*, which occurs mainly in the Appalachian and Ozark regions of the United States. It is absent from Canada and Mexico. The definitive range is from south-central New York southwest to southern Illinois, extreme northeastern Mississippi, and the northern parts of Alabama and Georgia. Disjunct populations occur in east-central Missouri and southeastern Missouri and adjacent Arkansas (Conant and Collins, 1998). The southeastern Missouri–Arkansas population has been given the subspecific name *Cryptobranchus alleganiensis bishopi.* Living North American *Cryptobranchus* may be distinguished from *Andrias* on the basis of having an open gill slit, having the frontal bone participate in the structure of the naris, and having a greater separation between the maxilla and the pterygoid. In the vertebrae, the angle between the neural spine and the centrum is 15–20 degrees in *Cryptobranchus.*

Cryptobranchus alleganiensis (Daudin, 1803)
Hellbender

Fig. 37

Fossil Localities. **Pleistocene (Rancholabrean NALMA):** Baker Bluff Cave, Sullivan County, Tennessee—Van Dam (1978), Fay (1988). Bell Cave, Colbert County, Alabama—Holman et al. (1990). Cheek Bend Cave, Maury County, Tennessee—Klippel and Parmalee (1982). Guy Wilson Cave, Sullivan County, Tennessee—Fay (written communication, 1993[1]), Holman (1995a). Saltville site, Saltville Valley, Virginia—Holman and McDonald (1986).

We have already discussed the range of this species and remarked on its uncomplimentary name. Conant and Collins (1998, p. 37) characterized the Hellbender as being "a huge, grotesque, thoroughly aquatic salamander." This animal has a flat head, and each side of the body has a large fold of wrinkled skin. The color ranges from yellowish brown to almost black. Adults have no external gills. The animals range from about 290 to 740 mm long, with the record length of males being 686 mm and females, 740 mm (Conant and Collins, 1998). They have a very slimy skin and are hard to hold once they are found. Many people that live within the range of this species think it is poisonous, but it is actually perfectly harmless. I am not aware of any of my herpetologist friends ever having been bitten by one. Hellbenders mainly live in clear shallow streams, where they hide under large flat rocks. They are known to eat crayfish, worms, and aquatic insects. In many places they are getting very

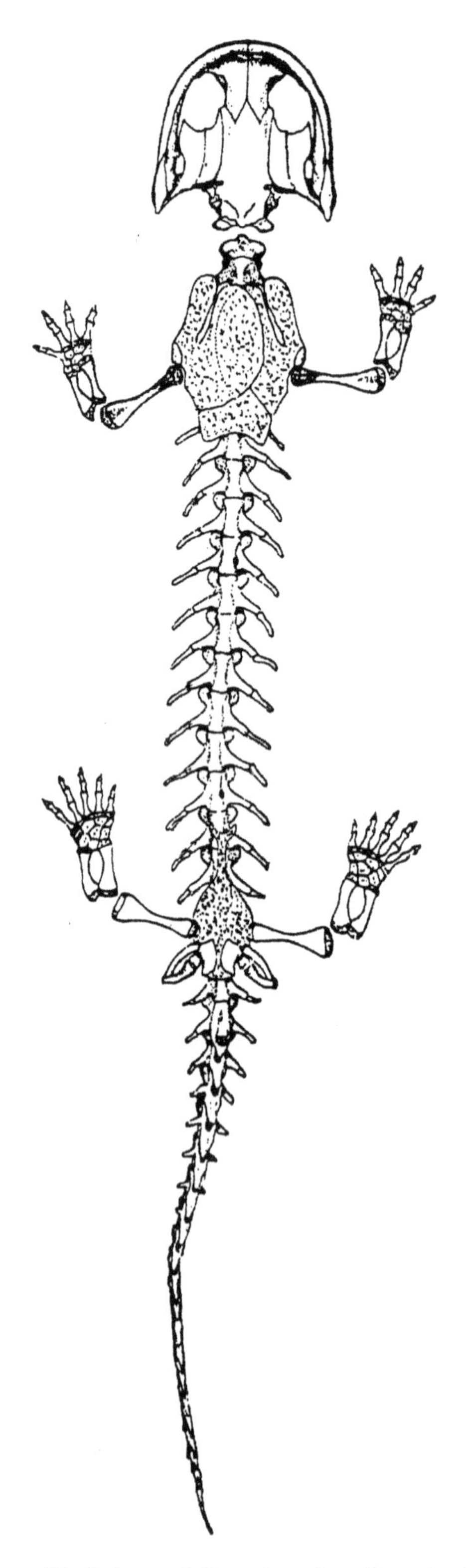

FIGURE 37. Skeleton of *Cryptobranchus alleganiensis* in dorsal view. After Cope (1889).

rare because of elimination or pollution of their stream habitats. We are privileged to have these living fossils around and should certainly provide for their welfare!

Identification of Fossils. This species is usually identified on the basis of vertebrae or pieces of maxillary or dentary bones. The large size of these elements in Hellbenders is usually the starting point for the identification process. (See the following account on how the extinct Irvingtonian Hellbender species differs from the Rancholabrean and modern ones.)

General Remarks. I would not advise anyone to keep a Hellbender in an aquarium as a pet. First, it is illegal to possess one in most states where it occurs. Second, it would need a huge tank to comfortably exist in captivity.

**CRYPTOBRANCHUS GUILDAYI* HOLMAN, 1977B

FIGS. 38, 39

Holotype. A left dentary, complete except for its anterior tip (CM 20470, Carnegie Museum of Natural History).

Type Locality. Cumberland Cave, Allegany County, Maryland.

Horizon. Pleistocene (Irvingtonian NALMA), Kansan glacial stage.

Other Material. Two right dentaries, one right epihyal, one atlas, one nearly complete and one fragmentary trunk vertebra, three fragmentary caudal vertebrae, two right femora, and one right scapula, all registered under CM 40416, were collected from Trout Cave, Pendleton County, West Virginia (Holman 1982a). This material also represents the Pleistocene (Irvingtonian NALMA), Kansan glacial stage.

Diagnosis. This emended diagnosis is directly from Holman (1982a, pp. 396–397): "(1) dentary differs from *C. alleganiensis* in having a longer Meckelian groove and in being more weakly curved; (2) epihyal bone differs from *C. alleganiensis* in having a strongly developed posterior process (this process is lacking in *C. alle-*

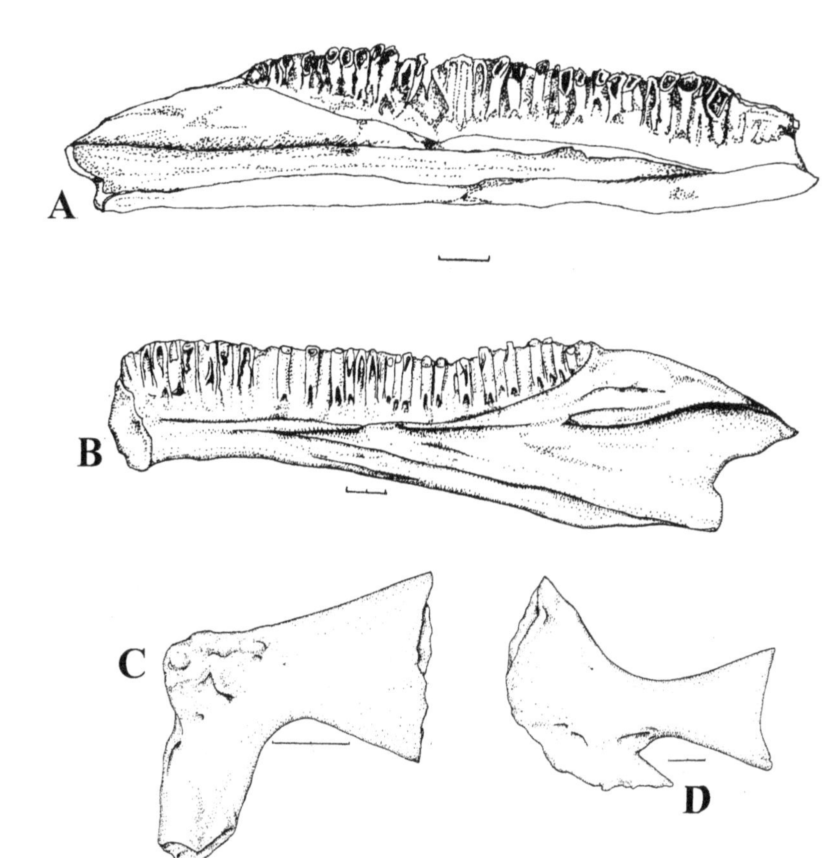

FIGURE 38. **Cryptobranchus guildayi.* (A) Holotype left dentary from the Pleistocene (Irvingtonian NALMA) of Cumberland Cave, Maryland. (B) Right dentary from the Pleistocene (Irvingtonian NALMA) of Trout Cave, West Virginia. (C) Epihyal from Trout Cave, West Virginia. (D) Scapula from Trout Cave, West Virginia. All scale bars = 2 mm.

ganiensis); (3) the available trunk vertebra [vertebrae] of *C. guildayi* is [are] shorter and wider than those of *C. alleganiensis*; (4) femur of *C. guildayi* has a much better developed and more extensive distal muscular line than in *C. alleganiensis*; (5) scapula of *C. guildayi* has its dorsal surface more rounded and with a greater angle between its posterior process and the shaft than in *C. alleganiensis*."

Description of the Holotype. This description is slightly modified from Holman (1977b). In dorsal view, the dentary is weakly curved. In lateral view, the Meckelian groove is long and occupies a distance of 34 teeth and tooth spaces. The Meckelian groove is narrowly V-shaped, is moderately deep, and has borders that are much thicker than those of *Cryptobranchus alleganiensis.* A total of at least 41 teeth and tooth spaces are present in the holotype, but the anterior tip of the bone is missing, so it is impossible to tell the total number of teeth and spaces. Because all the crowns are broken off, the teeth are represented only by remnants of their pedicels. The remains of at least 29 teeth are discernable. In labial view, a moderately deep groove runs along the length of the mental foramina. Four mental foramina are present. The first three are elongate; the last one is rounded. The height of the dentary through the pedicellar portions of the teeth at the level of the Meckelian groove is 5.4 mm. The length of the tooth row along the extent of the Meckelian groove is 18.4 mm.

Other Material. The additional material from the Pleistocene (Ir-

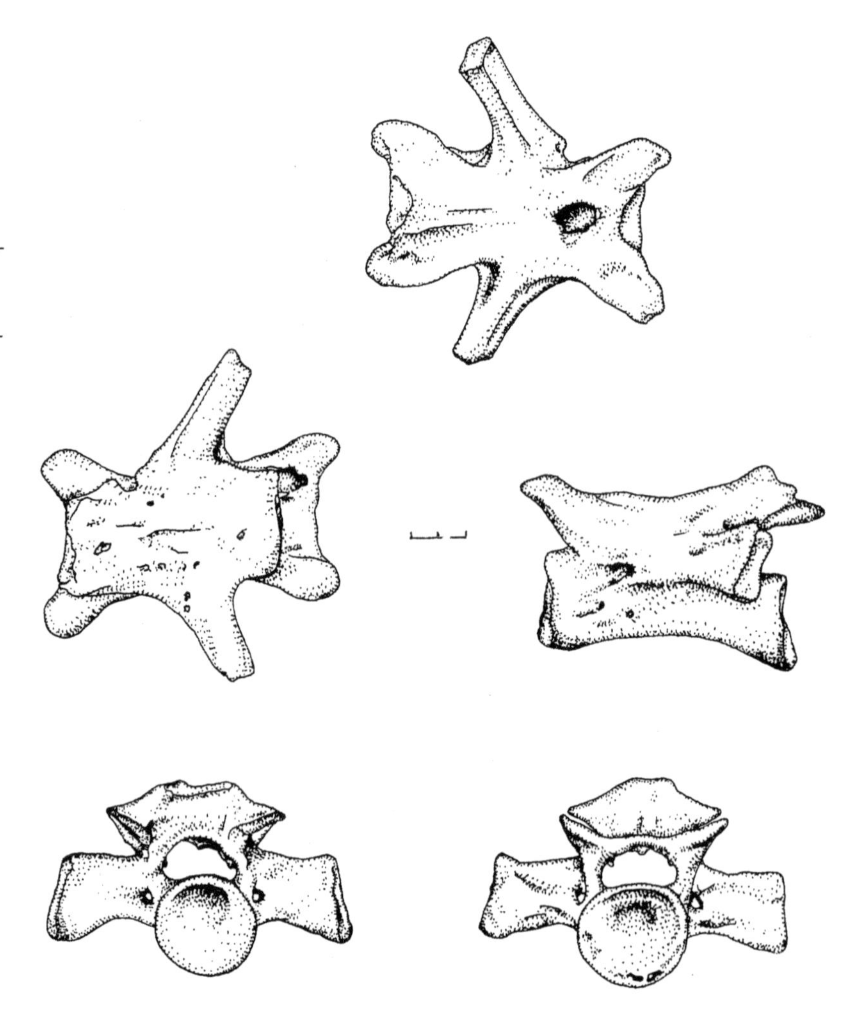

FIGURE 39. A trunk vertebra of **Cryptobranchus guildayi* from the Pleistocene (Irvingtonian NALMA) of Trout Cave, West Virginia. Top, dorsal view; middle left, ventral view; middle right, lateral view; bottom left, posterior view; bottom right, anterior view. Scale bar = 1 mm and applies to all figures.

vingtonian NALMA), Kansan glacial stage, from Trout Cave in West Virginia (Holman, 1982a), provided much more information about the extinct form. The diagnostic longer labial groove was present in the two new fossil dentaries. These have the Meckelian groove extending the length of 34 tooth-alveolar spaces in a well-preserved specimen and the length of 36 tooth-alveolar spaces in a second specimen that is eroded and worn. A second diagnostic character, the weaker curvature of the dentary, is also found in the new specimens. A third character, a deeper groove in the labial border of the dentary, is not supported by the new fossils, both of which have a groove similar in depth to that of *Cryptobranchus alleganiensis.*

The right epihyal in *Cryptobranchus guildayi* is quite different from that in *Cryptobranchus alleganiensis.* In the fossil there is a very strongly developed posterior process that is absent in *C. alleganiensis.* It seems that this process might be associated with some different feeding habits in the extinct species. Parenthetically, the epihyal does not ossify in the cryptobranchid genus *Andrias* (Meszoely, 1966).

The single, nearly complete trunk vertebra from the Trout Cave site is shorter and wider than trunk vertebrae from modern *Cryptobranchus alleganiensis*. In the fossil femur, the distal ridge of the femur is better developed and extends farther down the shaft than in *C. alleganiensis*. Moreover, in *Cryptobranchus guildayi* the distal muscular line extends two-thirds the length of the shaft, and in the modern form it extends only half the length of the shaft. The scapula of the fossil has a different shape from that in *C. alleganiensis*. In *C. guildayi* the scapula has a more rounded dorsal surface, and its posterior process makes a greater angle with the shaft than in *C. alleganiensis*.

General Comments. Someday, *Cryptobranchus guildayi* may be shown to be a Pleistocene (Irvingtonian NALMA) variant of the Pleistocene (Rancholabrean NALMA) and modern species *Cryptobranchus alleganiensis*—I was able to study only four modern adult *C. alleganiensis* skeletons at the time. Estes (1981) mentioned that all the characters used to define *C. guildayi* were variable in *C. alleganiensis* specimens he had seen, but he did not mention how many specimens he had observed; nor did he provide any data on the subject. He was aware of only the holotype dentary at that time.

CRYPTOBRANCHUS SP.

Fossil Localities. **Pleistocene (Irvingtonian NALMA):** Hamilton Cave, Pendleton County, West Virginia—Holman and Grady (1989). **Pleistocene (Rancholabrean NALMA):** New Trout Cave, Pendleton County, West Virginia—Holman and Grady (1987). Zoo Cave, Taney County, Missouri—Holman (1974) (this fauna may have encroached on the Early Holocene).

General Comments on the Cryptobranchidae. Milner (2000) pointed out that it is likely that cryptobranchids originated as neotenic "cryptobranchoids" in East Asia in the Cretaceous and extended to North America by way of the Bering land bridge in the Late Cretaceous or the Paleocene. This group is thought to have extended to Europe in the Middle Oligocene after the withdrawal of the Obik Sea across northern Asia. I would point out here that the major trend that occurred in cryptobranchids after *Andrias* arrived in North America was becoming smaller.

FAMILY PROTEIDAE HOGG, 1838

PROTEIDS

In North America, proteids are large, flattened, permanently aquatic salamanders with purple, plume-like external gills, the only thing "pretty" about them. There is only one genus, *Necturus*, in this continent, but it is composed of five species, one in the north and four in the south. Like Hellbenders, proteids have odd official common names. In the north they are Mudpuppies, and in the south they constitute four species of Waterdogs. This common name must have been derived in desperation, because these animals do not look like either puppies or dogs. Actually, most people (other than herpetologists) do not know what they are. Here in Michigan, I have heard them referred to as bullheads with legs.

The only other living proteid species (*Proteus anguinus*) is an elongated cave species that lives in underground streams and occurs only in northeast Italy, Slovenia, Croatia, and Bosnia (Arnold and Ovenden, 2002). The five living species of *Necturus* are *Necturus alabamensis* (Blackwarrior Waterdog), *Necturus beyeri* (Gulf Coast Waterdog), *Necturus lewisi* (Neuse River Waterdog), *Necturus maculosus* (Mudpuppies; two subspecies occur), and *Necturus punctatus* (Dwarf Waterdog).

The family Proteidae is unique in having (1) no maxillary bones; (2) two pairs of larval gill slits; (3) a periotic canal that is horizontal from its posterodorsal junction with the periotic cistern; and (4) a diploid number (set of chromosomes from each parent) of 38 (Duellman and Trueb, 1986). The origin of the proteids is somewhat obscure. Many think there is a relationship between the Proteidae and the odd, enigmatic extinct family #Batrachosauroididae, which will be discussed later in the book.

Definitive osteological characters of the Proteidae are slightly modified from Duellman and Trueb (1986) as follows. The premaxillary bones are paired. The septomaxillary and lacrimal bones and ypsiloid cartilage are absent, as is the basilaris complex of the inner ear. The opisthotic bone is not fused with the prootic or exoccipital. An internal carotid foramen is present. Proteids have pedicellate teeth, and the palatal teeth run parallel with the premaxillary teeth. Estes (1981) recognized the following, presumably derived, characters in the Proteidae. The columellar process of the squamosal has an ontogenetic ossification in the squamoso-columellar ligament. The parasphenoid almost reaches the occipital condyles. Two gill slits are present. The ypsiloid cartilage is absent.

Genus *Necturus* Rafinesque, 1819
Mudpuppies and Waterdogs

Genotype. Necturus maculosus (Rafinesque, 1818).

Diagnosis. The body is only moderately elongate. Four toes occur on each foot. There are 17–19 presacral vertebrae. Maxillary bones are absent. The posterior processes of the frontals are widely separated at the midline. Opisthotic bones are separate. The neural spines of the vertebrae are single and not split. The rib-bearers are double headed. The eyes are functional. The maximum snout to vent length (SVL) is 300 mm (modified from Estes, 1981).

**Necturus krausei* Naylor, 1978c

Fig. 40

Holotype. A single, incomplete trunk vertebra (UALVP 14310, University of Alberta Laboratory for Vertebrate Palaeontology).

Type Locality. Ravenscrag Formation, southern Saskatchewan site UAR 2g, near the town of Roche Percee.

Horizon. Late Paleocene.

Other Material. Referred specimens include an incomplete trunk vertebra (UALVP 14311), fragmentary trunk vertebrae (UALVP 14313–14315), and a centrum of a questionable second cervical vertebra (UALVP 14312).

Diagnosis. The diagnosis is from Naylor (1978c, p. 566): "A species of *Necturus*, differing from Recent species in the more elongate hypera-

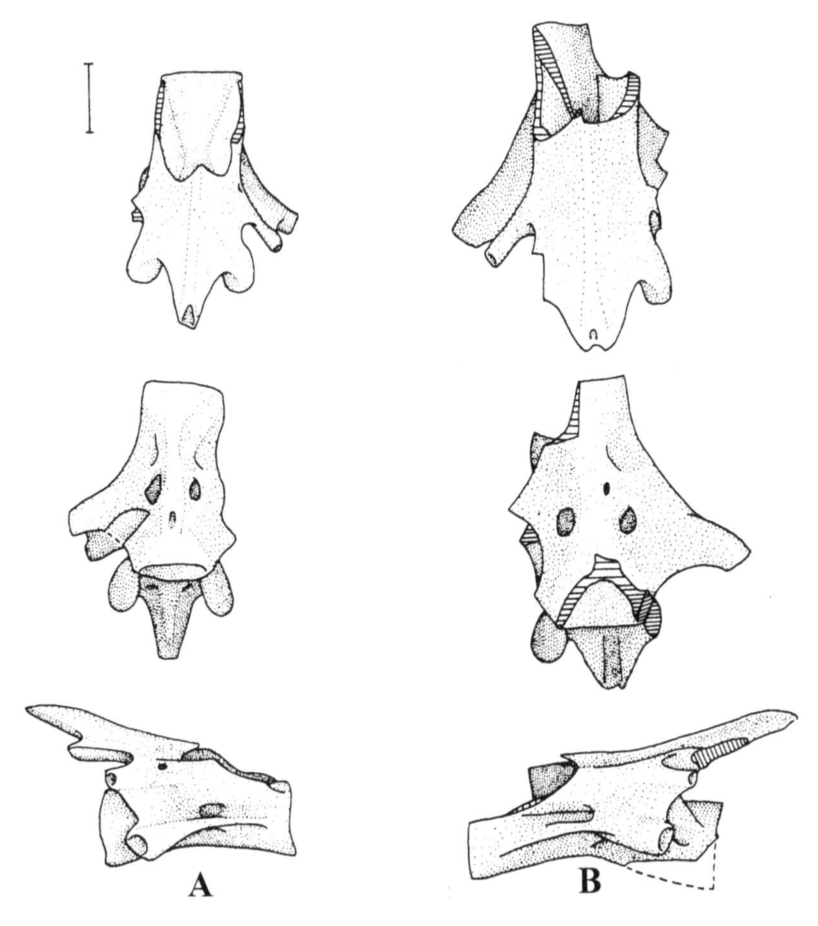

FIGURE 40. **Necturus krausei* from the Late Paleocene of Saskatchewan. (A) Holotype: top, dorsal view; middle, ventral view; bottom, lateral view. (B) Referred specimen: top, dorsal view; middle, ventral view; bottom, lateral view. Scale bar = 1 mm and applies to all figures.

pophysis, well-developed and hollow dorsal rib-bearers, and more anteroposteriorly elongate postzygapophyseal facets."

Description of the Material. The following description is modified from Naylor (1978c). The vertebrae are small and delicate. The anterior cotyle of the holotype is subcircular, with the dorsal rim somewhat flattened. The posterior cotyle has an irregular, subcircular shape. The centrum is deeply amphicoelous. Large subcentral foramina occur in both specimens, as well as faint subcentral keels. UALVP 14312, probably a second cervical, has a robust, divided subcentral keel resembling that on second cervicals of Recent *Necturus maculosus.* UALVP 14315, consisting of the anterior part of the centrum and subcentral keel from the anterior part of the trunk, is probably a third or fourth cervical. It has a robust subcentral keel, small ventral lamellae, and subcentral foramina resembling those in vertebrae from the pectoral region in *N. maculosus.* Basapophyses are absent from the specimens. **Author's note:** Today, vertebrae referred to by Naylor (1978c) as "cervical" vertebrae would be called anterior trunk vertebrae.

Except for being large, long, and low, the exact form of the neural canal cannot be determined, because of breakage. The roof of the neural arch is wide and flat, with a low neural crest extending farther forward than in living *Necturus*. The neural spine is elongate in comparison with *Necturus maculosus*, and it is more robust; nevertheless, it is a simple hollow tube. The holotype bears a ridge on the ventral surface of the neural arch as in *N. maculosus*, but UALVP 14311 has a groove in this area. Spinal nerve foramina are lacking in all specimens, but the postzygapophyseal articular facets resemble those of *N. maculosus*, except they are more elongate. The structure of the transverse process is *Necturus*-like. Well-developed lamellae, with large subcentral foramina, are present. The rib-bearers are divergently bicapitate and are directed posterolaterally. The dorsal rami extend from the level of the top of the neural arch, as in *N. maculosus*. These arms are relatively large and hollow and contrast with the smaller solid rods found in living *Necturus*.

General Remarks. All three of the characters in the diagnosis of *Necturus krausei* also occur in *Necturus maculosus*: (1) The posterior trunk vertebrae of *N. maculosus* match or even exceed those of *N. krausei* in the length of the posterior extension of the neural arch (hyperapophyses of Naylor, 1978c). (2) Well-developed and hollow dorsal rib-bearers also occur in most of the vertebrae of *N. maculosus* prepared as skeletons by the maceration process. These areas of the vertebrae are sometimes distally eroded away in specimens prepared with enzymes or chemicals. (3) Elongate, anteroposteriorly directed postzygapophyseal facets also occur in the posterior trunk vertebrae of *N. maculosus*. I have been unable to find other characters that do not vary between *N. krausei* and *N. maculosus* vertebrae. It is to be hoped that more material of *N. krausei* will be found.

Necturus maculosus (Rafinesque, 1818)

Mudpuppy

Fig. 41

Fossil Localities. **Pleistocene (Rancholabrean NALMA):** Baker Bluff Cave, Sullivan County, Tennessee—Van Dam (1978), Fay (1988), Holman (1995a). Cheek Bend Cave, Maury County, Tennessee—Miller (1992). Guy Wilson Cave, Sullivan County, Tennessee—Fay (written communication, 1993[1]), Holman (1995a). Peccary Cave, Newton County, Arkansas—Davis (1973), Holman (1995a, listed as *Necturus* cf. *Necturus maculosus*).

The Mudpuppy is much the larger of the five known species of living *Necturus*. It also has larger and darker spots on the belly than the other four species (called Waterdogs). The Mudpuppy consists of two subspecies: (1) *Necturus maculosus maculosus* occurs in southern Canada, mainly from Quebec to Manitoba; in the northeastern United States and Great Lakes region west of the Coastal Plain; then south as far as northern South Carolina, Georgia, and Mississippi. There are many odd extensions and gaps in this general distribution. (2) *Necturus maculosus louisianensis* occurs in the drainage system of the Arkansas River from southeastern Kansas and southern Missouri to north-central Louisiana.

Mudpuppies are mainly nocturnal in their habits. They are extremely aquatic; one was taken at a depth of 90 feet (27.4 m) in Green Bay, Wisconsin (Conant and Collins, 1998). Mudpuppies may eat almost any aquatic animal, and they will frequently take the hook of an angler. Their unusual appearance tends to bewilder the fishermen that catch them, and they are often treated badly. In Michigan, ice anglers often categorically throw Mudpuppies onto the ice to die, rather than returning them to the water. This is yet another living fossil that is persecuted because of its odd looks (see Holman, 1995b).

Identification of Fossils. Van Dam (1978) pointed out that it is sometimes difficult to separate fossil vertebrae of *Cryptobranchus alleganiensis* from those of *Necturus maculosus* because of their often fragmentary nature. He found the following criteria useful in differentiating the two large salamanders: "In *C. alleganiensis* the sides of the centrum are more sculptured than in *N. maculosus;* in *C. alleganiensis* the upper transverse process is heavy and cylindrical in shape (*N. maculosus* has a very winglike upper transverse process); *C. alleganiensis* has the articular facets of the transverse processes exhibiting a single opening, whereas in *N. maculosus* there are usually two distinct openings" (Van Dam, 1978, pp. 20–21).

NECTURUS SP. INDET.

Fossil Locality. **Pleistocene (Rancholabrean NALMA):** Arredondo site, Alachua County, Florida—Lynch (1965), Holman (1995a).

This genus does not currently occur in Florida (Conant and Collins, 1998, map, p. 420). The closest it occurs to Florida today is northern Georgia and Alabama. Lynch (1965) based his identification of this genus on the basis of two fragmentary vertebrae (both UF 9105, Florida Museum of Natural History), one of which consisted of the posterior half of the centrum and portions of the neural arch; the other lacked the neural arch and was well worn. He did not figure this material. It is odd that with the very abundant amphibian and reptile material that occurs in the Pleistocene sites of Florida (see Holman, 1995a) more *Necturus* material has not been found.

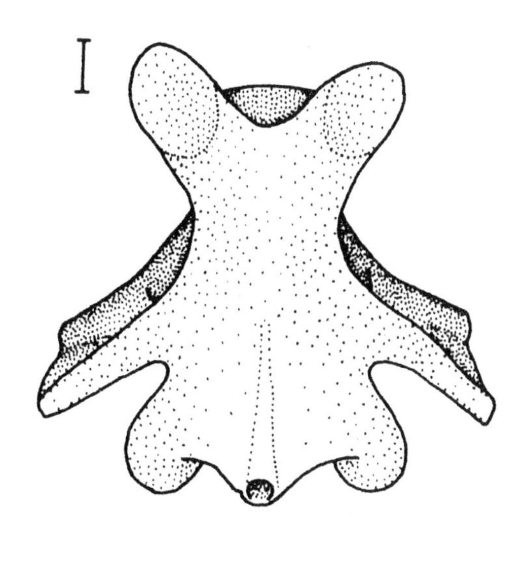

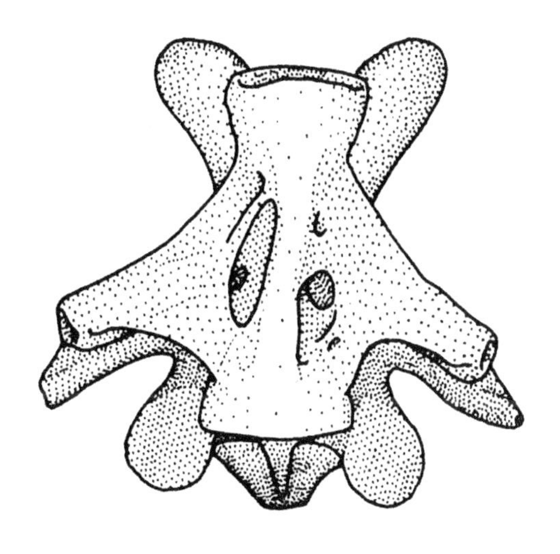

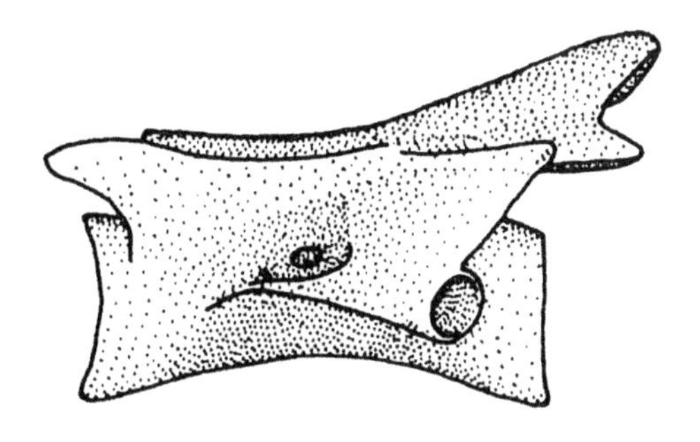

FIGURE 41. Trunk vertebra of modern *Necturus maculosus:* top, dorsal view; middle, ventral view; bottom, lateral view. Scale bar = 1 mm and applies to all figures.

General Remarks on Proteidae in North America. Necturus, the only member of the Proteidae in North America, appears to have remained virtually unchanged here from Paleocene to modern times.

Family Plethodontidae Gray, 1850

Lungless Salamanders

The Plethodontidae is by far the most successful family of living salamanders. This lungless but mainly terrestrial family contains more than half of the salamander species in the world and consists of about 28 genera. Even though an amazing radiation has occurred in the group, fossil records of plethodontids are relatively few. In North America, most fossil records are from cave faunas in Appalachia.

Plethodontids are not grotesque—in fact, far from it. Many are brightly colored, and even the dull-colored ones have an elfin charm. These creatures, having four good legs, are called lizards by some people. In the southern part of the United States several species are sold for fish bait under the semi-commercial name of Spring Lizards. The name comes from the fact that some plethodontids are common around springs and spring runs. The world range of these salamanders is from extreme southern Alaska and Nova Scotia, Canada, south to eastern Brazil and central Bolivia; they also occur in southern Europe (Arnold and Ovenden, 2002).

Some osteological characters used to define the members of this group are as follows (mainly from Duellman and Trueb, 1986). The premaxillae are paired and are fused in many genera. Both maxillae and premaxillae are present, but they may be secondarily reduced or absent in some taxa. The exoccipital, prootic, and opisthotic bones are fused. The teeth are pedicellate, and the palatal teeth extend posteriorly along the medial edges of the vomers. The nasals are ossified laterally from embryonic centers. Lacrimals, pterygoids, and the ypsiloid cartilages are absent, but pterygoids are present in the larvae. The vertebrae are opisthocoelous (condyle anteriorly, cotyle posteriorly) because nonskeletal plugs form in the cotylar area of the anterior part of the centrum. These plugs often part from the vertebrae of many plethodontids during the fossilization process. All but the first three spinal nerves exit intravertebrally.

Estes (1981) gave a list of both osteological and biological shared derived characters that define the family Plethodontidae; these characters are based on Wake (1966), Salthe (1967), and Edwards (1976) and are as follows. The angular, second epibranchial, lacrimal, and ypsiloid cartilages are absent. The palatal tooth row extends posteriorly onto the vomer. The replacement of vomerine teeth takes place posteriorly and laterally. The columella is fused. The spinal nerves exit intravertebrally. The lungs are lost. Fertilization is internal. Mating takes place on land. The haploid chromosome number is 14 and 13.

Because the family Plethodontidae comprises most of the salamander species in North America, we shall mainly follow Milner (2000) in discussing the subdivisions (subfamilies and tribes) of this huge group. The two subfamilies of the Plethodontidae are the Desmognathinae and the

Plethodontinae. The subfamily Desmognathinae contains two genera: *Desmognathus* and *Phaeognathus*. This group is most prevalent in the Appalachian region of North America, but it extends southward to Florida and westward to Texas and Oklahoma. Fossil *Desmognathus* have rather frequently been found in caves and fissures in the Pleistocene of the Appalachian region, but there are no earlier records.

The subfamily Plethodontinae is composed of three tribes: the Plethodontini, the Hemidactyliini, and the Bolitoglossini. The tribe Plethodontini is made up of only three genera: *Aneides*, *Ensatina*, and *Plethodon*, a genus with 53 species the last time I heard (see Crother, 2000). All these genera occur in North America. The tribe Hemidactyliini is made up of eight genera, including the large, colorful genera *Gyrinophilus* and *Pseudotriton*, as well as *Eurycea*, a genus with 22 species. The tribe Bolitoglossini is composed of 17 genera, mainly occurring in Mexico and southward into South America, but these genera include *Batrachoseps* and *Hydromantes*, which occur in the Pacific region of North America. *Batrachoseps*, a very slender form, has 15 species, and *Hydromantes* has only 3 (Crother et al., 2003).

Modern Plethodontini range across the extent of North America, except for very cold and very dry regions. *Aneides* was reported from the Miocene of North America by Tihen and Wake (1981) and Clark (1985). *Plethodon* was reported from the Miocene of North America by Tihen and Wake (1981). Both *Aneides* and *Plethodon* have additionally been reported from the Pleistocene: *Aneides* in California and *Plethodon* in the east (Holman, 1995a). The Hemidactyliini occur today in eastern North America as far west as Texas. Two genera, *Gyrinophilus* and *Pseudotriton*, are known from the Pleistocene. The living Bolitoglossini occur from western North America, south across Central America, and into the northern part of South America. This tribe has one highly disjunct genus, *Hydromantes*, which occurs in France, Italy, and Sardinia. Oddly, although the tribe Bolitoglossini contains more than half of the living salamander species in the world, the only true fossils of this group that are known from before the Pleistocene are vertebrae from the Miocene of California. These were attributed to *Batrachoseps* by Clark (1985). A set of footprints from the Late Miocene of California was also attributed to this genus (Brame and Murray, 1968).

Genus *Aneides* Baird, 1849

Climbing Salamanders

Genotype. *Aneides lugubris* (Hallowell, 1849).

Diagnosis. This diagnosis is modified from Estes (1981). *Aneides* is composed of derived Plethodontini that are similar to *Plethodon* but differ in that they have fused premaxillae and modified tarsal configurations. The more advanced species show both feeding and climbing specializations.

The living animals occur only in the Appalachian region and West Coast region of North America, but the fossil record shows they were previously more widespread.

Aneides lugubris (Hallowell, 1849)

Arboreal Salamander

Fig. 42

Fossil Localities. **Late Miocene (late Hemphillian NALMA):** Turlock Lake site 5, Stanislaus County, California—Clark (1985). **Pleistocene (Rancholabrean NALMA):** Costeau Pit, Orange County, California—Hudson and Brattstrom (1977), Holman (1995a). Newport Beach Mesa, Orange County, California—Hudson and Brattstrom (1977), Holman (1995a). Rancho La Brea, Los Angeles County, California—LaDuke (1991), Holman (1995a).

The Arboreal Salamander is a relatively large species of *Aneides.* It has a dark brown background color and cream to yellow spots on the head, trunk, tail, and limbs. These vary in brightness and distribution on the body. The belly is a creamy white, and the undersides of the tail and feet are yellow. The head is large compared with the body of this salamander and is widest behind the eyes. *Aneides lugubris* lives in yellow pine or black oak forests in the Sierra Nevada and in coastal oak forests from northern California to Baja California. Animals that are geographically isolated from the general range live in the foothills of Sierra Nevada and on South Farallon, Santa Catalina, Los Coronados, and Año Nuevo islands (Petranka, 1998).

These salamanders are climbers, as their common name suggests, and their climbing abilities are enhanced by expanded digits and a prehensile tail. Within this arboreal tendency they are remarkably ubiquitous in their microhabitats. On the other hand, large oaks are used for nesting and shelter during hot, dry spells. Arboreal Salamanders usually lay their eggs in June or July and guard them until they are hatched. These interesting animals mainly eat small invertebrates, such as snails, spiders, beetles, ants, and flies, but they occasionally prey on Slender Salamanders of the genus *Batrachoseps* (Miller, 1944).

Comments on the Miocene Fossil. The Turlock Lake very late Mio-

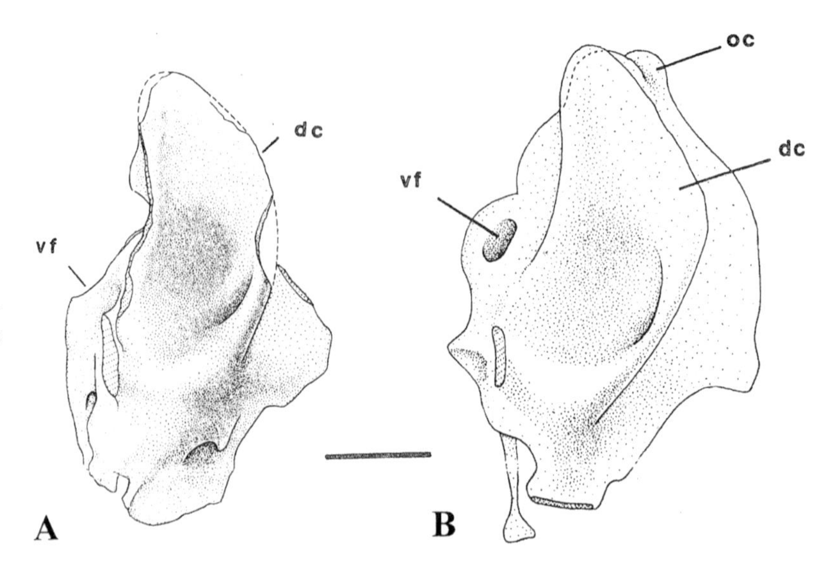

Figure 42. (A) Right occipito-otic unit of *Aneides lugubris* from the Late Miocene of California. (B) Same unit from a modern specimen of this species (female with snout to vent length of 74.1 mm). Abbreviations: dc, dorsal crest; oc, occipital crest; vf, vestibular foramen. Scale bar = 1 mm and applies to both figures.

cene site (ca 4.5 Ma BP) in Stanislaus County, California (Clark, 1985) yielded a very significant find: a right occipito-otic bone (UCMP 125852 A–C, University of California Museum of Paleontology). The bone was accidentally broken into three pieces during the removal of the matrix, but it yielded a large amount of information. This specimen is identical in all ways to the occipito-otic bone of *Aneides lugubris*, which is distinguished by the extreme development of a large dorsal crest. According to Clark (1985), a crest is developed only rarely in salamander occipito-otic bones, and only in *Aneides* is it found in the exact position seen on the fossil specimen. There is a morphocline in the development of this crest (Wake, 1963; Larson et al., 1991). It is least developed in the most primitive species (*Aneides aeneus*), well developed in *Aneides hardii* males, and highly developed in *A. lugubris*. Clark (1985) stated that the crest of *A. lugubris* is developed to the extent that this species can be distinguished from other species of the genus on the basis of this character alone. This is another case of a currently living herpetological species being found in the latest part of the Miocene (see Holman, 2000a, 2003).

Identification of the Pleistocene Fossils. Hudson and Brattstrom (1977) provided detailed information on how they identified Pleistocene fossils from the localities listed above. They identified three complete left dentary bones and one left dentary fragment on the basis of the robust size of the tooth pedicels and the triangular elevation that arises from the dorsal surface, just posterior to the midpoint of the bone. The complete dentaries all had nine tooth pedicels, which occupied less than one-half of the length of the ramus, a character that was reported for the species by Wake (1966).

Twenty-five trunk vertebrae of Arboreal Salamanders were described. Hudson and Brattstrom (1977) identified these on the basis that they are amphicoelous, constricted ventrally, and lacking a ventral spine. The neural spine is reduced in size, but it is present in all the vertebrae that were studied. The parapophyses of these vertebrae are supported by a dorsoventrally compressed, plate-like shelf that arises from the ventrolateral surfaces of the anterior and posterior margins of the centrum and extends to the distal tip of the parapophyses. The shelf gives the vertebra a diamond-shaped appearance in ventral view (Wake, 1966). The atlantes have a boss that is rounded anteriorly, becoming broader posteriorly and terminating at the posterior margin of the neural arch.

General Comments. This species of salamander currently occurs in Orange County, California, but it does not occur in the vicinity of the Newport Beach Mesa site. The fossil salamanders were found in a water-soluble, clay-like matrix at the Newport site, and the abundance and clumping of this matrix suggest that it could have been formed from the erosion of an owl pellet cache (Miller, 1971).

ANEIDES SP. INDET.

FIGS. 43D–43F

Fossil Locality. **Late Oligocene to Early Miocene (Arikareean NALMA).** Cabbage Patch Formation, Granite County, Wyoming—Tihen and Wake (1981).

FIGURE 43. Seventh trunk vertebra of *Aneides ferreus*, a living species from Oregon, in (A) dorsal, (B) ventral, and (C) lateral views. Seventh trunk vertebra from Early Miocene *Aneides* sp. from Montana in (D) dorsal, (E) ventral, and (F) lateral views. Vertebra of *Aneides flavipunctatus*, a living species from California, in (G) ventral view. Scale bar = 1 mm and applies to all figures.

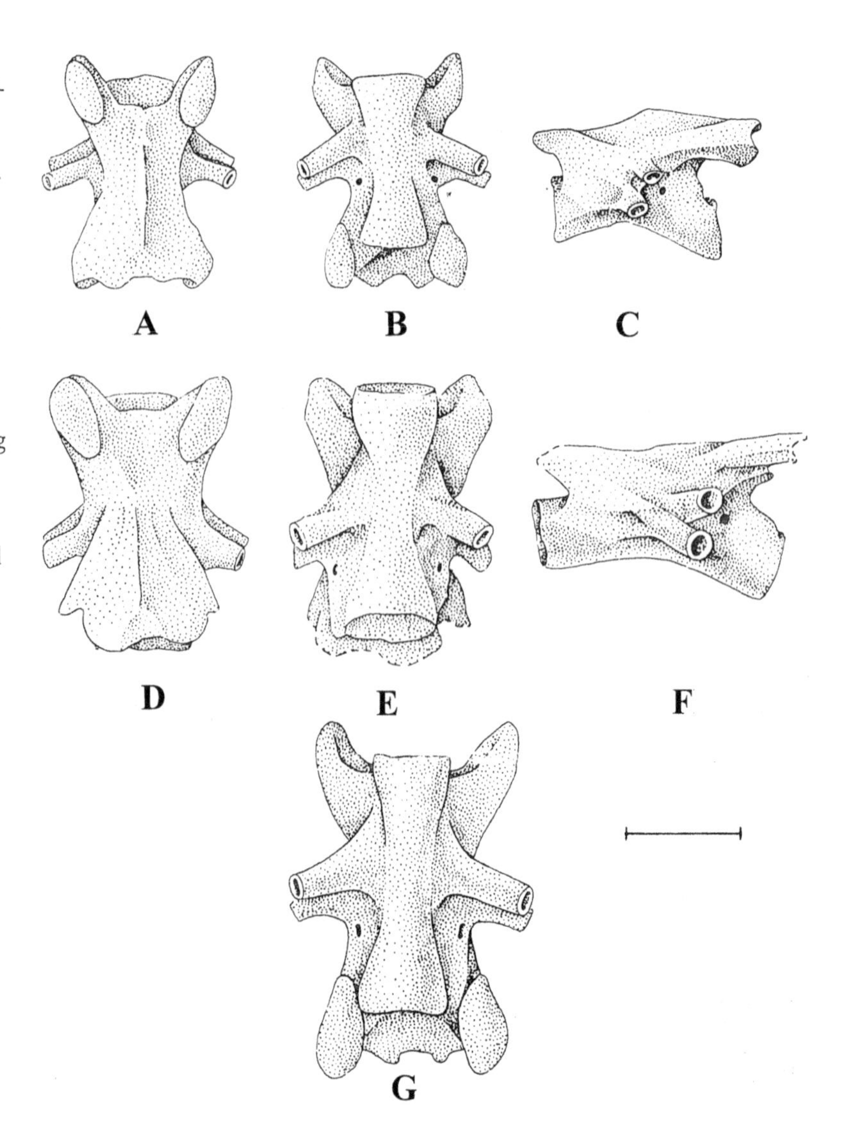

The Late Oligocene to Early Miocene Cabbage Patch Formation has produced the earliest records of the Plethodontidae in North America. The presence of two genera, *Aneides* and *Plethodon*, at this site corroborates indirect paleobiological records of these taxa being widely distributed in central North America by Late Paleocene or Early Miocene times.

Tihen and Wake (1981) so clearly pointed out the detailed process that must be followed in identifying fossil salamander taxa that I include it here in a modified version. It is hoped that this will help potential paleoherpetologists.

Vertebrae in salamanders vary regionally within an animal, ontogenetically within populations, as well as geologically within species (see Worthington and Wake, 1972). Fossil salamander vertebrae were examined under a binocular dissecting microscope and compared directly with vertebrae of living caudate species. When it was found that vertebrae were similar to those of modern genera of the Plethodontini, the vertebrae of living species (see *Aneides ferreus* in Figs. 43A–43C and *Aneides flavipunctatus* in Fig. 43G) were systematically studied, and detailed comparisons were made with the species and individuals that were most nearly identical.

The spinal nerve foramina in the vertebrae of salamanders have distinct patterns that are characteristic of major taxa (Edwards, 1976). All four of the fossil vertebrae from the Cabbage Patch Formation of Montana had distinct foramina posterior to the rib-bearers. If it is assumed that the vertebrae represented living families, the only possibilities for assignment to families were the Sirenidae, the Plethodontidae, the Ambystomatidae, and the Sirenidae. The vertebrae of sirenids are distinctive and differ from those of the Cabbage Patch fossils in size, proportions, rib-bearers, and dorsoventral dimension. The slender build of the fossil vertebrae, the long, narrow, husk-like centra, and the low neural arch with its weakly developed neural spine all pointed to a plethodontid rather than ambystomatid or salamandrid relationship. There was no sign of mineralized intervertebral cartilages; thus, the plethodontid subfamily Desmognathinae was eliminated. The authors (Tihen and Wake, 1981) then were able to limit their comparisons to the subfamily Plethodontinae.

The Cabbage Patch vertebrae turned out to be nearly identical in all structural details with those of the living members of the genera *Aneides* and *Plethodon*. The authors (Tihen and Wake, 1981) eliminated members of the tribe *Hemidactyliini* (Wake, 1966), in part on the grounds that many taxa in this group have double spinal nerve foramina (Edwards, 1976) and have transverse processes that are more anterior in position than in the fossils. However, many of the species in the large tribe Bolitoglossini were unavailable for study. Only two bolitoglossine genera, *Hydromantes* and *Batrachoseps*, occur in the United States today. *Hydromantes* has elongate, cylindrical transverse processes, and the shape of its vertebrae is different from that of the fossils. *Batrachoseps* has transverse processes that have a distinct angle along their length; these processes also have different proportions and are generally smaller than those of the fossils. Among the bolitoglossines in the New World tropics that were studied, *Thorius* and *Parvimolge* were eliminated on the basis of their very small size. *Lineatriton* has very elongated vertebrae, and both *Oedipina* and *Lineatriton* have fused transverse processes. Many species of *Chiropterotriton* and *Pseudoeurycea* were eliminated on the basis of having fused transverse processes, mineralized intervertebral cartilages, or both. It was found that many *Bolitoglossa* also have fused transverse processes and that these features differed strongly from those in the Cabbage Patch fossils. The only close relative of *Aneides* and *Plethodon* was found to be *Ensatina*, and Tihen and Wake (1981) found that this genus had very distinctive vertebrae, with proportions (centra of relatively large diameter) different from those of the fossils. At this point the authors were able to assign their fossils to *Aneides* and *Plethodon*.

A middle trunk vertebra (MSUMP 6504-2004, Montana State University Museum of Paleontology), one of the four Cabbage Patch Formation salamander vertebrae, was determined by the authors (Tihen and Wake, 1981) to be different from a Cabbage Patch vertebra (KU 18298, Museum of Natural History, University of Kansas) they had assigned to

Plethodon. They noticed that the centrum of MSUMP 6504-2004 was noticeably stouter and that the rib-bearers were positioned farther back on this vertebra than in KU 18298. Moreover, the dorsal and ventral rib-bearers were completely independent of one another, and the vertebral canal was observed to be dorsal to the bulk of the lower rib-bearer. All these characters indicated that it was not *Plethodon*. MSUMP 6504-2004 closely resembled the western species of *Aneides*. Another middle trunk vertebra (KU 18296) from the same locality was also assigned to *Aneides* by the same authors.

General Comments. Biochemical evidence (Maxson et al., 1979; Larson et al., 1981) indicated that the division of both *Aneides* and *Plethodon* (*Plethodon* from the Cabbage Patch Formation will be discussed later) into eastern and western groups predated the age of the fossils reported here. Tihen and Wake (1981, p. 39) stated that "discovery that both genera were in existence in western United States in early Miocene [possibly latest Oligocene] times is also consistent with paleogeographical, morphological, and biochemical evidences, all of which suggests that lineage divergence in the tribe Plethodontini was mainly an early Tertiary event."

Genus *Batrachoseps* Bonaparte, 1841

Slender Salamanders

Genotype. *Batrachoseps attenuatus* (Eschscholtz, 1833).

Diagnosis. This diagnosis is modified from the definition of the genus provided by Wake (1966). *Batrachoseps* is a plethodontine salamander, with maxillae that are either fused or separated. The prefrontals are either extremely reduced or absent. Posterolateral spurs occur on the parietals. The columellae are somewhat reduced. The tongue is highly modified and has extremely long and slender genioglossal muscles. Sixteen to 21 trunk vertebrae and 2 or 3 caudosacral vertebrae are present. There is a slight basal tail constriction. Four toes are present on each foot.

General Comments. These salamanders of the plethodontine tribe Bolitoglossini have a long body and very reduced limbs, and they tend to have very long tails. They are a markedly derived group of salamanders that are fossorial (burrowing) or semifossorial in their habits. They occur in western North America today. Several new species have been described in the last few years (see Crother, 2000; Crother et al., 2003), so at latest count there are 20. Today these animals occur mainly from western Oregon and California to northwestern Baja California, Mexico. As far as I am aware, there are no fossil populations known from outside the present general range of this genus.

Batrachoseps relictus Brame and Murray, 1968

Relictual Slender Salamander

Fossil Locality. **Late Miocene (late Hemphillian NALMA)**: Buchanan Tunnel, near Columbia, Tuolumne County, California—Peabody (1959), Wake (1966), Brame and Murray (1968).

The Relictual Slender Salamander has a relatively broad head and long legs, like the closely related *Batrachoseps pacificus*, but it is smaller and rarely exceeds 45 mm in SVL (Petranka, 1998). This tiny animal has a dorsal stripe that is reddish, yellowish brown, or dark brown, but in larger specimens this stripe is often faint or obscure. It has a background color that is usually dark gray to black that occurs both above and below. The Relictual Slender Salamander is distributed patchily along the western slope of the southern Sierra Nevada from the American River drainage to the Kern River canyon area (Petranka, 1998). This species typically occurs in mixed coniferous forests. Fifteen eggs of *Batrachoseps relictus* were found in a log lying in river-washed debris on a January day by Grinnell and Storer (1924).

Identification of Fossils. Two fossil trackways (UCMP 35400, 36411, University of California Museum of Paleontology) led to the identification of *Batrachoseps relictus* from the Late Miocene of California (see above). Fossil trackways may consist of crawl marks made by worms and clams and footprints made by mice, mastodonts, dinosaurs, birds, and even salamanders. Sometimes tails dragged along by vertebrate animals form part of the trackway as well. The Relictual Slender Salamander left a good set of footprints.

These fossil trackways were originally referred to the living *Batrachoseps pacificus* by Peabody (1959). However, Wake (1966) reported that the trackway fitted equally well with a then undescribed living species that occurs near the locality where the fossil trackways were found. This new species of living salamanders was described by Brame and Murray (1968) as *Batrachoseps relictus* on the basis of its primitive traits within the genus *Batrachoseps.* Brame and Murray (1968) then studied the fossil trackways and finally determined that the wider front and hind feet of the of the footprints indicated that *B. relictus* was the ancient walker rather than *B. pacificus* or other species of *Batrachoseps* examined by themselves and other workers.

BATRACHOSEPS SP. INDET.

FIGS. 44A, 44B

Fossil Locality. **Late Miocene (late Hemphillian NALMA):** Pinole Tuff, San Francisco Bay area, California (Clark, 1985).

Referred Material. A second trunk vertebra (UCMP 125362, University of California Museum of Paleontology); trunk vertebrae (LACMVP 121992–121994, 121996, Natural History Museum of Los Angeles County, Vertebrate Paleontology); and caudal vertebrae (LACMVP 62020, 121995; UCMP 82586).

The *Batrachoseps* trunk vertebra in Fig. 44A was described by Clark (1985). This description is slightly modified as follows. The vertebra is missing the right postzygapophysis and the tip of the right parapophysis but is otherwise complete. The dorsal surface is flattened, except for a low, anteriorly placed neural spine. The prezygapophyses are slightly bipartite, and a spinal nerve foramen occurs both anterior to and posterior to the diapophyses. The diapophyses themselves project posterolaterally, along with the parapophyses, to form an angle of about 45 degrees with the lateral edge of the centrum.

Ossifications occur between the ventral longitudinal midline and the ventral surface of the parapophyses, spanning longitudinal bilateral canals.

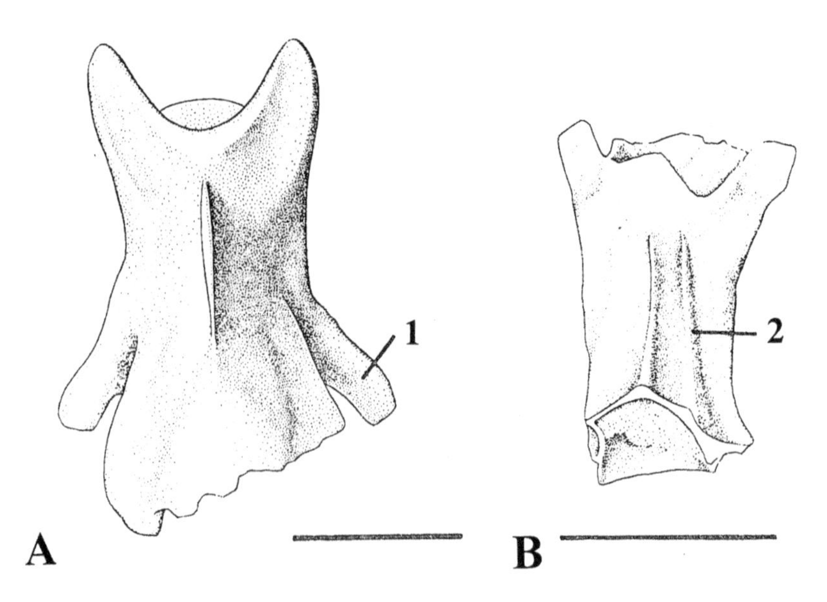

FIGURE 44. (A) Second trunk vertebra of *Batrachoseps* sp. from the Late Miocene of California in dorsal view. (B) Caudal vertebra of *Batrachoseps* sp. from the Late Miocene of California in dorsal view. Legend: 1, prezygapophyseal articular facet; 2, hyperapophysis. Scale bars = 1 mm.

An anterior process of the ventral surface of the right parapophysis can be seen, and a similar process probably extended from the left side before breakage occurred. Clark (1985) thought that the eroded texture of these ventral ossifications was probably natural.

Identification of Fossils. Clark (1985) explained the identification of UCMP 125362, the vertebra discussed immediately above, as follows. In salamanders, only plethodontid second trunk vertebrae (and the odd salamandrid first trunk vertebrae) have both anterior and posterior spinal nerve foramina (Edwards, 1976). Moreover, because of its gracile build, the specimen is clearly a plethodontid and not a salamandrid. Also, no salamandrid has a posteriorly deflected parapophysis or diapophysis as in UCMP 125362. This character is limited to the first two trunk vertebrae of *Batrachoseps* among the Plethodontidae. All other features of this specimen, including the eroded texture of the ventral surface, are characteristic of *Batrachoseps*. Nevertheless, it was not possible for Clark to refer this specimen beyond the generic level.

General Remarks. The very large size of the Pinole Tuff fossils relative to living *Batrachoseps* is noteworthy. As Clark (1985) pointed out, Peabody (1959) reported that the *Batrachoseps* trackways (discussed above) were made by a relatively large individual of this genus. The fossils reported here are within the known size range of the genus but at the highest end; the centra range in length from 2.2 to 2.6 mm (Clark, 1985). In fact, their size is comparable to that of the largest available modern skeleton known to D. B. Wake, LACMVP 36167, a specimen of *Batrachoseps pacificus* with an SVL of about 63 mm. It is not impossible that the Late Miocene *Batrachoseps* is very closely related to *Batrachoseps relictus* discussed above (my comment). However, Clark did point out that the living *Batrachoseps campi* and *Batrachoseps stebbensi* both reach a similar size. These specimens represent the first published remains of actual fossils of the tribe Bolitoglossini in North America.

Genus *Desmognathus* Baird, 1850

Dusky Salamanders

Genotype. Desmognathus fuscus (Rafinesque, 1820).

Diagnosis. This diagnosis is modified from the definition of the genus by Wake (1966). *Desmognathus* is a desmognathine salamander with 15 trunk vertebrae. The maxillary jugal processes are small. The internal nares open medially. The vomerine preorbital processes are distinctly separated from the vomer proper and do not extend laterally beyond the bodies of the vomer. An internasal fontanelle is present. The vomers and the premaxillae are joined by a septum-like process, which forms the fontanelle borders. Hypapophyseal keels are developed on the anterior trunk vertebrae. These either are not present or are only slightly developed on the posterior vertebrae. Neural crests are present only on the first two or three trunk vertebrae. The parapophyses and diapophyses are separated for most of their lengths. The bodies of these salamanders are stocky and stout, and the limbs are moderately long. They have aquatic larvae that undergo direct development.

General Comments. The salamanders of the subfamily Desmognathinae currently range well through eastern North America, except in cold regions, and they are especially abundant in the Appalachian region. As far as I am aware, all the fossils of this species are from the Pleistocene, and none extend outside the limits of their present distribution. The vertebrae of *Desmognathus* are very difficult to separate from a former member of the Desmognathinae, *Leurognathus* Moore, 1899, which has recently been included with the genus *Desmognathus* to "render *Desmognathus* monophyletic" (Crother, 2000, p. 21). This means, in effect, that it is so similar to other members of the genus *Desmognathus* that it should be included within it, thus indicating that *Desmognathus* and the former *Leurognathus* evolved from a common ancestor.

Desmognathus fuscus (Rafinesque, 1820)

Northern Dusky Salamander

Fossil Localities. **Pleistocene (Irvingtonian NALMA), Kansan glacial stage:** Hamilton Cave, Pendleton County, West Virginia—Holman and Grady (1989), Holman (1995a). **Pleistocene (Rancholabrean NALMA):** New Trout Cave, Pendleton County, West Virginia—Holman and Grady (1987), Holman (1995a) (remains were found in all three stratigraphic units of this cave [see Holman and Grady, 1987]). Strait Canyon Fissure, Highland County, Virginia—Fay (1984) (these remains were referred to by Fay as *Desmognathus* (?) *fuscus*).

The Northern Dusky Salamander is a medium-sized member of this large genus and is rather plain, with a gray or brown coloration. It has a keeled tail, which is a good field mark. The base of the tail is usually lighter than the rest of the upper part of the body. This animal tends to be abundant within its range and occurs in brooks and near springs. It is seldom found far from water. This salamander has a wide range in the Appalachian area, occurring from southern New Brunswick and Quebec to southeastern Indiana and the Carolinas (Conant and Collins, 1998).

Identification of Fossils. This section deals with the identification of the vertebrae of the three species of *Desmognathus* (*Desmognathus fuscus, Desmognathus monticola,* and *Desmognathus ochrophaeus*) that have been found in the Pleistocene of North America. All three of these species are from the Appalachian region. The term *falsely opisthocoelous* (sometimes the term *pseudo-opisthocoelous* is used instead) below refers to the independently ossified, forward-extending plug that occurs in the cotyle area of the genus *Desmognathus.* This plug does not usually separate from the vertebrae during the fossilization process. The following paragraph is directly from Holman and Grady (1987, p. 311): "The small, falsely opisthocoelous vertebrae of this genus are easily distinguished from other North American salamander genera. The structure of the posterior end of the neural arch appears to separate most vertebrae of *D. fuscus, D. monticola,* and *D. ochrophaeus.* In *D. fuscus* and *D. ochrophaeus* the posterior end of the neural arch is notched, but the notching is never as deep in *D. fuscus* as it is in some *D. ochrophaeus.* Deeply notched vertebrae (Figure 5) [in Holman and Grady, 1987] are assigned to *D.* ochrophaeus, moderately notched ones to *D. fuscus.* In *D. monticola* the posterior end of the neural arch is slightly convex, flattened, or very weakly notched. Very weakly notched vertebrae are assigned to *Desmognathus* sp., as very weak notches are found in some *D. fuscus* and *D. ochrophaeus.* Vertebrae with convex or flattened posterior ends are assigned to *D. monticola.*"

It turned out that 13 vertebrae of *Desmognathus fuscus,* 18 vertebrae of *Desmognathus monticola,* and 9 vertebrae of *Desmognathus ochrophaeus* from the New Trout Cave site in West Virginia were complete enough to be identified in this manner (Holman and Grady, 1987).

Desmognathus monticola Dunn, 1916
Seal Salamander

Fossil Localities. **Pleistocene (Irvingtonian NALMA), Kansan glacial stage:** Hamilton Cave, Pendleton County, West Virginia—Holman and Grady (1989), Holman (1995a). **Pleistocene (Rancholabrean NALMA):** New Trout Cave, Pendleton County, West Virginia—Holman and Grady (1987), Holman (1995a) (remains were found in two of three stratigraphic units of this cave [see Holman and Grady, 1987]).

The Seal Salamander is an impressive, stout salamander that has a bold pattern on its upper body but is plain and pale below. Thus, there is usually a distinct line of separation between the colors of the upper and lower parts of the body. The color pattern on the upper part of the body is variable. The background color may be buff, gray, or light brown. The markings on this coloration range from netlike patterns, to circular areas, to dark or light spots or streaks. The tail is very compressed and has thin edges (Conant and Collins, 1998). This animal prefers damp areas and outright wet spots near shaded streams or torrential mountain brooks. Conant and Collins (1998) pointed out that this large salamander looks like a miniature seal when captured in the light of a flashlight.

Identification of Fossils. See the previous account.

Desmognathus ochrophaeus Cope, 1859
Allegheny Mountain Dusky Salamander
Fig. 45

Fossil Localities. **Pleistocene (Irvingtonian NALMA), Kansan glacial stage:** Hamilton Cave, Pendleton County, West Virginia—Holman and Grady (1989), Holman (1995a). **Pleistocene (Rancholabrean NALMA):** Bell Cave, Colbert County, Alabama—Holman et al. (1990), Holman (1995a) (remains were found in one of three stratigraphic units in this cave [see Holman et al., 1990]). New Trout Cave, Pendleton County, West Virginia—Holman and Grady (1987), Holman (1995a) (remains were found in all three stratigraphic units in this cave [see Holman and Grady, 1987]).

The Allegheny Mountain Dusky Salamander can be immediately distinguished from the two other *Desmognathus* species we have discussed here on the basis of its round, rather than sharp or keeled tail. It can be readily distinguished from the Northern Dusky Salamander (*Desmognathus fuscus*) on the basis of its very sinuous mouth line, which makes its mouth look shriveled from the side, especially in specimens preserved in alcohol. It is exceedingly variable in coloration, but one good field mark is that it has a light-colored line extending from the eye to the angle of the jaw. This animal wanders farther from water than most species of *Desmognathus*. This salamander occurs mainly in upland areas from the

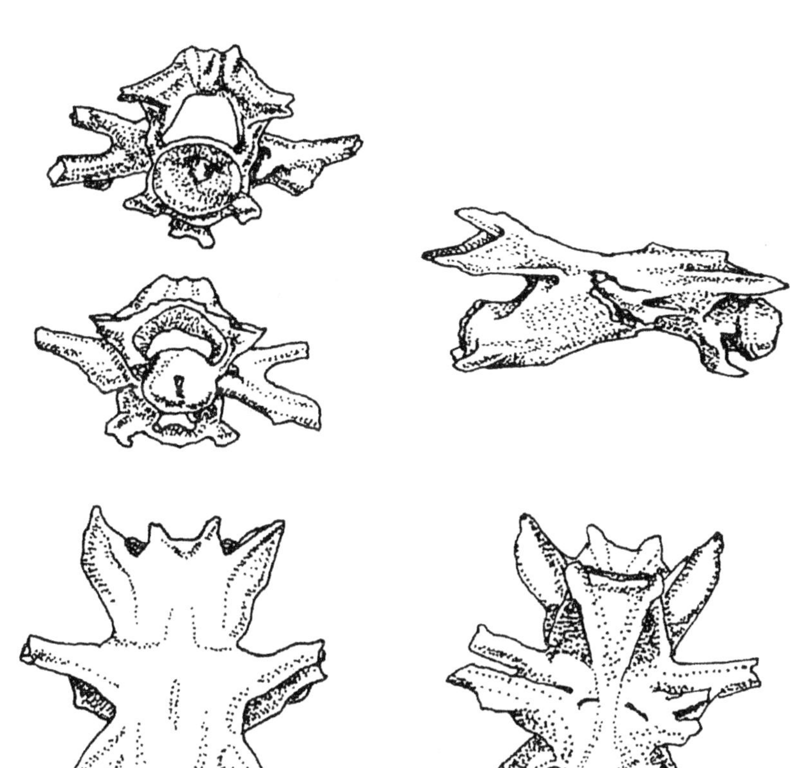

Figure 45. Trunk vertebra of *Desmognathus ochrophaeus* from the Pleistocene (Rancholabrean NALMA) of New Trout Cave, West Virginia. Left column: top, posterior view; middle, anterior view; bottom, dorsal view. Right column: top, lateral view; bottom, ventral view. No scale provided.

Adirondack Mountains in New York to northern Georgia and northeast Alabama. Isolated colonies exist in west-central Georgia and northeastern Kentucky. It reaches the coniferous forests on the very highest peaks in the southern Appalachians (Conant and Collins, 1998).

Desmognathus sp. indet.

Fossil Localities. **Pleistocene (Rancholabrean NALMA):** Baker Bluff Cave, Sullivan County, Tennessee—Van Dam (1978), Fay (1988), Holman (1995a). Bell Cave, Colbert County, Alabama—Holman et al. (1990), Holman (1995a) (these remains occurred in one of three stratigraphic units in the cave [see Holman et al., 1990]). Clark's Cave, Bath County, Virginia—Fay (1988), Holman (1995a). Guy Wilson Cave, Sullivan County, Tennessee—Fay (written communication, 1993[1]), Holman (1995a). Kingston Saltpeter Cave, Bartow County, Georgia—Fay (1988), Holman (1995a). Ladds Quarry site, Bartow County, Georgia—Holman (1985a, 1985b, 1995a). Natural Chimneys site, Augusta County, Virginia—Fay (1988), Holman (1995a). New Paris 4 site, Bedford County, Pennsylvania—Fay (1988), Holman (1995a).

General Comments. Except for the Bell Cave fossils, all these identifications were originally recorded as *Desmognathus* or *Leurognathus*. The two genera have been recently combined, as mentioned above.

Genus *Eurycea* Rafinesque, 1822
Brook Salamanders

Genotype. Eurycea lucifuga Rafinesque, 1822.

Diagnosis. The diagnosis presented here is modified from the definition of the genus provided by Wake (1966). The adults of the genus *Eurycea* are either neotenic or fully transformed. The eyes are well developed. They have either parietal–otic or squamosal–otic crests or no crests at all. Orbitosphenoids are well developed. The anterior and posterior vomerine teeth are discontinuous. Fifteen to 21 trunk vertebrae are present, and the transverse processes of the trunk vertebrae do not extend beyond the zygapophyses. There is no basal tail constriction. Four fingers and five toes are present.

Brook Salamanders range from small to medium sized. They are brightly colored yellowish, orangish, or reddish animals with dark stripes or spots. Most of them live near small streams, but some live in caves. Others are neotenic, some to the extreme.

Identification of Fossils. Miller (1992, pp. 49, 51) gave criteria for identifying vertebrae of the genus *Eurycea*. "*Eurycea* vertebrae have widely separated transverse processes not greatly exceeding lateral margins of zygapophyses (Figure 14A) [in Miller, 1992]. Also, parapophyses are well anterior to diapophyses. Neural ridges are weak-to-well developed. Hyperapophyses are separate and hypapophyses are absent. Posteriorly placed basapophyses are well-developed, especially on anterior vertebrae (Figure 16A) [in Miller, 1992]. Vertebrae of most *Eurycea* species are at least slightly opisthocoelous due to the ring of calcified cartilage on the anterior end of the centrum."

General Comments. This genus of the tribe Hemidactyliini is currently composed of 22 species. They range through eastern North America, except for very cold regions. In 1985, only 11 species of *Eurycea* were recognized (Frost, 1985). This means that the number of recognized species of this genus has literally doubled in about the last 20 years.

Eurycea cirrigera (Green, 1830)
Southern Two-lined Salamander

Fig. 46A

Fossil Locality. **Pleistocene (Rancholabrean NALMA):** Cheek Bend Cave, Maury County, Tennessee—Miller (1992), Holman (1995a) (this record was listed by Miller [1992] as *Eurycea* cf. *Eurycea bislineata;* later, Holman [1995a] referred it to *Eurycea bislineata*).

The Southern Two-lined Salamander and the Northern Two-lined Salamander are similar, except that *Eurycea cirrigera* has 15 or 16 costal grooves (conspicuous "rib grooves") on the body and *Eurycea bislineata* has 14. The common little Southern Two-lined Salamander is yellow, with a distinct dark line on either side of the upper body. In between these two lines is a broader but less distinct line. This salamander normally lives near brooks, where it hides under many kinds of objects. In wet weather it may wander into the woods. This salamander occurs from southern Virginia west to eastern Illinois and south to northern Florida and eastern Louisiana, with an isolated colony occurring in northeastern Illinois (Conant and Collins, 1998).

Identification of Fossils. Miller (1992) gave features of adult vertebrae of *Eurycea cirrigera* that are helpful in the identification of fossils of this species. His description is modified as follows. The vertebrae of adults (Fig. 46A) are very small, with prezygapophyseal to postzygapophyseal lengths ranging approximately from 1.5 to 2.5 mm. They are distinct in being pseudo-opisthocoelous—that is, having a ring of calcified cartilage on the anterior cotylar end of the centrum (Fig. 46A). The neural ridge (spine) is often weakly developed on the posterior trunk vertebrae. Basapophyses are well developed, especially on anterior trunk vertebrae. The Dwarf Salamander (*Eurycea quadridigitata* Holbrook, 1842) has vertebrae that appear to be indistinguishable from those of *E. cirrigera.* The Cheek Bend Cave fossils above, however, came from central Tennessee, in the heart of the present distribution of *E. cirrigera*, whereas *E. quadridigitata* is found well out of the range of the former species and occurs on the Atlantic and Gulf coastal plains (see Conant and Collins, 1998, map, p. 494).

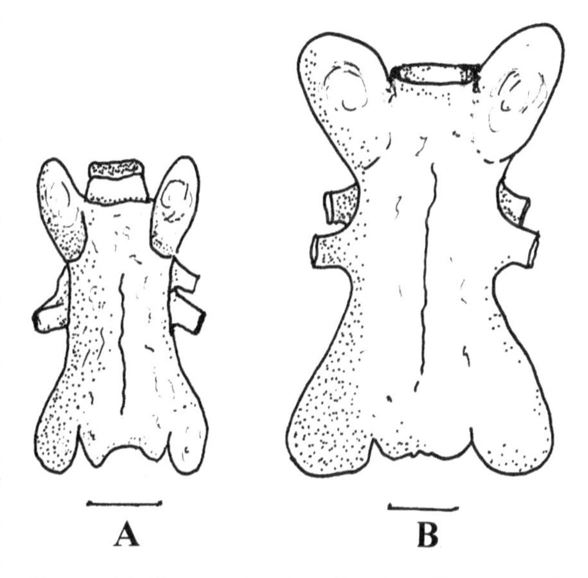

Figure 46. Comparative size of modern *Eurycea* trunk vertebrae in dorsal view. (A) *Eurycea cirrigera.* (B) *Eurycea lucifuga.* Scale bars = 0.5 mm.

General Remarks. Here we are including *Eurycea wilderae* Dunn, 1920 as a subspecies of *Eurycea cirrigera* rather than as a full species, because some gene exchange occurs where their ranges are contiguous but not overlapping (see discussion in Crother et al., 2003). **Author's note: Eurycea longicauda** (Green, 1818), the Long-tailed Salamander, was identified from lower Holocene strata of the Cheek Bend Cave site. The vertebrae of this species are generally similar to those of the species to follow in the next account, *Eurycea lucifuga*. Thus, it is not impossible that a few of the large *Eurycea* vertebrae in the Pleistocene strata are actually those of *E. longicauda*.

In Field and Lab. I once entered a cave in southern Appalachia by crawling through a long, very low crack in the rock structure. Pushing my aging body along with my toes for minutes on end caused me to be embarrassed and very nervous; which the experienced cavers accompanying me seemed to find irritating and funny at the same time. My tension was rewarded, however, when we entered a beautiful grotto. The colors of the walls were a magnificent reddish hue, and Cave Salamanders and Cave Crickets (the insects waving their unbelievably long antennae slowly about) were abundant.

Eurycea lucifuga Rafinesque, 1822

Cave Salamander

Fig. 46B

Fossil Locality. **Pleistocene (Rancholabrean NALMA):** Cheek Bend Cave, Maury County, Tennessee—Miller (1992).

The Cave Salamander is a reddish or pinkish salamander with a long tail. Dark spots occur on the background color. Conant and Collins (1998) reported that a favorite habitat is in the twilight zone of caves, near entrances where the light is weak. These salamanders use their prehensile tails as aids for climbing about on ledges and on piles of fallen limestone or shale. Cave Salamanders occur in areas of limestone, from Virginia and West Virginia southwest to central Alabama and northwest to the Ozarks.

Identification of Fossils. Miller (1992) made useful comments about the identification of fossil Cave Salamander remains in Cheek Bend Cave, Tennessee. The atlas of *Eurycea lucifuga* can be distinguished from that of *Eurycea longicauda* (a species that was found in a lower Holocene stratum of the Cheek Bend Cave) on the basis of the neural arch. In *E. longicauda* the dorsal ridge of the neural arch becomes well ossified at a smaller size and has an anteriorly raised crest. This crest is relatively straight in *E. lucifuga*. All adult *E. lucifuga* vertebrae are much larger than those of *Eurycea cirrigera*.

The vertebrae of populations of all (including Holocene) *Eurycea lucifuga* in the cave consisted of two morphological populations, with 114 "robust" and 539 "normal" vertebrae identified. Robust Cave Salamander vertebrae were characterized as always being larger, better ossified, and broader than the normal vertebrae. Only one modern comparative skeleton, a 62 mm (SVL) male, was found to be identical to the robust cave population fossils. Other comparative skeletons of males showed the normal form. However, whether the condition observed was due to sexual dimorphism or age-related factors is unknown because of the lack of modern comparative specimens (Miller, 1992).

EURYCEA SP. INDET.

Fossil Localities. **Pleistocene (Rancholabrean NALMA):** Bell Cave, Colbert County, Alabama—Holman et al. (1990), Holman (1995a) (these fossils occurred in two of three strata in the cave [see Holman et al., 1990]). Cheek Bend Cave, Maury County, Tennessee—Klippel and Parmalee (1982), Miller (1992), Holman (1995a). Clark's Cave, Bath County, Virginia—Fay (1988), Holman (1995a). Kingston Saltpeter Cave, Bartow County, Georgia—Fay (1988), Holman (1995a). Natural Chimneys site, Augusta County, Virginia—Fay (1988), Holman (1995a). New Paris 4 site, Bedford County, Pennsylvania—Lynch (1966), Fay (1988), Holman 1995a).

GENUS *GYRINOPHILUS* COPE, 1869

SPRING SALAMANDERS

Genotype. Gyrinophilus porphyriticus (Green, 1827).

Diagnosis. This diagnosis is modified from the definition of the genus provided by Wake (1966). The adults may be neotenic or fully transformed. The eyes are well developed and functional. The premaxillary bones are separate in fully transformed species but fused in neotenic species. Parietal–otic and squamosal–otic crests are present. Orbitosphenoids are well developed. Anterior and posterior vomerine teeth are continuous. Eighteen to 20 trunk vertebrae are present. The transverse processes of the trunk vertebrae extend beyond the zygapophyses. The tails are not basally constricted. Four fingers and five toes are present.

Spring Salamanders are medium-sized caudates that are usually reddish or salmon colored. Species of this genus are partial to cool, clear water in situations ranging from seepage areas to mountain brooks. This genus is currently composed of four species. The Spring Salamanders range from southwestern Maine and southern Quebec to northern Alabama and extreme northeastern Mississippi, with an isolated population occurring in Ohio near Cincinnati. They are in the tribe Hemidactyliini of the subfamily Plethodontinae.

GYRINOPHILUS PORPHYRITICUS (GREEN, 1827)

SPRING SALAMANDER

FIG. 47B

Fossil Localities. **Pleistocene (Irvingtonian NALMA), Kansan glacial stage:** Hamilton Cave, Pendleton County, West Virginia—Holman and Grady (1987), Holman (1995a). **Pleistocene (Rancholabrean NALMA):** Clark's Cave, Bath County, Virginia—Fay (1988), Holman (1995a). Guy Wilson Cave, Sullivan County, Tennessee—Fay (written communication, 1993[1]), Holman (1995a). Natural Chimneys site, Augusta County, Virginia—Fay (1988), Holman (1995a). New Trout Cave, Pendleton County, West Virginia—Holman and Grady (1987), Holman (1995a).

The Spring Salamander is a vividly colored caudate. The upper part

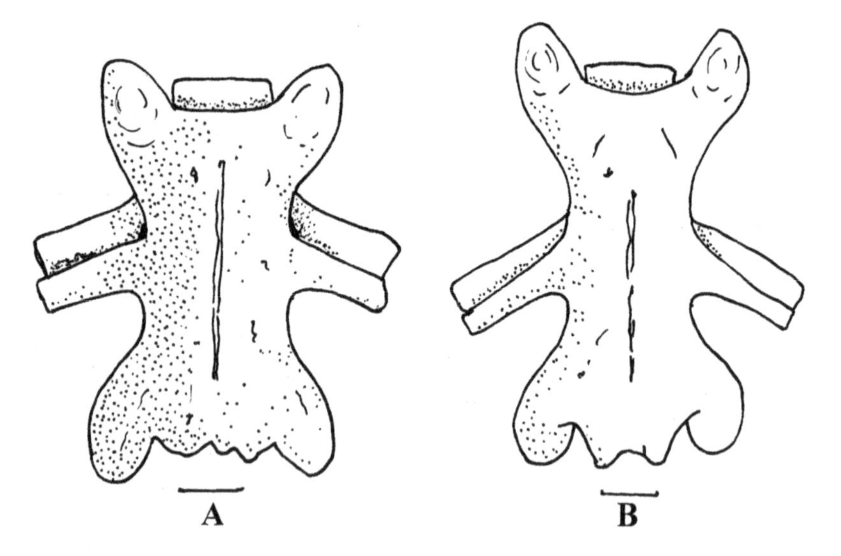

FIGURE 47. (A) Trunk vertebra of *Pseudotriton ruber* in dorsal view. (B) Vertebra of *Gyrinophilus porphyriticus* in dorsal view. Scale bars = 0.5 mm.

of the body is reddish, salmon, or dark yellow; these colors are marked with black or brown spots or flecks. The best field mark is a white line from the eye to the nostril that is bordered below by a distinctive dark line. These animals frequent habitats such as mountain springs, seepages, and edges of brooks and streams, as well as wet forest areas near all these habitats. The present range of this salamander is mainly in Appalachia, from southern Maine and Quebec to northern Georgia and Alabama.

In Field and Lab. This is a beautiful salamander that normally lives in lovely forested areas. A person never forgets seeing the first one in the wild.

Identification of Fossils. The trunk vertebrae have obsolete, very low neural spines, with no raised central portion. The rib-bearers are widely separated, and the end of the neural arch is unnotched. When the notochordal canals of the vertebral centra in fossils of these salamanders are closed or tiny, the indication is that metamorphosed adults are represented.

GYRINOPHILUS—POSSIBLE UNNAMED NEOTENIC SPECIES

Fossil Locality. **Pleistocene (Irvingtonian NALMA), Kansan glacial stage:** Hamilton Cave, Pendleton County, West Virginia—Holman and Grady (1989), Holman (1995a).

Two vertebrae (USNM 421564, United States National Museum) from the Cheetah locality at Hamilton Cave have widely open notochordal canals and are much larger than vertebrae of *Gyrinophilus porphyriticus* identified from the *Smilodon* locality at the cave. These two vertebrae could, therefore, represent an unnamed neotenic, troglodytic (cave) species of *Gyrinophilus*. The closest living troglodytic species in the area is the West Virginia Spring Salamander (*Gyrinophilus subterraneus* Besharse and Holsinger, 1977), which lives microsympatrically (side by side, without interbreeding) with *G. porphyriticus* at the type locality. Among other differences between *G. subterraneus* and *G. porphyriticus* (e.g., a wider head and a paler skin in the troglodytic species) is the fact that G. *subterraneus* is larger. This species is restricted to General Davis Cave, Greenbrier County, in the Ridge and Valley province of West Virginia (Green and Pauley, 1987). Hamilton Cave is located in Pendleton County, also in the Ridge and Valley province, but about 85 miles (about 137 km) to the northeast of the West Virginian Spring Salamander, as far as I can determine. It thus appears unlikely that the Hamilton Cave fossils are G. *subterraneus*.

The other troglodytic species in the *Gyrinophilus* complex are the Berry Cave Salamander (*Gyrinophilus gulolineatus* Brandon, 1965) and the Tennessee Cave Salamander (*Gyrinophilus palleucus* McCrady, 1954). Together, these species live in the Tennessee–northern Georgia–northern Alabama region. This would seem to eliminate them as species to be considered as taxonomic candidates for the unidentified neotenic Hamilton Cave Spring Salamander.

General Comments. Holman (1975a) pointed out that widely open notochordal canals are characteristic of fossil neotenic salamanders. In metamorphosed salamanders the notochordal canals are normally constricted or closed.

Gyrinophilus cf. *Gyrinophilus porphyriticus* (Green, 1827)

Probable Spring Salamander

Fossil Locality. **Pleistocene (Irvingtonian NALMA), Kansan glacial stage:** Cumberland Cave, Allegany County, Maryland—Holman (1977b, 1995a).

Gyrinophilus sp. indet.

Fossil Localities. **Pleistocene (Rancholabrean NALMA):** Kingston Saltpeter Cave, Bartow County, Georgia—Fay (1988), Holman (1995a). Ladds Quarry site, Bartow County, Georgia—Holman (1967, 1985a, 1985b, 1995a).

Gyrinophilus or *Pseudotriton*

Fossil Locality. **Pleistocene (Rancholabrean NALMA):** Guy Wilson Cave, Sullivan County, Tennessee—Fay (written communication, 1993[1]), Holman (1995a).

Fay (written communication, 1993[1]) was able to separate these vertebrae to the taxonomic level of either *Gyrinophilus* or *Pseudotriton* (discussed later in the book) but could go no farther in the identification process.

Genus *Hydromantes* Gistel, 1848

Web-toed Salamanders

Genotype. Hydromantes genei (Temminck and Schlegel, 1838).

Diagnosis. This definition is modified from the definition of the genus by Wake (1966). *Hydromantes* comprises a plethodontine salamander group. The premaxillae are separated, and the prefrontals are absent. Posterolateral parietal spurs are absent. Well-developed columellae are present. Fourteen trunk and three caudosacral vertebrae occur. There is no basal tail constriction. Five toes are present.

General Comments. These salamanders of the plethodontine tribe Bolitoglossini have webbed toes (as in some of their tropical relatives), a very long, mushroom-like tongue, and a flat body. Males have projecting teeth on the upper jaw. Three species are currently recognized in North America (Crother, 2000; Crother et al., 2003), and all occur only in California.

Hydromantes sp. indet.

Fossil Locality. **Pleistocene (Rancholabrean NALMA):** Kings Canyon Biota, south-central Sierra Nevada, California—Mead et al. (1985).

Identification of Fossils. The identification of *Hydromantes* sp. was based on vertebral characters. Mead et al. (1985, p. 108) stated, "The identification of the web-toes salamander was based on the shape of the transverse process and the height and division of the crests on the posterior portion of the neural arch. The keel on the anterior part of the arch is like that of *Hydromantes*, and unlike the other plethodontid salamanders in the area."

General Comments. The vertebral characters matched those of both the Mount Lyell Salamander (*Hydromantes platycephalus*) and the Limestone Salamander (*Hydromantes brunus*). Both species live in the Sierra Nevada range today. Neither of these species occurs in the immediate vicinity of the fossil locality at present, but the Limestone Salamander occurs at about the same elevation in mixed chaparral–oak woodland along the Merced River north of the site. As far as I am aware, this is the only fossil record of *Hydromantes*.

Genus *Plethodon* Tschudi, 1838
Woodland Salamanders

Genotype. Plethodon glutinosus (Green, 1818).

Diagnosis. This genus of the tribe Plethodontini of the subfamily Plethodontinae differs from the other two genera in the tribe in the following ways. Adults differ from *Aneides* on the basis of having two premaxillae and from *Ensatina* in having tibial spurs and in lacking a basal constriction of the tail (Wake, 1966).

Woodland Salamanders (*Plethodon*) are completely North American in their distribution. They are very widespread in the woodlands of eastern North America. Others occur in forested parts of the far west, from the northwestern part of California to British Columbia, in the mountains of northern Idaho and adjacent Montana, and in north-central New Mexico (Conant and Collins, 1998). There is no aquatic larval stage in any of the species of *Plethodon*, and their eggs are laid in small clusters in logs, on leaves on the woodland floor, or in other such situations.

In Field and Lab. Systematic zoologists talk (sometimes approvingly and sometimes hysterically) about "lumpers" and "splitters." Lincoln et al. (1982, p. 144) defined a *lumper* as one who engages in lumping: "The taxonomic process of ignoring minor variation in the definition or recognition of species." The same authors (p. 232) defined a *splitter* as one who engages in splitting: "The taxonomic process of subdividing species on the basis of more or less minor differences." I want to clear that all up before I discuss the number of species in the genus *Plethodon*.

As I write, there are 53 officially recognized species of the genus *Plethodon;* most are confined to the Appalachian region, but some can be found in the far west (Crother, 2000). Yesterday, as I wrote, I had an e-mail message warning me that there was yet another new species of *Plethodon* to be reckoned with. One species, *Plethodon glutinosus*, which had four subspecies in 1953 (Schmidt, 1953), has been divided (split) into 13 species, and no subspecies are recognized. The point for any potential paleosalamander people reading this book is that we used to be able to identify individual *P. glutinosus* vertebrae in Pleistocene fossil sites (or at least we thought that we could). It is literally impossible to find definitive characters in vertebrae from any of the recently described species of "*Plethodon glutinosus*"; thus, we shall have to list these as *Plethodon glutinosus* species complex.

Plethodon glutinosus (Green, 1818) species complex

Slimy Salamander species complex

Fossil Localities. **Pleistocene (Irvingtonian NALMA), Kansan glacial stage:** Cumberland Cave, Allegany County, Maryland—Holman (1977a, 1995a). **Pleistocene (Rancholabrean NALMA):** Bell Cave, Colbert County, Alabama: Holman et al. (1990), Holman (1995a) (these remains occurred in one of three strata [see Holman 1995a]). Cave Without a Name Fauna, Kendall County, Texas—Holman (1968a, 1995a). Ladds Quarry site, Bartow County, Georgia—Wilson (1975), Holman (1967, 1985a, 1985b). New Trout Cave, Pendleton County, West Virginia—Holman and Grady (1987), Holman (1995a) (these remains occurred in one of three strata [see Holman 1995a]). Orange Lake site, Marion County, Florida—Holman (1959b, 1995a). Peccary Cave, Newton County, Arkansas—Davis (1973), Holman (1995a). Sabertooth Cave, Citrus County, Florida—Holman (1958, 1995a). Strait Canyon Fissure, Highland County, Virginia—Fay (1984), Holman (1995a). Williston IIIA site, Levy County, Florida—Holman (1959a, 1995a, 1996a).

Identification of Fossils. Wilson (1975, p. 48) identified *Plethodon* cf. *Plethodon glutinosus* from the Ladds Quarry site, Bartow County, Georgia (2.2 miles [about 3.54 km] west-southwest of Cartersville), as follows: "*Plethodon glutinosus* vertebrae are amphicoelous. They can be separated from the vertebrae of *Leurognathus* and *Desmognathus* which are opisthocoelous, and from *Pseudotriton* and *Gyrinophilus* vertebrae which are falsely opisthocoelous (Soler, 1950). Holman (1969) [*sic;* is actually 1968a in reference list, this volume] separates *Plethodon glutinosus* vertebrae from those of *Eurycea bislineata* (Green), *E. longicauda* (Green), and *E. lucifuga* (Rafinesque) on the basis of the shorter and wider dimensions of the former. He also notes that adult *P. glutinosus* vertebrae are larger than those of *Plethodon cinereus* (Green), *E. bislineata* and the troglodyte species of *Eurycea.* Using these criteria, the fossil vertebrae are assigned to *Plethodon glutinosus* from which they cannot be separated."

Holman (1967, pp. 155–156) was able to identify four salamander femora from the Ladds Quarry site as *Plethodon* cf. *Plethodon glutinosus* as follows: "The femur of the several large species of *Plethodon* available is stouter and has more robust processes than that of *Eurycea*. The femur of *Pseudotriton* and *Gyrinophilus* is as large as or larger than that of the large species of *Plethodon*, but it has a much more flattened head and a wider trochanter than in *Plethodon*. Distinguishing species of *Plethodon* on the basis of the femur is difficult. The femur of *Plethodon glutinosus* is much larger than *P. cinereus* (Green), *P. dorsalis* (Cope), *P. richmondi* Netting and Mittleman, and *P. welleri* Walker. The other *Plethodon* species have the femur similar in size to *P. glutinosus*, but the head of *P. jordani* Blatchley is flatter than in *P. glutinosus*, and in a single specimen of *P. yonahlossee* Dunn the bone is less stout and the processes less massive than in *P. glutinosus*. On the basis of the above characters the fossil femora are tentatively assigned to the species *P. glutinosus*."

PLETHODON SP. INDET.

FIGS. 48D–48F (LATE OLIGOCENE TO EARLY MIOCENE *PLETHODON*)

INDETERMINATE WOODLAND SALAMANDERS (*PLETHODON*)

Fossil Localities. **Late Oligocene to Early Miocene (Arikareean NALMA):** Cabbage Patch Formation, Granite County, Wyoming—Tihen and Wake (1981). **Pleistocene (Rancholabrean NALMA):** Clark's Cave, Bath County, Virginia—Fay (1988), Holman (1995a). Guy Wilson Cave, Sullivan County, Tennessee—Fay (written communication, 1993[1]), Holman (1995a). Kingston Saltpeter Cave, Bartow County, Georgia—Fay (1988), Holman (1995a). Natural Chimneys site, Augusta County, Virginia—Fay (1988), Holman (1995a). New Paris 4 site, Bedford County, Pennsylvania—Fay (1988), Holman (1995a).

Identification of Fossils. The process by which Tihen and Wake (1981) identified *Aneides* from the Late Oligocene to Early Miocene Cab-

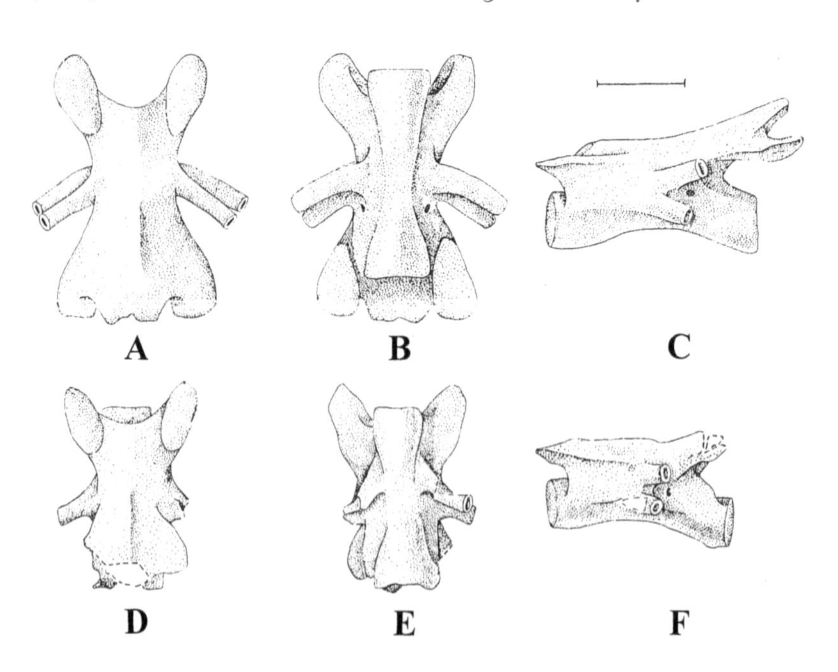

FIGURE 48. Seventh trunk vertebra of modern *Plethodon dunni* female (snout to vent length, 60.5 mm) from Oregon in (A) dorsal, (B) ventral, and (C) lateral views. (D)–(F) Trunk vertebra of *Plethodon* sp. from the Early Miocene of Montana in (D) dorsal, (E) ventral, and (F) lateral views. Scale bar = 0.5 mm and applies to all figures.

bage Patch Formation was given in the account of *Aneides* sp. indet. (see above). These authors also confidently identified a middle trunk vertebra (MSUMP 65042003, Montana State University Museum of Paleontology) (Figs. 48D–48F) from the Cabbage Patch Formation as belonging to the genus *Plethodon*, by the following process (modified).

Identification of Fossils. The centrum of 65042003 is relatively long and narrow. The rib-bearers are situated relatively far forward on the vertebra (compared with *Aneides* sp., MSUMP 65042004), and the dorsal and ventral rib-bearers are connected with a thin web of bone to their tips. The impression of the vertebral canal forms a depression along the dorsolateral side of the centrum, and the canal is almost entirely ventral to the rib-bearer. This feature usually distinguishes western species of *Plethodon* from *Aneides.* Tihen and Wake (1981) decided that 65042003 is very similar to middle trunk vertebrae of modern species such as *Plethodon dunni* (Figs. 48A–48C) and *Plethodon vehiculum.* An anterior caudal vertebra (KU 18298, Museum of Natural History, University of Kansas) from the Cabbage Patch Formation that Tihen and Wake assigned to *Plethodon* has a mid-vertebral spinal foramen and a long and slender centrum. It also has a thin keel on one side of the ventral portion of the centrum. The position of the transverse processes—slightly anterior to the midpoint of the vertebra—indicates that this is from about the sixth or seventh vertebral point behind the sacrum. Tihen and Wake suggested that this vertebra could be from the same species as 65042003.

Genus *Pseudotriton* Tschudi, 1838

Red and Mud Salamanders

Genotype. Pseudotriton ruber (Latreille, 1801).

Diagnosis. This definition is modified from the definition of the genus provided by Wake (1966). The adults are fully transformed. The eyes are well developed and functional. The premaxillae are fused. Parietal–otic and squamosal–otic crests are present. The anterior and posterior vomerine teeth are continuous. Seventeen or 18 trunk vertebrae are present. The transverse processes of the trunk vertebrae extend beyond the zygapophyses. There is no basal constriction in the tail. Four fingers and five toes are present.

Red and Mud Salamanders are rather large, reddish or bright red salamanders with small, dark spots. They are members of the subfamily Plethodontinae and the tribe Hemidactyliini. They occur mainly in moist woodland habitats in the eastern United States, especially the southeastern portion. The genus has two species, the Mud Salamander (*Pseudotriton montanus*) and the Red Salamander (*Pseudotriton ruber*), each of which has four subspecies. Only *P. ruber* is known in the fossil record.

Pseudotriton ruber (Latreille, 1801)

Red Salamander

See Fig. 47A

Fossil Localities. **Pleistocene (Irvingtonian NALMA), Kansan glacial stage:** Cumberland Cave, Allegany County, Maryland—Holman (1977b, 1995a). **Pleistocene (Rancholabrean NALMA):** Kingston Saltpeter Cave, Bartow County, Georgia—Fay (1988), Holman (1995a). Ladds Quarry site, Bartow County, Georgia—(Holman (1967, 1985a,b, 1995a)

The modern Red Salamander, in regions where these fossils have been found, are mainly red or reddish orange and are covered above with irregularly rounded black spots. They are moderately large salamanders that occur in or near clear, cool water situations of many kinds. The Red Salamander ranges from southern New York and Pennsylvania south, through the Appalachians, to northernmost Florida; then west to Louisiana and north to north-central Kentucky. A small population occurs in southeastern Indiana.

In Field and Lab. I suspect that these robust salamanders will grab at any small creatures that come nearby. I found a beautiful Red Salamander under a rotting log on an April day in the Howard College Natural Area in greater Birmingham, Alabama (Holman 1961). The salamander was 116 mm long and disgorged the tail of a large Ground Skink (*Scincella lateralis*), a lizard that prowls the leaf litter on the woodland floor.

Identification of Fossils. Holman (1967, p. 156) remarked about the identification of *Pseudotriton ruber* from the Pleistocene (Rancholabrean NALMA) of the Ladds Quarry site, Bartow County, Georgia, as follows: "The presacral vertebrae of *Pseudotriton* and *Gyrinophilus* may be distinguished from those of *Desmognathus* and *Leurognathus* in that they are amphicoelous (more precisely falsely opisthocoelous, Soler, op. cit.) rather than truly opisthocoelous as in the latter two genera. Presacral vertebrae of *Pseudotriton* and *Gyrinophilus* may be distinguished from those of *Plethodon* in that they are larger and have more restricted notochordal canals. *Pseudotriton* vertebrae may be separated from those of the two available specimens of *Gyrinophilus porphyriticus* (Green) in that the neural spines are well developed in the former and obsolete in the latter form. On the basis of three skeletons of *P. ruber* and two of *P. montanus* Baird the neural spines are higher in the former than in the latter species. The fossil agrees with *P. ruber* in this character."

Pseudotriton cf. *Pseudotriton ruber*

Probable Red Salamander

Fossil Locality. **Pleistocene (Rancholabrean NALMA):** New Trout Cave, Pendleton County, West Virginia—Holman and Grady (1987), Holman (1995a) (this species occurred in two of the three strata in this cave [see Holman, 1995a]).

Pseudotriton sp. indet.

Fossil Localities. **Pleistocene (Rancholabrean NALMA):** Clark's Cave, Bath County, Virginia—Fay (1988), Holman (1995a). Guy Wilson Cave, Sullivan County, Tennessee—Fay (written communication, 1993[1]), Holman (1995a). Natural Chimneys site, Augusta County, Virginia—Fay (1988), Holman (1995a). New Paris 4 site, Bedford County, Pennsylvania—Fay (1988), Holman (1995a) (this species was found in one of two stratigraphic units [see Holman, 1995a]).

Family Amphiumidae Gray, 1825

Amphiumas

Fig. 49

If there is any salamander family that can rival the family Sirenidae in its ability to be misunderstood, it's the Amphiumidae. These animals may be called Congo eel, conger eel, lamper eel, or ditch eel; yet I have never heard anyone but herpetologists call it by its real name, amphiuma, which is probably just as odd as the other names people use! Moreover, this very large eel-like critter can bite like the devil if you grab it, and then it squirms through your hands before you can render the necessary oaths. The leg situation in these eel-like animals is more than ridiculous. Each amphiuma has two sets of absolutely tiny legs. One species has three toes on each tiny leg; another species has two toes; and a third has only one toe, but it doesn't make any difference, because, as far as I know, the toes are not used for anything anyway.

Three living species occur. *Amphiuma means* (Two-toed Amphiuma) occurs in the coastal plain from Virginia to the southern tip of Florida and west to southern Mississippi. *Amphiuma pholeter* (One-toed Amphiuma) is known only from Levy, Calhoun, and Liberty counties, Florida. *Amphiuma tridactylum* (Three-toed Amphiuma) occurs from southeastern Missouri and extreme southeastern Oklahoma to the Gulf of Mexico in tributaries of the Mississippi River (Frost, 1985).

Some osteological characters used to define the living members of

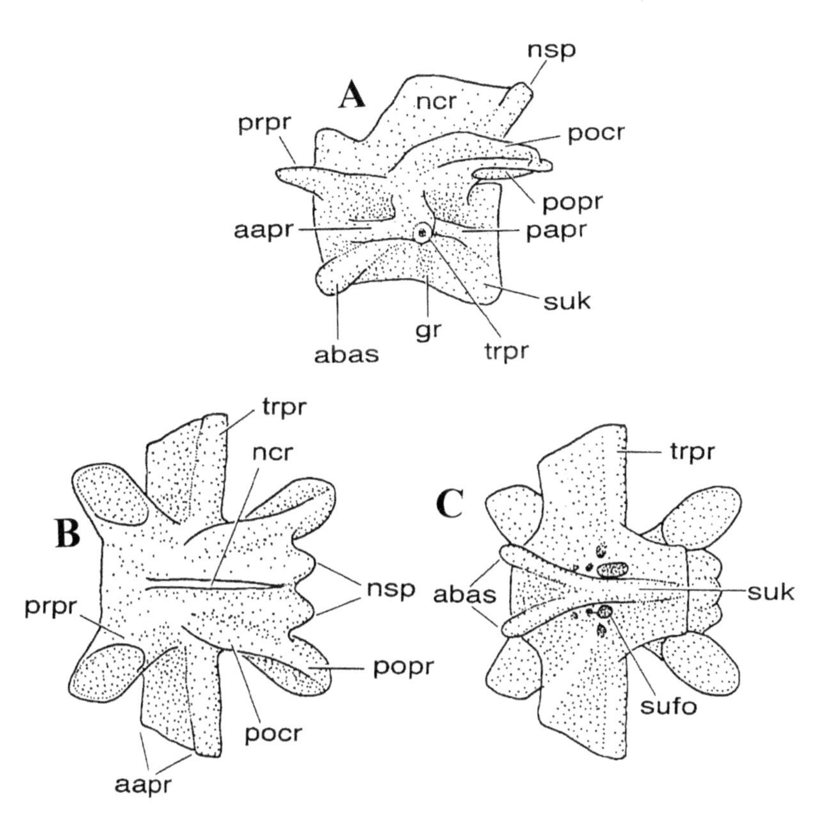

Figure 49. Structures on a generalized vertebra of the Amphiumidae in (A) lateral, (B) dorsal, and (C) ventral views. Abbreviations: aapr, anterior alar process; abas, anterior basapophyses; gr, groove for blood vessel from subcentral foramen; ncr, neural crest; nsp, neural spine; papr, posterior alar process; pocr, postzygapophyseal crest; popr, postzygapophyseal process; prpr, prezygapophyseal process; sufo, subcentral foramen; suk, subcentral keel; trpr, transverse process.

the family (Duellman and Trueb, 1986) are modified as follows. The premaxillae are fused. The septomaxillae, lacrimals, and pterygoids are reduced. The ypsiloid cartilage is absent. The exoccipital, prootic, and opisthotic are not fused. An internal carotid foramen is present. The nasals ossify from lateral developmental centers. The teeth are pedicellate. The palatal teeth parallel the maxillary and premaxillary teeth. The columella is fused with the operculum. The vertebrae are amphicoelous. Only the posterior caudal nerves exit through foramina in the vertebrae. The pectoral and pelvic girdles are highly reduced.

Diagnosis. The following general diagnosis of the family is from Estes (1981, p. 41): "Elongated aquatic permanently larval salamanders, which may reach 90 cm. Front and hind limbs extremely reduced; adults lacking gills but a single persistent gill slit present; three gill slits present in larvae; ypsiloid cartilage, articular bone, septomaxilla, lacrimal, second epibranchial absent; columella fused to skull; spinal nerves intervertebral except in posterior caudals, where they exit through the posterior half of the neural arch; chromosomes *n* 14."

Vertebral Diagnosis. Gardner (2003b) reported that vertebrae (Fig. 49) are diagnostic for Amphiumids and are easily differentiated from those of other salamanders. Diagnostic characters of the atlas vertebra are as follows. (1) The notochordal pit is retained. (2) Anterior cotyles are large and hemispherical, slightly compressed, and deeply concave. (3) A scoop-shaped odontoid process bears a raised subcircular condyle on either side of the ventral midline. (4) The neural arch is robust, has a posteriorly short roof, and bears low neural and accessory crests. Diagnostic characters of the post-atlantal vertebrae (vertebrae posterior to the atlas) include the following. (1) All these vertebrae are amphicoelous and have deeply concave cotyles. (2) All but the posteriormost caudals bear a pair of postzygapophyseal crests. (3) The trunk vertebrae lack spinal foramina but bear a subcentral keel, a pair of well-developed anterior basapophyses, and an elongate and moderately high neural crest. (4) Paired, posteriorly diverging neural spines often occur on the trunk vertebrae. (5) Weakly bicapitate transverse processes occur on the anteriormost trunk vertebrae, but unicapitate ones occur on the others.

In Field and Lab. In the spring of 1959 I was driving west through the eastern Florida Panhandle with my fellow student and herpetological colleague Howard (Duke) Campbell. It was 4 A.M. and perfectly dark, and we decided to shine our headlights on a roadside ditch to see whether anything of note was moving about. As we approached the ditch we heard several splashes. When we focused our lights on the water, we saw both swirls and splashes. I suspected the turmoil was from small bass or crappies feeding. But we finally saw about a dozen large amphiuma heads; all barely at the surface, some stationary and others moving through the water. We finally confirmed that each swirl or splash was an attempt by an amphiuma (*Amphiuma means*) to catch one of the scores of top minnows that were also at the surface looking for prey. When we tried to catch an amphiuma with a long-handled net, both the amphiumas and top minnows moved quickly out of range, and the swirls and splashes stopped.

Genus *Amphiuma* Garden, 1821

Amphiumas

Genotype. Amphiuma means Garden, 1821.

**Amphiuma antica* Holman, 1977a

Holotype. A trunk vertebra (SMPSMU 63676, Shuler Museum of Paleontology, Southern Methodist University).

Type Locality. Moscow site in the Fleming Formation near the town of Moscow, in northern Polk County, Texas.

Horizon. Early Middle Miocene (early Barstovian NALMA).

Other Material. No other material is available.

Diagnosis. The diagnosis is taken directly from Holman (1977a, p. 392): "The vertebra of this form differs from modern species of *Amphiuma* in having a lower, thicker, more rounded ventral keel on the centrum and in having the neural arch more depressed posteriorly. It differs from *Amphiuma jepseni* Estes of the Late Paleocene of Wyoming in having a much more broadly V-shaped posterior portion of the neural arch in dorsal view, and a convex rather than a concave ventral keel of the centrum in lateral view."

Description of the Holotype. This description is slightly modified from Holman (1977a). In anterior view the cotyle is round and is much larger than the neural canal, which is subrounded. The centrum is perforated by the canal of a consistent notochord, a not inconsistent occurrence in neotenically evolved salamander species. In dorsal view, the anterior part of the neural arch is broken off just above the level of the top of the vertebra. The posterior part of the neural arch lacks the V-shaped aliform processes of the Sirenidae that form by a splitting of the neural spine. The posterior border of the neural arch is very broadly V-shaped. In lateral view, the transverse processes are broken. The posterior part of the neural arch is strongly upswept. The keel on the ventral part of the centrum is low and straight posteriorly. In posterior view, the border of the neural arch is very thick, and the dorsal part of the neural arch is depressed. The round cotyle is larger than the subrounded neural canal. In ventral view, the keel of the centrum is thick and rounded. A large foramen is present of each side of the keel.

General Remarks. Holman (1996b) referred again to *Amphiuma antica* from the Moscow site but did not add any additional information. Estes (1981) reviewed this species and remarked that the shape of the neural arch and the straight to convex border of the subcentral keel resemble the situation in most anterior vertebrae of the living *Amphiuma means*, although it differs from *A. means* in being thick and rounded. Estes (1981, p. 42) went on to say, "As it is the first late Tertiary *Amphiuma* described it is likely to be a new species but the available material does not offer clear cut differences. It seems to differ from the Paleocene *Amphiuma jepseni* in having a higher neural arch like the living species." Unfortunately, the figure of *A. antica* in Holman (1977a, fig. 2, p. 393) was very poorly rendered.

*AMPHIUMA JEPSENI ESTES, 1969

FIG. 50A

Holotype. At least 14 associated but dislocated vertebrae in a block of matrix and an associated left quadrate and left dentary (PU 14666, Princeton University).

Type Locality. Polecat Bench Formation (Fort Union Formation), Silver Coulee Quarry, Park County, Wyoming.

Horizon. Late Paleocene.

Other Material. Referred material consists of a partial skull and associated mandibles (PU 14668) from the type locality.

Diagnosis. The diagnosis is directly from Estes (1981, p. 42): "Vertebrae relatively narrow as in *Proamphiuma,* crests and basapophyses well developed as in modern *Amphiuma;* snout short and blunt in contrast to modern *Amphiuma;* vomers less larval, with fewer and relatively more pointed teeth than in modern *Amphiuma.*"

Description of the Holotype. This description is modified from Estes (1981). The restoration of this fossil indicates that the snout is wider and shorter than in modern *Amphiuma.* The premaxillae are unpaired, flattened dorsally, and elongated posteriorly. A strong median septum occurs. Thirteen teeth are present. The ethmoid process of the frontals is visible in section posteriorly. The prefrontals are sculptured and elongated and have a long suture with the maxillae, posterior to the orbit. Strong sculpturing also is present on the frontal. The maxilla is bluntly curved and is sculptured on its nasal process. A row of foramina is present exteriorly on the maxilla. The teeth are placed in a deep, channeled dental sulcus. The teeth are pedicellate and sharply pointed, and they have recurved crowns.

The vomer is expanded anteriorly with a small, hollowed-out dental sulcus present near the midline. Ten or 11 teeth are present here. A choanal excavation is present laterally on the vomer, and it reaches the

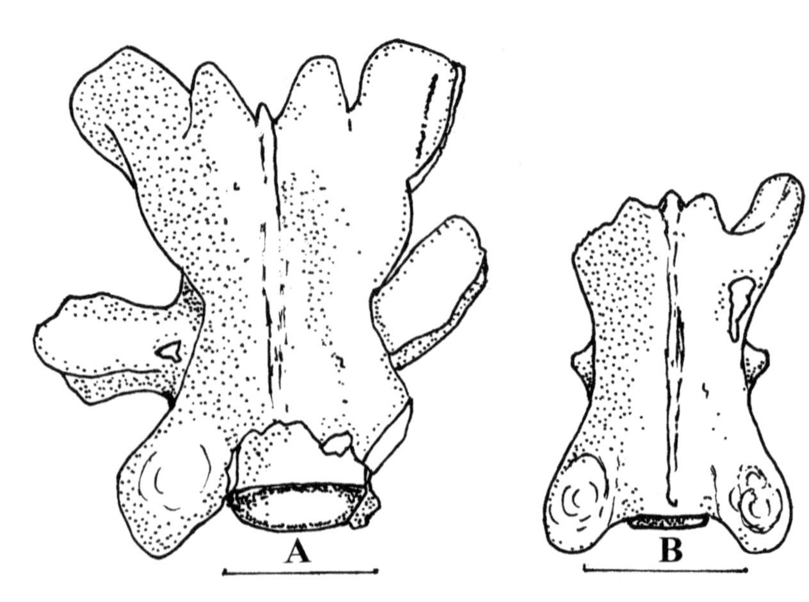

FIGURE 50. (A) Holotype trunk vertebra of *Amphiuma jepseni* from the Paleocene of Wyoming in dorsal view. (B) Holotype trunk vertebra of #*Proamphiuma cretacea* of the Late Cretaceous of Montana in dorsal view. Scale bars = 2 mm.

seventh tooth from the front. The ethmoid is visible medial to the vomer, with an anterior ethmoid foramen visible on the left. The parasphenoid is blunt and extends anteriorly, covering the ethmoids and separating the vomers to the level of the anterior end of the dental sulcus. The orbitosphenoids are missing, but grooves for them are present on the ventral surfaces of the ethmoid. A flange of the vomer covers the orbitosphenoid area ventrally.

The dentary is strongly curved and has a deep dental sulcus. The teeth are similar to those of the upper jaw. The dentary symphysis is flat and prominent. About 18 teeth are indicated. The prearticular is robust and medially crested, reaching anteriorly to about the sixth tooth from the symphysis. A prominent fossa for the adductor muscle attachment occurs posterior to the last tooth. A blunt coronoid process is present. The quadrate is blunt and robust, and its articular surface is gently concave, with the palatoquadrate and hyoid connections separated.

The vertebrae of *Amphiuma jepseni* are amphicoelous, with the cotyles being suboval anteriorly and teardrop shaped posteriorly. The anterior basapophyses are strong, ventrally flattened, and crested on their posterior surfaces. The subcentral keel is well defined, and the subcentral foramina are simple and elongate. The rib-bearers of the posterior vertebrae are single headed and blunt distally and have no rib articulation facet. The anterior vertebrae have double-headed rib-bearers, but they lie close together. Ventral lamina of the transverse processes are prominent in the posterior vertebrae, but they are absent or reduced in the anterior vertebrae. Zygapophyseal ridges are poorly developed near the midpoint of the vertebrae. The neural arch is narrow at the level of the transverse processes. Well-developed keels are present on the posterior zygapophyses; these reach anteriorly to the level of the transverse processes. The neural spine is prominent but thin; it is squared off dorsally. The posterior border of the neural arch is forked and has a median septum that bisects the fork, making a three-pronged structure.

General Comments. In the skull of *Amphiuma jepseni* the olfactory tracts are enclosed by the frontal, which is considered a derived state unique to the amphiumids (Estes, 1981). The snout is shorter than in the Holocene species. The vomer is more like that of adult salamanders, a condition indicating a less neotenic skull than in modern amphiumas.

Estes (1981) went on to say that it would take only slight evolutionary changes, probably all related to neoteny, to derive modern species from the Paleocene *Amphiuma jepseni.* On the other hand, living amphiumids are more robust and have chisel-shaped teeth, unlike the simple conical teeth of *A. jepseni.* This indicates that a change in diet may have occurred during the evolution of modern species of *Amphiuma.* The teeth of modern *Amphiuma,* with their aligned and chisel-shaped crowns, form an effective shearing device when coupled with the sharp twisting movements of the body during predatory attacks. Estes (1981, p. 42) stated, "It may be that this feeding behavior evolved since the Paleocene, and the continued paedomorphosis [neoteny] that seems to have characterized post-Paleocene amphiumid evolution would have resulted in two parallel rows of premaxillary, maxillary and vomerine teeth that provide a double slicing adaptation in the Holocene species."

Estes (1981) continued in his evolutionary speculation by pointing out that the ancient adaptation for long bodies and reduced limbs that is often seen in North American salamanders must be related to the development of ancient continent-encroaching seas during the Late Cretaceous in North America; and that these shallow seas provided extensive and persistent coastal plains, with slow drainage and abundant shallow-water habitats.

A close parallel with this proposed amphiumid evolution is seen in the previously discussed sirenid salamanders. The Cretaceous and Paleocene sirenids already had the vertebral modifications necessary for an elongate body, but the skull was less paedomorphic (neotenic) than in living sirens, a situation that parallels that in *Amphiuma jepseni.*

Rieppel and Grande (1998) proposed that the name *Amphiuma jepseni* is a nomen dubium (dubious name). Gardner (2003b), however, stated that the published drawings of the crushed and incomplete topotypic skull of *A. jepseni* indicated that the species differed from the extant members of the genus in having a shorter and broader snout and a shorter maxilla and vomer, with both bones having a correspondingly shorter tooth row; and that this extinct species evidently had a broader parasphenoid. Gardner disagreed with Rieppel and Grande's proposal that *A. jepseni* is a nomen dubium, and I do so as well.

Amphiuma means Garden, 1821

Two-toed Amphiuma

Fossil Localities. **Pleistocene (Rancholabrean NALMA):** Seminole Field, Pinellas County, Florida—Brattstrom (1953), Holman (1995a). Sims Bayou Fauna, Harris County, Texas—Holman (1965a, 1995a). Vero Beach Strata 2 and 3, Indian River County, Florida—Weigel (1962), Holman (1995a). Wakulla Spring site, Wakulla County, Florida—Brattstrom (1953), Holman (1995a).

This is the largest and probably the most well known member of the three living species of *Amphiuma,* growing to a length of 1162 mm (about 45.75 inches) (Conant and Collins 1998). This animal is almost completely aquatic, but it sometimes may be seen slipping through very shallow swamps. The upper part of the body of this salamander ranges from dark brown to black, and the belly is dark gray. As I found out in my younger years, this species is generally very active at night.

Identification of Fossils. Holman (1965a, p. 419) discussed criteria he used to identify 11 vertebrae and a left dentary from the Sims Bayou Fauna, Harris County, Texas: "The vertebrae of the monotypic family Amphiumidae are quite characteristic and may be diagnosed by the following combination of characters: vertebrae large, amphicoelous; cotyla higher than broad; centrum strongly keeled, perforated ventrally; transverse processes well developed, flared; neural arch vaulted, not modified posteriorly by a V-shaped notch [arising from the neural spine]; neural spine well-developed, elongate; strong ridges beginning on posterolateral parts of neural arches and extending onto dorsal parts of postzygapophyses present." About the dentary Holman (1965a, p. 419) stated, "Moreover, the fossil dentary is identical to that of Recent A. *means* of the same size, and both have a tooth number of 20."

General Comments. *Amphiuma* is one of the most abundant fossils in the Vero Beach deposits and is represented by several hundred vertebrae. It is odd that at the Reddick I, Florida, Pleistocene deposit, probably the richest herpetological deposit in the southeast, no *Amphiuma* are present. Moreover, favorable *Amphiuma* habitat in the Reddick fossil fauna is indicated by the presence of two eel-shaped, swamp-dwelling salamanders, *Pseudobranchus* and *Siren* (Holman 1995a).

Genus #*Proamphiuma*

Genotype. *Proamphiuma cretacea* Estes, 1969.

Diagnosis. The diagnosis is the same as for the genotype and only known species, *Proamphiuma cretacea.*

#*Proamphiuma cretacea* Estes, 1969

Fig. 50B

Holotype. A vertebra (MCZ 3504, Museum of Comparative Zoology, Harvard University).

Type Locality. Bug Creek Anthills, McCone County, Montana. At present, this genus is known only from the type locality (Gardner, 2003b).

Horizon. Late Cretaceous or Early Paleocene.

Other Material. Referred specimens (from Gardner, 2003b, p. 772) are as follows: "Atlantes: MCZ 3505, 3637, UALVP 40045, 43813–43816; middle trunk vertebrae: MCZ 3506, 3507, UALVP 43818, 43820–43824, 43827, 43828, 43830–43833, 43837–43839; anterior trunk vertebrae: MCZ 3508, 3509, 3632, UALVP 43817, 43825, 43826, 43834; posterior trunk vertebrae: MCZ 3629, 3631, 3634, 3636, UALVP 43819, 43829, 43835, 43836; caudal vertebrae: MCZ 3630, UALVP 43830." All these referred specimens are from the holotype locality. The above inventory of specimens excludes four specimens listed by Estes (1969c, 1981). These are MCZ 3627 (trunk vertebra of *Habrosaurus*), MCZ 3635 (trunk vertebra of *Opisthotriton*), and MCZ 3633 and 3628 (Caudata indet.).

Revised Diagnosis. The diagnosis is directly from Gardner (2003b, pp. 772–773): "Genus and species of Amphiumidae primitively differing from *Amphiuma* in having trunk vertebrae with neural crest relatively lower, subcentral keel relatively shallower, and anterior basapophyses relatively shorter with distal end in line with or behind anterior cotylar rim and in having trunk vertebrae and anterior caudals with postzygapophyseal crests relatively lower (see Fig. 2 [of Gardner, 2003b]). Differs further from extant species of *Amphiuma* (conditions uncertain for late Paleocene A. *jepseni*) as follows: more derived in having atlas with indistinct postzygapophyseal processes, trunk vertebrae with vertebrarterial canal closed posteriorly, and anterior caudals pierced by spinal foramen; more primitive in having atlas with ventral rims and of anterior and posterior cotyles approximately in line, postatlantal vertebrae with neural crest extending anteriorly almost to leading edge of neural arch roof, and trunk vertebrae with leading edge of neural crest inclined posteriorly at shallower angle;

and in one character state of uncertain polarity, namely neural crest on inferred middle trunk vertebrae broadly rounded anteriorly and dorsal edge horizontal in lateral profile."

Description. The following is modified from the description in Estes (1981) of his holotype vertebra (MCZ 3504) and referred atlantes from the type locality. The vertebral centra are amphicoelous. The cotyles are teardrop shaped and have a thin internal coating of calcified cartilage. The subcentral keel is prominent or low and with or without channels for the segmented blood vessels. There are either prominent or flattened anterior basapophyses, which usually project beyond the anterior margin of the centrum. Indications of two closely appressed rib-bearers appear on the four anterior vertebrae; the others have only one. Ventral lamina of the rib-bearers are present, but they are apparently not well developed anteriorly.

The zygapophyses are elongate in an anterior–posterior direction. The posterior zygapophyses are prominent and widely separated. They have keeled crests dorsally that extend forward to the level of the root of the transverse process. These crests are in a more medial position in presumed anterior vertebrae and in a more lateral position in presumed posterior vertebrae. The neural spine is relatively low, but it is prominent, thin, keeled, and squared off in lateral view. The posterior border of the spine is slightly forked, often in a three-pronged fashion.

The atlantes have rounded anterior cotyles and prominent interglenoid tubercles. The centra of the atlantes are short, and they have their neural arches relatively high and blunt, with their dorsal surfaces horizontal. The atlantal neural spines are represented by only a faint ridge.

General Comments. The vertebrae of *Proamphiuma cretacea* closely resemble those of modern *Amphiuma* in having posterior zygapophyseal crests that indicate the presence of intervertebral muscles found uniquely in *Amphiuma* (Auffenberg, 1959). Main differences between *Proamphiuma* and *Amphiuma* are the less-developed crests, narrower centra, and less-prominent zygapophyses in the fossil form when compared with *Amphiuma* specimens of comparable size (Estes, 1981).

Rieppel and Grande (1998) considered *Proamphiuma cretacea* a nomen dubium. The thorough restudy of this taxon by Gardner (2003b) disputes this. Actually, Gardner has shown that *P. cretacea* is a good structural ancestor for *Amphiuma*, perhaps even the actual ancestor of this genus. As shown by the fossil record, the characteristic joint between the head and the atlas, the muscular complex associated with the vertebrae, and the elongate trunk region occurred before the elongation of the snout and other cranial modifications in living amphiumids. The limitations of the fossil record, however, have not yet reflected on the sequence of these latter modifications.

Amphiumid indet.

Albright (1994) reported fragmentary "amphiumoid" remains from the Toledo Bend Local Fauna (Early Miocene: Arikareean NALMA) of Newton County in eastern Texas.

Family Dicamptodontidae Tihen, 1958
Pacific Giant Salamanders

See Fig. 11

The modern terrestrial adults of the family Dicamptodontidae are large, stout salamanders that may reach up to 340 mm in total length (Petranka, 1998). They have compressed, blade-like teeth that can deliver a nasty bite to the human hand. Dicamptodontids lay large, unpigmented eggs. Adults with gills are common in many populations and may sometimes outnumber the transformed individuals in a given population. Until fairly recently, the Dicamptodontidae were included in the family Ambystomatidae (Mole Salamanders). At present, four species of dicamptodontids are recognized (Crother, 2000).

Some characters used to define this group were given by Tihen (1958). I present a slightly modified account of that definition here. The lacrimal bone is independent, as are the exoccipitals, prootics, and columella. The skull is exceptionally solid. The individual teeth are compressed and have somewhat the shape of a curved, double-edged blade. Nasal bones are present. The premaxillary spines are short and broad, and they have no fontanelle, or at most a small one. There is a decided linear variation in the proportions of the trunk vertebrae. The lungs, eyes, and ypsiloid cartilage are normal.

Tihen (1958) went on to report that the most distinctive feature of the skeleton of *Dicamptodon* is the solidity and rigidity of the skull (see Fig. 11). In the evolution of the salamanders the tendency has been toward a less rigid skull. It is uncertain whether this condition in *Dicamptodon* represents the retention of a primitive condition or actually represents a secondary development from a skull structure nearly typical of the closely related ambystomatids. Tihen tended to believe that the rigid skull is, in fact, a secondary development, possibly related to habitat or habitats.

The features that contribute strongly to the skull solidity of *Dicamptodon*, particularly in the palatal region, are the firm, extensive sutural connections of the prevomers with the premaxillae and the maxillae. This has led to the presence of a distinct palatal shelf, not found in other salamander taxa; and a more extensive palatal portion of the maxilla, usually not found in other salamander taxa. Moreover, the pterygoid is very heavy and extensive. In all the individuals of *Dicamptodon* that Tihen (1958) was able to see, the pterygoid actually abutted against the posterior end of the maxillae; some figures he had seen indicated that the pterygoid approached the maxillary very closely, without actually being in contact with it.

The shape of the individual teeth of *Dicamptodon* is unique and is undoubtedly a very specialized feature of the family. There is no question that *Dicamptodon* is an especially voracious salamander, and it is very possible that the tooth shape is, at least to some extent, correlated with its feeding habits.

The range of *Dicamptodon* is along the Pacific Coast from southwestern British Columbia to Santa Cruz, California, and it is also known from northern Idaho (Tihen, 1958).

In Field and Lab. A herpetologist friend of mine kept a *Dicamptodon* for a pet for several years and mentioned that it not only ate a very wide variety of diced vertebrate muscle, but would attack just about any small creatures put in the terrarium with it.

GENUS #*CHRYSOTRITON* ESTES, 1981

Genotype. Chrysotriton tiheni Estes, 1981.

Diagnosis. The diagnosis is the same as for the genotype and only known species, *Chrysotriton tiheni.*

#*CHRYSOTRITON TIHENI* ESTES, 1981

FIG. 51

Holotype. A series of vertebrae, most of which are articulated or only slightly disassociated from a natural position. All the vertebrae have the same registration No. (PU 17314, Princeton University).

Type Locality. Golden Valley Formation, Stark County, North Dakota.

Horizon. Early Eocene.

Diagnosis. The diagnosis is taken directly from Estes (1981, p. 47): "Vertebrae differ from those of all other known dicamptodontids in having subcentral keel; winglike ventral lamina and single rib-bearer of the transverse process; more well-developed basapophyses, elongated body form, the vertebral column having more than twenty-six vertebrae."

Additional Preparation of the Holotype. Estes (1988) reported that the original specimen was further prepared so that it consisted of two separate, overlapping blocks of matrix mounted in plaster, one of which exhibited 17 trunk vertebrae, and the other of which showed 1 caudal and 9 trunk vertebrae. Other modifications of the type material allowed better viewing of the holotype.

Revised Description of the Holotype. Estes (1988) gave a succinct revised description of the holotype that appears to better characterize the specimen than his previous one (Estes, 1981). This new description is presented here in a slightly modified form. The vertebrae are all fairly compact in form. The centrum is simple and amphicoelous and has a prominent ventral keel. The cotyles are rounded anteriorly, but they have

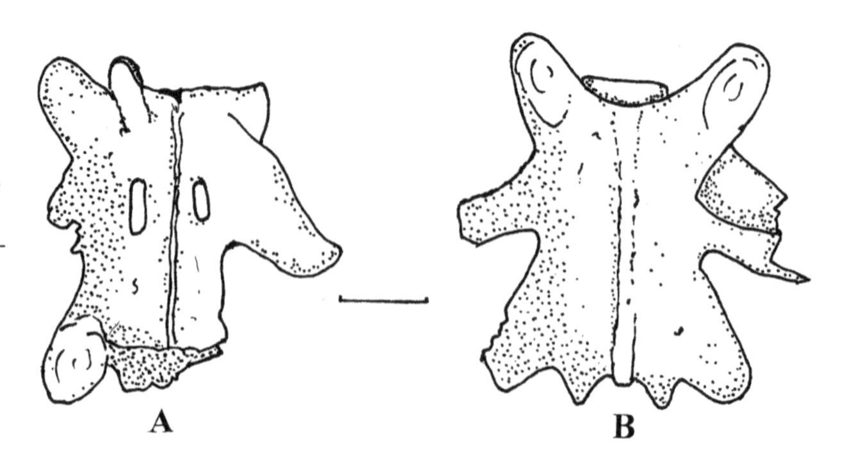

FIGURE 51. Views of two vertebrae from the holotype series of vertebrae that represent #*Chrysotriton tiheni* from the Eocene of South Dakota. (A) Trunk vertebra in ventral view. (B) Another trunk vertebra in dorsal view. Scale bar = 1 mm.

a slightly elongated, inverted teardrop shape in posterior view. A pair of basapophyses protrudes anteriorly beyond the border of the centrum. There is usually a prominent pair of subcentral foramina, one on either side of the central keel, near the bottom of the basapophyses. The transverse processes are single headed and arise at about the midpoint of the centrum. On all the vertebrae in which these processes are not broken, there is a thin, wing-like ventral lamina that curves forward to meet the anterior border of the centrum.

The neural arch has a low medial neural ridge, and the posterior border has a spinous projection on each side of the midline. The zygapophyses are robust and project strongly from the centrum. The medial edges of the zygapophyses are usually concave; and the lateral edges, usually convex. The posterior border of the neural arch projects about as far as, or very slightly farther than, the posterior borders of the postzygapophyses. In a specially prepared vertebra, the total length of the neural arch from the anterior border of the prezygapophyses to the posterior border of the postzygapophyses is 3.7 mm. The maximum width across the anterior zygapophyses in this specimen is 2.9 mm. The width of the neural arch at its narrowest point is 1.3 mm.

General Remarks. Chrysotriton may be separated from other dicamptodontids and from ambystomatids in having a prominent subcentral keel and a single rib-bearer on the transverse process that is closely attached to the wing-like ventral lamina of the transverse process. The elongate body form of *Chrysotriton* was common among the salamanders represented in the primarily slow stream, pond, or marshy deposits of the Late Cretaceous and Early Cenozoic. Neither the related dicamptodontids nor the ambystomatids currently have this habitus.

Genus *Dicamptodon* Strauch, 1870

Genotype. Dicamptodon ensatus (Eschscholtz, 1883).

Diagnosis. See diagnosis of the family Dicamptodontidae, above.

**Dicamptodon antiquus* Naylor and Fox, 1993

Fig. 52

Holotype. An incomplete skeleton that has a skull, left pectoral limb, and 14 vertebrae (UALVP 32387, University of Alberta Laboratory for Vertebrate Palaeontology).

Type Locality. Paskapoo Formation (location 1 of Wilson, 1980), Alberta, Canada.

Horizon. Late Paleocene.

Other Material. No other material is known.

Diagnosis. The diagnosis is directly from Naylor and Fox (1993, p. 816): "An extinct species of *Dicamptodon* differing from extant species in the presence of palatal teeth on the anterolateral portion of the pterygoid, a more oblique orientation of the vomerine tooth row, and well-developed alar processes on the ventral arms of the rib-bearers."

Description of the Holotype. The description here is modified from that of Naylor and Fox (1993). The skull is fairly well preserved, but it is flattened because of the compaction of the sediments that enclosed it. In

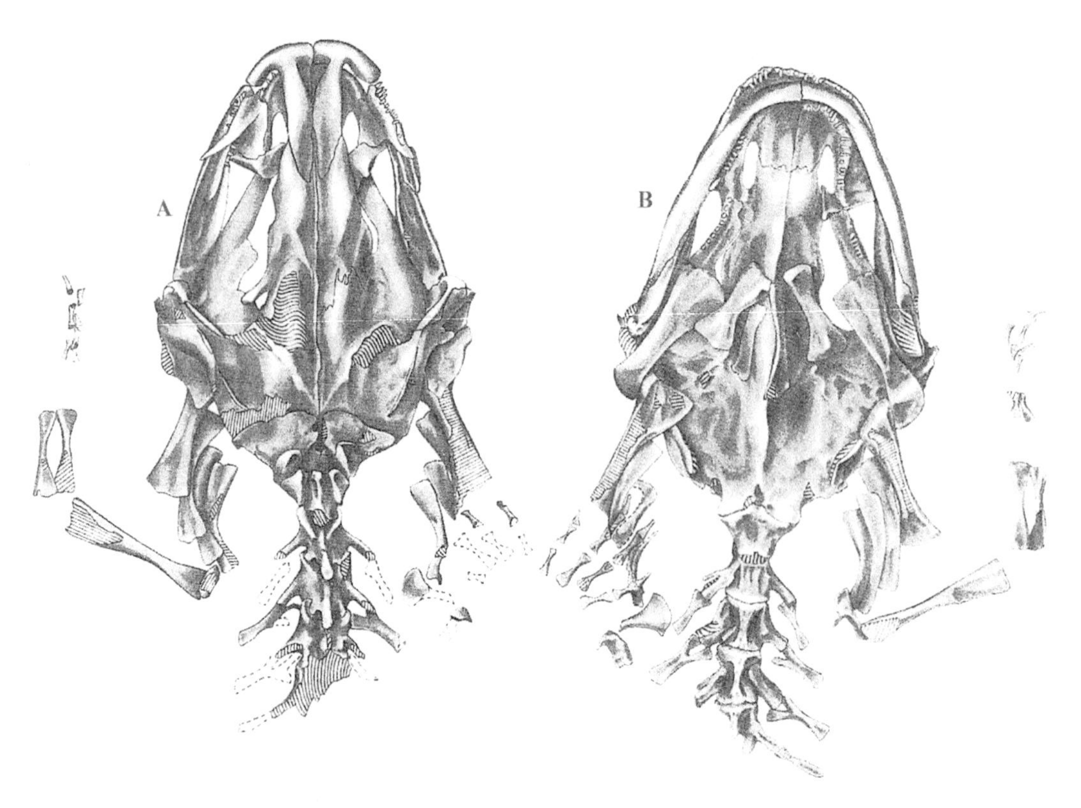

FIGURE 52. Holotype of **Dicamptodon antiquus* from the Paleocene of Alberta, Canada, consisting of the skull, pectoral girdle, and a portion of the vertebral column. (A) Dorsal view. (B) Ventral view. The skull measures about 150 mm at its widest portion.

dorsal view, the skull is approximately triangular from the tip of the snout to the quadrates. In these proportions the skull resembles neotenic salamanders such as *Necturus*, but it more closely resembles the skull of *Dicamptodon ensatus*, which is somewhat broader. The fossil skull is well ossified, as in *D. ensatus*. This contrasts with the weak ossification of the skull of *Rhyacotriton*, which is probably the nearest living relative to *Dicamptodon*.

Continuing with the fossil skull, posterior to the quadrate condyles, the outline of the skull tapers gradually toward the midline and then turns inward more sharply at the occiput. The occipital condyles are distinctly stalked, as in the larvae of extant *Dicamptodon*. The snout of UALVP 32387 is bluntly rounded and the premaxillae are paired, the primitive state in salamanders. For example, they differ from the fused premaxillary in *Amphiuma*. The dental portion of the premaxillaries and the premaxillary spine are robust and have the larval character of joining one another at the midline to enclose a central triangular fontanelle. The dental portion of the premaxillary extends from the midline along a broad curve to its articulation with the maxillary, rather than being restricted to the end of the snout as in *Amphiuma*, sirenids, or *Opisthotriton* of the Batrachosauroididae. It is uncertain whether a single row of teeth is present on the premaxillary (a feature of larval *Dicamptodon*), but only a single row appears to have been present on the fossil. The facial portion of the premaxillary is narrow and elongate, differing from the short, broad area in cryptobranchids. Distal to the fontanelle, the premaxillary spines join along a midline suture that begins at about one-third of their length posteriorly and then diverge slightly from one another at their distal tips. *Necturus* lacks a fontanelle, and the spines are joined for a short distance proximally and then diverge through their remaining distal lengths. In the fossil, the premaxillary spines broadly meet the frontals.

The maxillary bones of UALVP 32387 are somewhat displaced, as they have been rotated outward. This exposes the teeth laterally. In contrast, the obligate neotenes *Necturus, Proteus,* and *Siren* have lost the maxillary bones entirely. The dental portion of the maxillary in the fossil tapers posteriorly. The facial portion is crushed, and the outline of these bones cannot be seen on either side of the skull. The maxillary teeth are short, slender, and conical; Naylor and Fox (1993) thought they were probably non-pedicellate, although this condition cannot be clearly seen. The shape of these teeth is similar to that in larval *Dicamptodon ensatus*, which contrasts with the blade-like teeth of the metamorphosed adults.

Moving on to the more posterior bones of the midline in UALVP 32387, we find that the frontals are crushed. The frontals are said to extensively overlap the parietals, although this is difficult to determine from the figure of the fossil (Naylor and Fox, 1993, fig. 1, p. 815). The parietals are of the "normal shape" in salamanders, they articulate anteriorly with the frontals, and they send a broad process laterally to the squamosals (although the suture with the squamosal cannot be seen on the specimen). Back to the more lateral bones in the fossil skull, we find that because of the postmortem crushing of the skull, the quadrate and the squamosal have been displaced to the extent that they lie in the same horizontal plane as the remainder of the skull roof. A prominent crest extends down the squamosal. In dorsal view, the quadrates are visible on either side of the skull and are directed anterolaterally, as in the squamosal. The quadrate is shaped like a somewhat compressed, slender pillar that distally flares into a concave cotyle, which is intercepted by the condyle of the articular.

Turning to the bones of the underside of the skull, we find that parts of the vomers and pterygoids are actually exposed dorsally because of the compression of the skull roof down onto the palate. However, no significant features of these bones are evident in dorsal view. The vomerine tooth row extends posteriorly in a more oblique way than in modern *Dicamptodon*. The pterygoid itself is a long, narrow bone that extends anteromedially from the juncture of the upper and lower jaws to the posterolateral border of the vomer. The vomerine tooth row extends onto the lateral margin of the pterygoid as a single line of small teeth. This differs from some (but not all) modern *Dicamptodon* in which the pterygoid in both larval and neotenic individuals remains toothless. Neither the lacrimal, which develops as a tube associated with the prefrontal in larval *Dicamptodon ensatus* (Larson, 1963), nor the nasal and septomaxillary, which develop only at metamorphosis in living *Dicamptodon*, can be seen on the Late Paleocene fossil.

With regard to the lower jaw, Naylor and Fox (1993) stated that the dentary may have been the only toothed bone in this structure. Nevertheless, a toothed bone on the left side of the fossil is in the position of a coronoid. However, a bone corresponding to that tooth-bearing bone is not seen on the right side; thus, the bone on the left may actually be a broken part of the dentary that was displaced posteriorly during the fossilization process. The Meckelian groove is long and wide open dorsally. This grove is bounded medially by the prearticular, which rises well above the margin of the jaw lateral to the groove. These features of the groove and its surrounding bones in the fossil are observable in larval *Dicamptodon*, but not in *Cryptobranchus, Necturus, Amphiuma,* or sirenids.

Other parts of the skeleton of the Late Paleocene fossil salamander include the hyobranchium, a partial vertebral column, and a left pectoral limb. The hyobranchium is well ossified and robust, and Naylor and Fox (1993) described it well, but they did not compare it with that of other salamander species. Unfortunately, although 14 vertebrae are preserved, postmortem crushing has made a taxonomic analysis very difficult. Crushing has obscured the neural arch of the atlas. Nevertheless, the apparent lack of an odontoid process of the atlas contrasts markedly with the situation in modern *Dicamptodon*, where this process is notably robust. The neural spines on the next 13 vertebrae are tubular in appearance, with their tips having been finished in cartilage. The rib-bearers are long and double headed, as in modern *Dicamptodon* (and various other salamanders). The ventral rib-bearers have well-developed alar processes, which is not the case in modern members of the genus, but it sometimes occurs in other caudates of aquatic habits.

A humerus, an ulna, a radius, and phalanges with two digits each represent a left pectoral limb of *Dicamptodon antiquus*. The humerus is long and slender. Both ends are wider than the central part of the shaft. A small tubercle is located proximally on the anterior edge of the humerus, and a faint ridge occurs distal and posterior to the tubercle. The ulna and radius occur side by side in a presumably natural position; both elements are well ossified throughout their lengths, and both are expanded at either end. The carpus is missing. The phalanges are arranged in two somewhat parallel rows that appear not to have been displaced from their normal position relative to the distal ends of the ulna and radius. Three apparently well-ossified phalanges occur in each row.

General Comments. Regarding the assignment of this exciting fossil to *Dicamptodon*, Naylor and Fox (1993, p. 818) stated, "The separated premaxillaries, with elongate spines forbid assignment to the Hynobiidae, Cryptobranchidae, Amphiumidae, Plethodontidae, Proteidae, Batrachosauroididae, Sirenidae, or Salamandridae. Within the Ambystomatidae [and as currently construed the Dicamptodontidae], the separated premaxillaries, with spines converging medially towards [the] rear, as well as the association of the vomers and pterygoids, show clear assignment to *Dicamptodon*." I would only pick at one point, and that would be that additional modern skeletal material of *Dicamptodon* could have been compared with the fossil specimen. Only two cleared and stained skeletons of *Dicamptodon ensatus* were studied.

Dicamptodon sp. indet.

Fossil Locality. **Late Miocene (late Hemphillian NALMA):** Buchanan Tunnel, near Columbia, Tuolumne, California—Peabody (1959), Estes (1981). These trackways indicated the genus *Dicamptodon*, but the species was not determined.

Dicamptodon sp. indet.

Fossil Locality. **Paleocene:** Bear Butte locality, Sweet Grass County, Montana—Peabody (1954), Estes (1981).

Comments. These dicamptodontid tracks were carefully studied and analyzed by Peabody (1954). The palm impressions were bilobate, which is a condition found only in *Dicamptodon* among living salamanders. However, the trackways were twice as large as those in living *Dicamptodon*. It is of interest that the floral associates of these trackways included dawn redwoods (*Metasequoia*), which shows the ancient association of dicamptodontids with a redwood-type forest (Estes, 1981).

Family Ambystomatidae Oppel, 1811
Mole Salamanders

Mole Salamanders consist of a single living and fossil genus, *Ambystoma*, which occurs in North America south through Mexico. The genus *Amphitriton* Rogers, 1976 from the Late Pliocene (late Blancan NALMA) of Scurry County, Texas, falls within the structural range of *Ambystoma* (see Estes, 1981) and thus is not considered a valid genus. Fourteen living species currently occur in North America north of Mexico (Crother, 2000).

Although this family is much less diversified than the Plethodontidae, its occurrence in the fossil record in North America is much more common. In part, this may be due to their burrowing habit and their generally more robust skeleton. Mole Salamanders have robust bodies and short, rounded heads. They also have prominent costal grooves. Nasolabial grooves are absent, and transformed adults have lungs. Incompletely metamorphosed aquatic populations of *Ambystoma*, especially *Ambystoma tigrinum*, are not uncommon today or in the fossil record, and some individuals in these populations become very large.

Some osteological characters used to define the Ambystomatidae are modified from Duellman and Trueb (1986). The premaxillary bones are paired. Septomaxillae, pterygoids, and the ypsiloid cartilage are present. Lacrimals are absent. The exoccipital, the prootic, and the opisthotic are fused. The teeth are pedicellate, and the palatal teeth are transverse. The nasal bones are ossified from lateral developmental precursors. The columella and operculum are fused. The vertebrae are amphicoelous. All but the first three spinal nerves exit intravertebrally.

Genus *Ambystoma* Tschudi, 1838
Mole Salamanders

Genotype. Ambystoma maculatum (Shaw, 1802).

Diagnosis. This diagnosis is modified from the definition of the genus by Tihen (1958). Metamorphosed adults have the prevomerine teeth in a typically adult position. The premaxillary fontanelle is nearly or completely obliterated in adults. The maxillae are of normal extent. The hyobranchium is almost completely cartilaginous. The tips of the teeth are definitely bifid (with two cusps) and often very blunt or even expanded. Occasionally the teeth are pointed but very sharply hooked inward. The sides of the parasphenoid are either parallel or concave, but they do not diverge. There are four phalanges in the fourth toes in most taxa. Metamorphosis is customary, but neoteny may occur in some species and is apparently obligatory in a few.

AMBYSTOMA ALAMOSENSIS ROGERS, 1987

FIG. 53

Holotype. One presacral vertebra, with the distal ends of the rib-bearers partially broken (UCM 54087, University of Colorado Museum of Natural History).

Type Locality. Alamosa Local Fauna, CT3 locality, Alamosa Formation, Alamosa County, Colorado.

Horizon. Middle Pleistocene (late Irvingtonian NALMA).

Other Material. Paratype material consists of a parietal (UCM 50475) from the CT2 locality and a vertebra (UCM 50522) from the B-80-3 locality, both in the Alamosa Local Fauna.

Diagnosis. The diagnosis is taken directly from Rogers (1987, p. 84): "A species of *Ambystoma* with the dorsolateral surfaces of the neural arch convex as a result of the upward projection of the posterolateral corners of the neural arch. Differs from all other ambystomatids in that the posterolateral corner of the vertebrae of all other ambystomatids projects ventrally."

Description of the Holotype and Referred Parietal. These descriptions are slightly modified from Rogers (1987). In dorsal view, the prezygapophyses are laterally angled, and the vertebra is not constricted posterior to the zygapophyses. The neural arch has a raised rib-bearer medially and is concave laterally. The posterolateral corners of the neural arch project upward. In ventral view, the centrum is generally spool shaped, except that it is ventrally flattened. Webs of bone connect the anterior and posterior ends of the centrum to the rib-bearers, resulting in the presence of pits on either side of the flat part of the centrum. The postzygapophyseal articular facets are completely visible, because the postzygapophyseal area overhangs the posterior end of the centrum. In anterior view, the centrum is almost round, the notochordal canal has a tiny perforation at its center, and the prezygapophyseal articular facets are flat, rather than tilted upward. In posterior view, the centrum is round, except for ventral flattening. The posterior end of the centrum appears to be imperforate. The postzygapophyseal area is bow tie shaped, with the knot of the bow tie being the area for insertion of the interneural muscle. Rib-bearers appear to be single on each side of the centrum, with webs of bone extending from the dorsal surface of the vertebra and the centrum as well.

The parietal referred to *Ambystoma alamosensis* is similar in general shape to that of *Ambystoma tigrinum*, but it is medially thicker than in *A. tigrinum*. Alamosa Local Fauna *A. tigrinum* fossil parietals and those

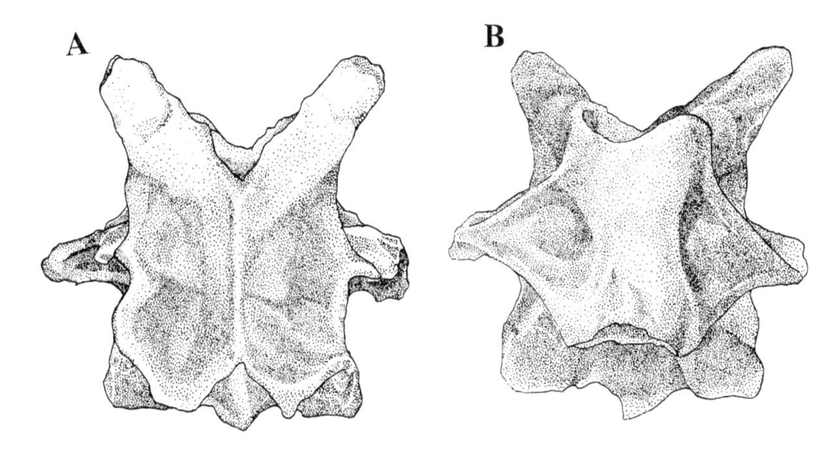

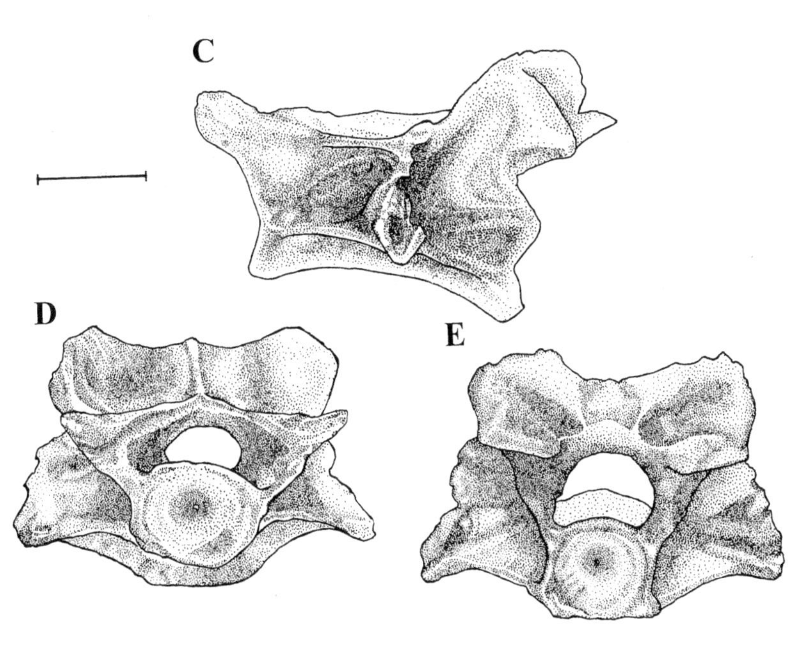

FIGURE 53. Holotype vertebra of **Ambystoma alamosensis* from the Pleistocene (Irvingtonian NALMA) of Colorado in (A) dorsal, (B) ventral, (C) lateral, (D) anterior, and (E) posterior views. Scale bar = 2 mm and applies to all figures.

from the Recent San Luis Valley parietals vary in thickness between 12 and 15 mm, whereas the parietal referred to A. *alamosensis* is 19 mm thick.

General Comments. I am not aware of any normal salamander species in the world that has the posterior part of the vertebra pushed upward as in *Ambystoma alamosensis.* This, in itself, makes one think that this odd vertebral shape among the hundreds of *Ambystoma tigrinum* fossils from the Alamosa Local Fauna is due to some genetic or pathological condition.

*AMBYSTOMA HIBBARDI TIHEN, 1955

FIG. 54

Holotype. A posterolateral portion of a left vomer (UMMPV 31440, University of Michigan Museum of Paleontology, Vertebrate Collection).

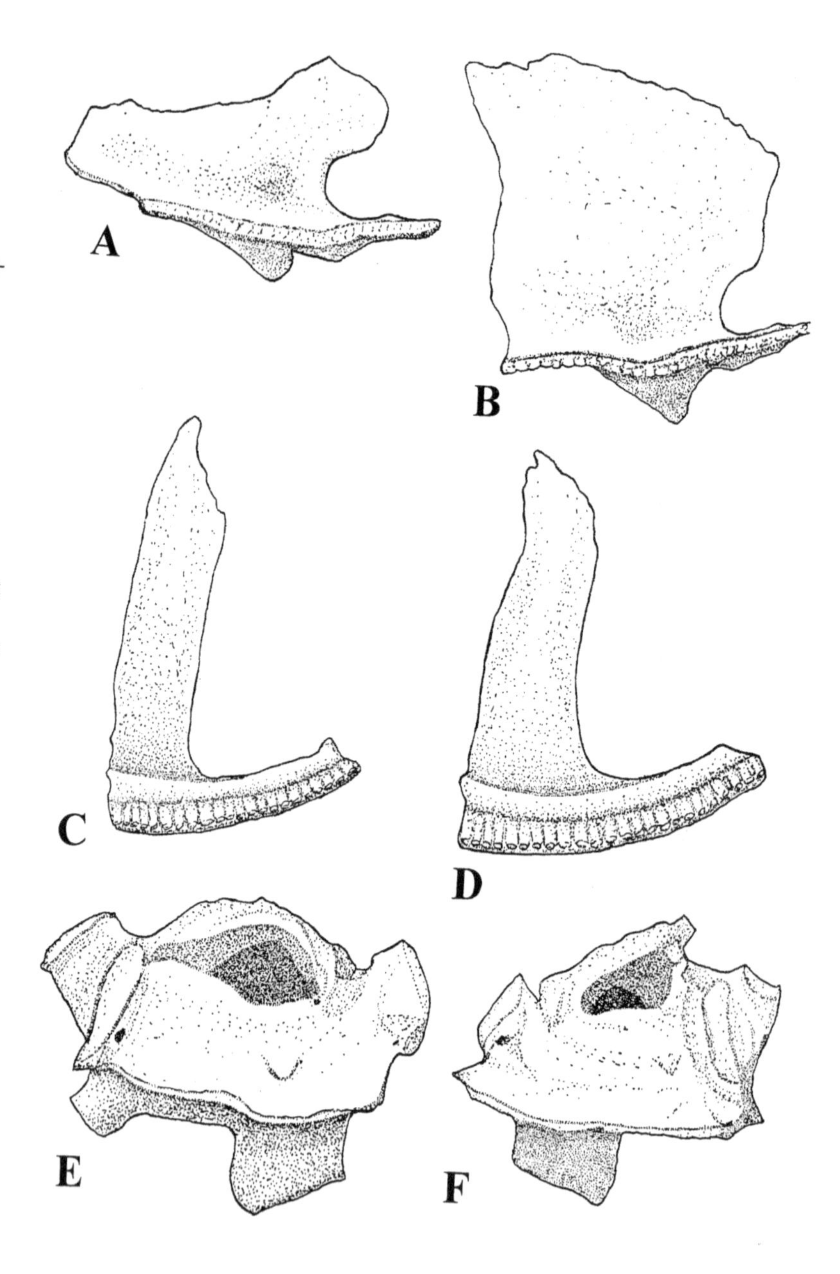

FIGURE 54. **Ambystoma hibbardi* from the Late Pliocene of Kansas and modern *Ambystoma tigrinum mavortium* for comparison: (A) A. *hibbardi* holotype, posterolateral part of left prevomer; (B) A. *tigrinum mavortium*, left prevomer; (C) A. *hibbardi* paratype, right premaxilla in lingual aspect; (D) A. *tigrinum mavortium*, right premaxilla in lingual aspect; (E) A. *hibbardi* paratype, left otic capsule; (F) ?A. *hibbardi*, atypical otic capsule. Scale not given.

Type Locality. Rexroad Formation, Fox Canyon, Meade County, Kansas.

Horizon. Pliocene (Blancan NALMA).

Other Material. Additional topotypic material includes a complete right premaxilla (UMMPV 31616), a left otic capsule (UMMPV 31617), and numerous cranial and postcranial elements (UMMPV 27187, 27189, 27190, 27195, 31234, 31235, 31261, 31618, 31619).

Diagnosis. The diagnosis is directly from Estes (1981, p. 54): "A transformed species of *Ambystoma* closely related to extant A. *tigrinum* in size, proportions, and general features; estimated snout–vent length 100 mm; agrees with A. *tigrinum* in absence of diastema in vomerine tooth series at level of choana, in possession of nearly straight-sided parasphenoid, and in similarity of form and proportions of vertebrae. Differs from *A. tigrinum* in having a narrow premaxillary spine, in (apparently) usual failure of columellar footplate to fuse with otic capsule, and in probable weak ossification of orbitosphenoids, quadrates, and ischia."

Description. This description is modified from Estes (1981). The vomer has at least 35 and probably 45 teeth. The choanal border of the vomer is slightly less than semicircular. The premaxilla has 23 tooth bases and a spine arising lateral to the medial edge of its tooth-bearing portion. The premaxillary fontanelle is very small or absent. The otic capsule is one that one would find in an advanced or neotenic larva of *Ambystoma tigrinum*, except for the absence of a fused columellar footplate. The maxilla bears at least 33 teeth. The parietals vary in the age class of the individuals represented. The frontal and squamosal are as in *A. tigrinum.* The parasphenoid has its sides nearly parallel or slightly expanded; and its underlying otic region, in its shape and its extent of variation, is similar to the *A. tigrinum* group of *Ambystoma.* The dentary bears at least 55 and probably between 60 and 65 teeth. The prearticular is long and low and lacks a strong coronoid process, or it is shorter and higher with a distinct coronoid process. The articular is most often fused with the prearticular.

None of the vertebrae are distinguishable from those of *Ambystoma tigrinum.* Both larval specimens with continuous notochordal canals and adult specimens with interrupted notochords are present in the fossil specimens.

Turning to the limb girdle material, we find that the scapulocoracoid is less fully ossified than in corresponding stages of *Ambystoma tigrinum.* The humerus, radius, and ulna are not distinguishable from those of *A. tigrinum.* The ilium, femur, tibia, and fibula are also indistinguishable from those of *A. tigrinum.*

General Comments. Ambystoma hibbardi is represented by much more material than are the majority of fossil ambystomatids and appears to be a member of the *Ambystoma tigrinum* species group of Tihen (1958). In fact, it is distinguished from *A. tigrinum* itself only by a few features in the diagnosis above. There is clear evidence that neoteny occurred in this fossil population, but other members of *A. hibbardi* underwent metamorphosis, which Estes (1981) considered the principal character state differentiating them from the neotenic *Ambystoma kansense,* which follows. Estes (1981, p. 55) stated, "Unfortunately Tihen (1955) chose as the type specimen a vomer that is indistinguishable from that of *A. tigrinum* and the species is technically a synonym of the latter; the minute differences in premaxillary spine and sound-transmitting apparatus may serve provisionally to maintain this species."

**AMBYSTOMA KANSENSE* (ADAMS AND MARTIN, 1929) (*PLIOAMBYSTOMA KANSENSE* ADAMS AND MARTIN, 1930; *LANEBATRACHUS MARTINI* TAYLOR, 1941; *OGALLALABATRACHUS HORARIUM* TAYLOR, 1941)

FIG. 55

Holotype. A disarticulated skeleton (KU 5250, Museum of Natural History, University of Kansas).

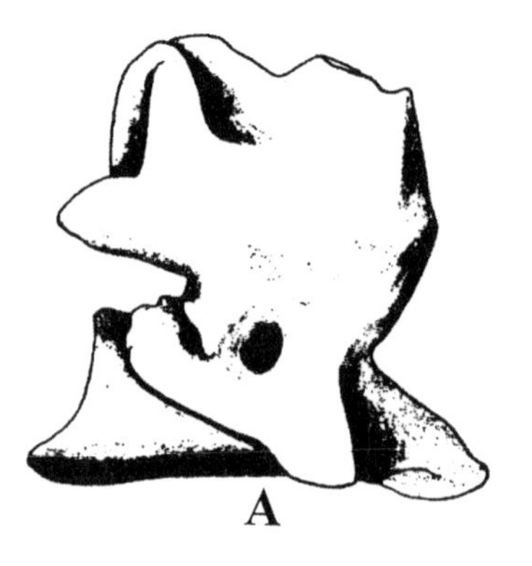

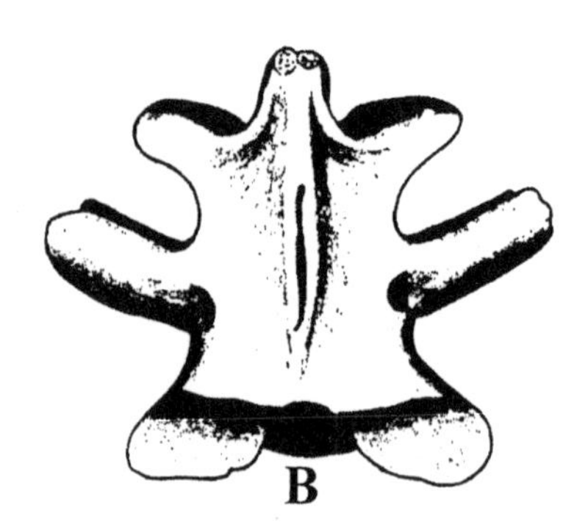

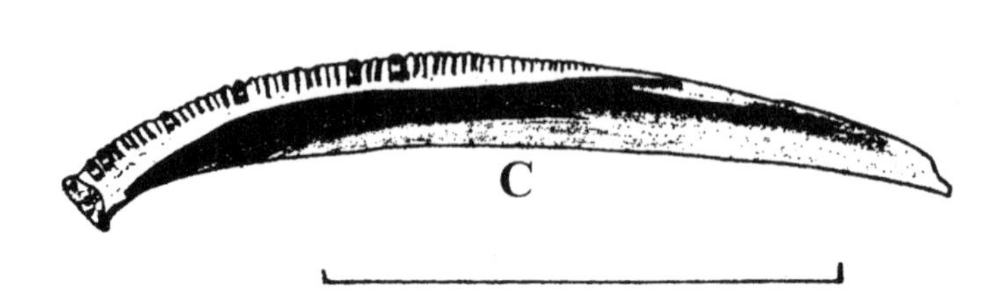

FIGURE 55. **Ambystoma kansense* from the Pliocene of Kansas. (A) Atlas in lateral view. (B) Trunk vertebra in dorsal view. (C) Right dentary in lingual view. Scale bar = 10 mm and applies only to (C). From Adams and Martin (1929).

Type Locality. Edson Quarry, Ogallala Formation, Sherman County, Kansas.

Horizon. Late Miocene (Hemphillian NALMA).

Other Material. Other material includes topotypic specimens (KU 1468, 1470, 5251, 5252, 5253) and many other specimens. The minimum number of individuals represented by these topotypic specimens was an amazing 150! There were several more or less complete skeletons (KU 9919) from Wallace County, Kansas, also representing the Late Miocene. In addition, the species was represented by several elements from the Rexroad Formation (Pliocene: late Hemphillian NALMA) of Meade County, Kansas.

Diagnosis. The diagnosis is directly from Estes (1981, p. 55): "Palatine joined to pterygoid and extending anteriorly, probably to articulate with the vomer, a condition suggesting the larval condition in modern *Ambystoma.* Palatine end of pterygoid with 6–8 teeth; maxilla twice as long as wide, while in A. *tigrinum* the width of the maxilla is contained in the length three times. Tips of the teeth cone-shaped, as opposed to the peg-like teeth of most modern *Ambystoma.*"

Description. This description is modified from Estes (1981). The premaxillae are fairly small, averaging 7 mm in length and 5 mm in width. The maxillary teeth number less than 30. The maxilla is short and tapers abruptly to a sharp point posteriorly; 23–28 teeth are present. The prefrontal is thin and fragile and narrows abruptly at its proximal end. The posterior border of the frontal is convex. The squamosal articulation is roughened by a conspicuous ridge. The parasphenoid usually lacks parallel ridges. The palatine is small and bears six to eight teeth. The pterygoid is a thin plate that is firmly articulated with the otic capsule and in contact with the squamosal, quadrate, and palatine. The squamosal is a fairly large, thin, spatulate bone. The dentary is short and bears 50–55 teeth. These teeth are cone shaped and gradually taper from their junction with the bases to the tip. The articular cartilage is large and calcified.

The vertebrae are amphicoelous. The ventral surface of the centrum of the atlas is rounded on its posterior half. The other vertebrae are not distinctive. The distal end of the scapula is slightly forked. The humerus, radius, ulna, tibia, and fibula are fairly short.

General Comments. The above description includes the main elements that show differences from those of modern species of *Ambystoma*. Tihen (1958) examined many larval skeletons of modern *Ambystoma* and came to the conclusion that *Ambystoma kansense* did not differ significantly from Recent larval members of his *Ambystoma tigrinum–Ambystoma mexicanum* species groups. Thus, *A. kansense* is definitely a neotenic species that was principally assigned, on that basis, specifically to the *A. mexicanum* species group of Tihen (1958). In fact, that group differs mainly in its obligate neotenic habit and somewhat larger size. The living *A. mexicanum* group occurs in the ancient lakes of the southern Mexican plateau, but it is possible that if *A. kansense* is properly construed, it may link the two species groups. However, this is highly speculative.

Ambystoma laterale Hallowell, 1856
Blue-spotted Salamander—
Ambystoma jeffersonianum (Green, 1827)
Jefferson Salamander

Author's note: *Ambystoma laterale* and *Ambystoma jeffersonianum*, both widespread species in eastern North America, are very similar. The composite range of the two species is from northern Canada to southern Kentucky and Virginia, with a huge zone of hybridization in between. "Pure" Blue-spotted Salamanders occur mainly north of Michigan; "pure" Jefferson Salamanders occur mainly in southern Indiana, north-central Kentucky, southern Ohio, Pennsylvania, and southern New York. In between these areas, frequent hybridization between the two species occurs. Paleoherpetologists thus far have not been able to distinguish between the skeletons of these species, so the term "*Ambystoma laterale* complex" used in this account means that the fossil salamander could be either *A. laterale*, *A. jeffersonianum*, or a hybrid of the two.

Fossil Localities. **Pleistocene (Irvingtonian NALMA), Kansan glacial stage:** Albert Ahrens Local Fauna, Nuckolls County, Nebraska—Ford (1992), Holman (1995a). Hamilton Cave, Pendleton County, West Virginia—Holman and Grady (1989), Holman (1995a). **Pleistocene (Rancholabrean NALMA):** New Trout Cave, Pendleton County, West Virginia—Holman and Grady (1987), Holman (1995a) (found in all strata of the cave). Sheridan Pit Cave, Wyandot County, Ohio—Holman (1997). Worm Hole Cave, Pendleton County, West Virginia—Holman and Grady (1994).

Ambystoma laterale and *Ambystoma jeffersonianum* are both rather small, somewhat nondescript salamanders with blackish or grayish bodies flecked with bluish spots. Both species and their hybrids are mainly woodland species that may be found, sometimes abundantly, under rotting logs and rotting leaves during moist conditions. These salamanders move underground with the onset of dry weather, often using the burrows of other small animals.

Identification of Fossils. Rogers (1984), Holman and Grady (1987), and Ford (1992) separated the *Ambystoma laterale* complex from *Ambystoma maculatum*, the other member of the *A. maculatum* group of Tihen (1958), on the basis of two characters. In *A. maculatum* the neural arch often extends beyond the postzygapophyses, and the postzygapophyseal area is relatively wide. In the *A. laterale* complex the neural arch ends anterior to the most posterior extent of the postzygapophyses, and the postzygapophyseal area is relatively narrow. Additionally, the vertebrae of *A. maculatum* are larger than those of the *A. laterale* complex.

Comment. The *A. laterale* complex does not currently occur near the Irvingtonian Albert Ahrens Local Fauna in Nebraska. The nearest occurrence of that complex at present is about 600 km to the northeast of Nuckolls County, Nebraska (Ford, 1992).

Ambystoma maculatum (Shaw, 1802)
Spotted Salamander

Fig. 56A

Fossil Localities. **Late Miocene (Clarendonian NALMA):** WaKeeney Local Fauna, Trego County, Kansas—Holman (1975b). **Pliocene (late Blancan NALMA):** Hornets Nest Quarry, Knox County, Nebraska—Rogers (1984). **Pleistocene (Irvingtonian NALMA), Kansan glacial stage:** Cumberland Cave, Allegany County, Maryland—Holman (1977b, 1995a). Hall Ash Local Fauna, Jewell County, Kansas—Rogers (1982). Hamilton Cave, Pendleton County, West Virginia—Holman and Grady (1989), Holman (1995a). Vera Local Fauna, Knox County, Texas—Parmley (1988). **Pleistocene (Rancholabrean NALMA):** Bell Cave, Colbert County, Alabama—Holman et al. (1990), Holman (1995a) (these remains occurred in two of three strata [see Holman, 1995a]). Cheek Bend Cave, Maury County, Tennessee—Miller (1992), Holman (1995a). Crankshaft Pit, Jefferson County, Missouri—Holman (1965b, 1995a). New Trout Cave, Pendleton County, West Virginia—Holman and Grady (1987), Holman (1995a). Peccary Cave, Newton County, Arkansas—Davis (1973), Holman (1995a). Strait Canyon Fissure, Highland County, Virginia—Fay (1984), Holman (1995a). Worm Hole Cave, Pendleton County, West Virginia—Holman and Grady (1994).

References to "*Ambystoma maculatum* group" salamanders without pinning down the exact species represented (probably *Ambystoma maculatum*) are as follows. **Pleistocene (Rancholabrean NALMA):** Baker Bluff Cave, Sullivan County, Tennessee—Van Dam (1978), Fay (1988), Holman (1995a). Clark's Cave, Bath County, Virginia—Fay (1988), Holman (1995a). Guy Wilson Cave, Sullivan County, Tennessee—Fay (written communication, 1993[1]), Holman (1995a). Kingston Saltpeter Cave, Bartow County, Georgia—Fay (1988), Holman (1995a). Natural Chimneys site, Augusta County, Virginia—Fay (1988), Holman (1995a). New

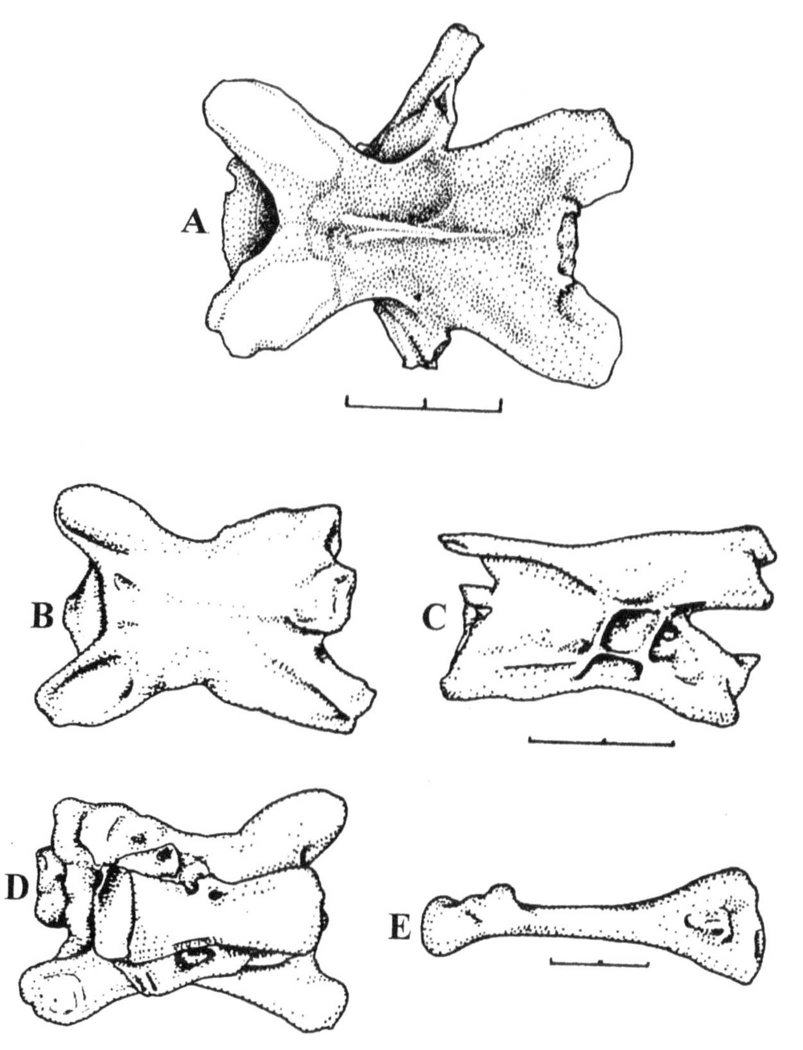

FIGURE 56. Trunk vertebra of *Ambystoma maculatum* from the Late Miocene of Kansas in (A) dorsal view. Caudal vertebra of **Ambystoma minshalli* from the Middle Miocene of Nebraska in (B) dorsal, (C) lateral, and (D) ventral views. Humerus of *A. minshalli* in (E) lateral view. Scale bars = 2 mm. Scale bar for (C) also applies to (B) and (D).

Paris 4 site, Bedford County, Pennsylvania—Fay (1988), Holman (1995a).

The Spotted Salamander is a very attractive species of medium size that ranges from about 110 to 200 mm in total length. It may easily be recognized in the field, as it has up to about 50 round, bright yellow or orange spots extending in an irregular row on each side of the back from the eye to the tip of the tail (Conant and Collins, 1998). The background color of the back and sides ranges from black to bluish black. The Spotted Salamander is mainly a denizen of moist woodlands and breeds early in the spring, mainly in small, seasonal woodland ponds. This species has a wide range in eastern North America, occurring from Nova Scotia and the Gaspé Peninsula to south-central Ontario and then south to Georgia and eastern Texas. It is absent from the prairie region of Illinois and from most of southern New Jersey and the Delmarva Peninsula. Some isolated populations occur in the northern part of Illinois and in North Carolina.

Identification of Fossils. Tihen (1958) measured vertebrae of individual modern ambystomatid vertebrae to establish vertebral ratios that help in identifying the various species groups, as follows.

Length of Centrum Divided by Width of Centrum at Anterior End	
Ambystoma mexicanum group	1.9–2.2
Ambystoma tigrinum group	1.8–2.3
Ambystoma opacum group	2.0–2.6
Ambystoma maculatum group	2.2–2.9

Combined Zygapophyseal Width Divided by Zygapophyseal Length	
Ambystoma mexicanum group	1.3–1.6
Ambystoma tigrinum group	1.3–1.7
Ambystoma opacum group	1.3–1.5
Ambystoma maculatum group	1.1–1.4

The groups above do not include all of the North American ambystomatid salamander species (see Tihen 1958, table 1, p. 19). Salamander species in the four above groups contain the following species specifically referred to in this volume (except for *Ambystoma texanum*).

Ambystoma mexicanum group:
    *_Ambystoma kansense_
Ambystoma tigrinum group:
    *_Ambystoma hibbardi_
    Ambystoma tigrinum
Ambystoma opacum group:
    Ambystoma opacum
    *_Ambystoma tiheni_
Ambystoma maculatum group:
    Ambystoma jeffersonianum
    Ambystoma laterale
    Ambystoma maculatum
    *_Ambystoma minshalli_
    *_Ambystoma priscum_

Ambystoma texanum is not in any of the above groups in Tihen (1958). It has a "centrum" ratio of 1.9–1.3 and a "zygapophyseal" ratio of 1.0–1.3.

Using the ratios to help determine the proper species group, Holman and Grady (1989, p. 36) made the following comments about their identification of *Ambystoma maculatum* from the Pleistocene (Irvingtonian NALMA) Hamilton Cave, Pendleton County, West Virginia. "Vertebrae of *A. maculatum* are shorter than in the *A. jeffersonianum* complex [*A. jeffersonianum* and *A. laterale*] and have a wider, more flared postzygapophyseal area. The neural arch usually extends posterior to the posterior extent of the postzygapophyses, and the posterior end of the vertebra is depressed."

General Comments. The Spotted Salamander has been around since the Late Miocene and apparently has changed very little since. It apparently evolved from *Ambystoma minshalli*, earlier in the Miocene.

*AMBYSTOMA MINSHALLI TIHEN AND CHANTELL, 1963

FIGS. 56B–56E

Holotype. A trunk vertebra (UNSM 61002, University of Nebraska State Museum), complete except that the transverse processes are represented only by their bases.

Type Locality. Norden Bridge Local Fauna, Brown County, Nebraska.

Horizon. Middle Miocene (medial Barstovian NALMA).

Other Material. Topotypic referred specimens listed by Tihen and Chantell (1963) included a cervical vertebra (UNSM 61003), a right tibia (UNSM 61004), several vertebrae and fragments of vertebrae (UNSM 61005), and two trunk vertebrae (UMMPV 42190). Other topotypic specimens reported by Estes and Tihen (1964) included two humeri, proximal portions of three femora, a dental fragment, and several vertebral fragments, including a cervical, all under one number (UNSM 61012); a complete humerus (UNSM 61013) was also listed from the type locality. Holman (1973) added two vertebrae (MSUVP 720, Michigan State University Museum, Vertebrate Paleontology Collection) from the type locality; and Holman (1982b) added six vertebrae—three trunk vertebrae, one caudal vertebra (Figs. 56B–56D), and two vertebrae of uncertain position—and one right humerus (Fig. 56E) from the type locality, all under one number (MSUVP 944).

Ambystoma minshalli has also been reported from other localities, as follows. **Middle Miocene (medial Barstovian NALMA):** Achilles Quarry, Brown County, Nebraska—Voorhies (1990). Carrot Top Quarry, Brown County, Nebraska—Voorhies (1990). Egelhoff Local Fauna, Keya Paha County, Nebraska—Chantell (1971), Holman (1976a, 1987), Voorhies (1990). Hottell Ranch rhino quarries, Banner County, Nebraska—Voorhies et al. (1987). Lost Duckling site, Brown County, Nebraska—Voorhies (1990). Quarry Without a Name, Brown County, Nebraska—Voorhies (1990). **Middle Miocene (late Barstovian NALMA):** Annies Geese Cross Local Fauna, Knox County, Nebraska—Holman (1995b). Bijou Hills Local Fauna, Charles Mix County, South Dakota—Holman (1978a). Glad Tidings Prospect, Knox County, Nebraska—Holman (1996c). Glenn Olson Quarry, Charles Mix County, South Dakota—Green and Holman (1977). University of Notre Dame locality V-377, Cherry County, Nebraska—Estes and Tihen (1964), Holman (1976a). Valentine Formation, type locality, Cherry County, Nebraska—Holman and Sullivan (1981). *Ambystoma* cf. *Ambystoma minshalli* was also identified, as follows. **Late Miocene (medial Hemphillian NALMA):** Coffee Ranch Local Fauna of Hemphill County, Texas—Parmley (1984).

Diagnosis. The diagnosis is directly from Tihen and Chantell (1963, p. 509): "An *Ambystoma* of the *maculatum* group, differing from most members of that group in its smaller size, and perhaps from all in the more extensive development of the flange, or crest, continuous with the spine, along the posterodorsal surface of the tibia."

Description. This description is slightly modified from Tihen and Chantell (1963). The general form of all the vertebrae (see Tihen and Chantell, 1963, fig. 4, p. 509) is completely typical of the genus *Ambys-*

toma, and specifically of the *Ambystoma maculatum* group. The posterior zygapophyses always extend farther posteriorly than the neural spine, usually very much so. All the vertebrae appear to represent mature individuals: a septum interrupting the notochord is present in those specimens where its presence or absence could be observed.

The size variation in these vertebrae is not great. In those vertebrae with the centrum length complete enough to be measured, the length of the centrum varies only between 2.4 and 2.8 mm, the latter figure representing the holotype. Other measurements of the holotype are as follows: width of centrum at anterior end, 1.1 mm; width of centrum at posterior end, 1.2 mm; length at zygapophyses, 3.9 mm; width at anterior zygapophyses, 2.2 mm; width at posterior zygapophyses 2.3.

A total of seven vertebrae allowed for the determination of proportions of the centrum and the zygapophyses. Values for the ratio of centrum length divided by the width at the anterior end are 2.3 (three vertebrae), 2.4 (one vertebra), 2.5 (one vertebra), and 2.6 (two vertebrae), with an average value of 2.4. Values for the ratio of combined zygapophyseal width divided by zygapophyseal length are 1.1 (two vertebrae), 1.2 (three vertebrae), and 1.4 (two vertebrae), with an average of 1.2.

The referred atlas is broken posteriorly, but the anterior part is nearly intact. The total width of that specimen is 2.8 mm. The odontoid process (Tihen and Chantell, 1963) was somewhat narrower than in most of the modern *Ambystoma* skeletons examined.

The referred tibia is 3.0 mm in length and is nearly complete, except that the tibial spine and one corner of the distal end have been broken off. A relatively extensive crest occurs along the posterior edge of the tibia, which is continuous with the remnant of the spine proximally and extends distally almost to the full extent of the bone. The outer (actually the posterior) edge of the crest forms nearly a straight line and is not as concave as in most modern specimens that have a crest present.

Chantell (1971) further enhanced the credibility and taxonomic stability of *Ambystoma minshalli* by collecting and reporting on additional material of *A. minshalli* from the Egelhoff Local Fauna (Middle Miocene: medial Barstovian NALMA). This material included 67 trunk vertebrae and three atlantes (UMMPV 56555, University of Michigan Museum of Paleontology, Vertebrate Collection), three left dentaries (UMMPV 57355), two left and one right humeri (UMMPV 57354), and two left and one right femora (UMMPV 56542).

The ratio of centrum length to anterior centrum width and the ratio of combined zygapophyseal width to zygapophyseal length for 30 measurable vertebrae were within the ranges given for the *Ambystoma maculatum* group of Tihen (1958) (see Chantell 1971, table 1, p. 240). It was also found that the atlantes had plate-like odontoid processes that are flatter than those from extant *Ambystoma* specimens.

Caudal vertebrae of *Ambystoma minshalli* (Figs. 56B–56D) were not discussed by Tihen and Chantell (1963). Holman (1982b) recovered this element at the type locality. This specimen is described here as follows. In dorsal view, the vertebra is somewhat longer than it is wide. The remaining left prezygapophyseal articular process is ovoid. The medially indented neural arch extends posteriorly beyond the postzygapophyseal articular surfaces. The neural spine is poorly developed and exists in the form of a faint keel. The left postzygapophysis is missing.

In lateral view, the rib-bearing processes are broken. The bottom of the centrum is moderately concave. The top of the neural arch is sinuate. In ventral view, the centrum is amphicoelous and is moderately constricted at its middle portion; its postzygapophyses are ovoid, the right one being somewhat less ovoid than the left one. In posterior view (not shown in Figs. 56B–56D) the top of the neural arch is slightly convex. The round cotyle is larger than the neural canal. In anterior view (not shown in Figs. 56B–56D) the top of the neural arch is slightly convex, and the subrounded cotyle is much larger than the neural canal.

Measurements and ratios of the caudal vertebra are as follows: length of centrum, 3.6 mm; width of centrum anteriorly, 1.1 mm; width through postzygapophyses, 3.0 mm; length through zygapophyses, 4.8 mm; ratio of centrum length to anterior centrum width, 3.27.

General Comments. Ambystoma minshalli is restricted in time and space, being known only in Nebraska, South Dakota, and probably Texas, where it was tentatively identified. This species is known from Nebraska and South Dakota only in the Middle Miocene (medial to late Barstovian NALMA). In Texas it was tentatively identified from the Late Miocene (Parmley, 1984). If the Texas record is correct, this species existed from about 16 to about 8 Ma BP. In Nebraska and South Dakota this species is very abundant in comparison with the other salamanders of the time. Unspecified material of "*Ambystoma* cf. *A. minshalli*" from the Middle Miocene Marsland Formation of Box Butte County, Nebraska, was mentioned but not documented by Yatkola (1976). In pages to follow, I shall discuss the less well-known *Ambystoma priscum,* another member of the *Ambystoma maculatum* species group that lived in the Miocene (Barstovian NALMA) of Nebraska.

Ambystoma opacum (Gravenhorst, 1807)

Marbled Salamander

Fig. 57A

Fossil Localities. **Pliocene (Blancan NALMA):** Beck Ranch Local Fauna, Scurry County, Texas—Rogers (1976). **Pleistocene (Irvingtonian NALMA), Kansan glacial stage:** Cumberland Cave, Allegany County, Maryland—Holman (1977b, 1995a). Hamilton Cave, Pendleton County, West Virginia—Holman and Grady (1989), Holman (1995a). **Pleistocene (Rancholabrean NALMA):** Baker Bluff, Sullivan County, Tennessee—Van Dam (1978). Boney Spring, Benton County, Missouri—Saunders (1977), Holman (1995a). Frankstown Cave, Blair County, Pennsylvania—Fay (1988), Holman (1995a). New Trout Cave, Pendleton County, West Virginia—Holman and Grady (1987), Holman (1995a) (two of three strata in this cave yielded this species [see Holman, 1995a]). Worm Hole Cave, Pendleton County, West Virginia—Holman and Grady (1994).

References to "*Ambystoma opacum* group" salamanders without pinning down the exact species represented (probably *Ambystoma opacum*) are as follows. **Pleistocene (Rancholabrean NALMA):** Baker Bluff Cave, Sullivan County, Tennessee—Van Dam (1978), Fay (1988), Holman

FIGURE 57. Trunk vertebra of *Ambystoma opacum* from the Late Pliocene of Texas in (A) dorsal view. Holotype trunk vertebra of **Ambystoma priscum* from the Middle Miocene of Nebraska in (B) ventral, (C) dorsal, (D) anterior, (E) posterior, and (F) lateral views. Upper scale bar at (A) = 1 mm; lower scale bar = 2 mm and applies to (B)–(F).

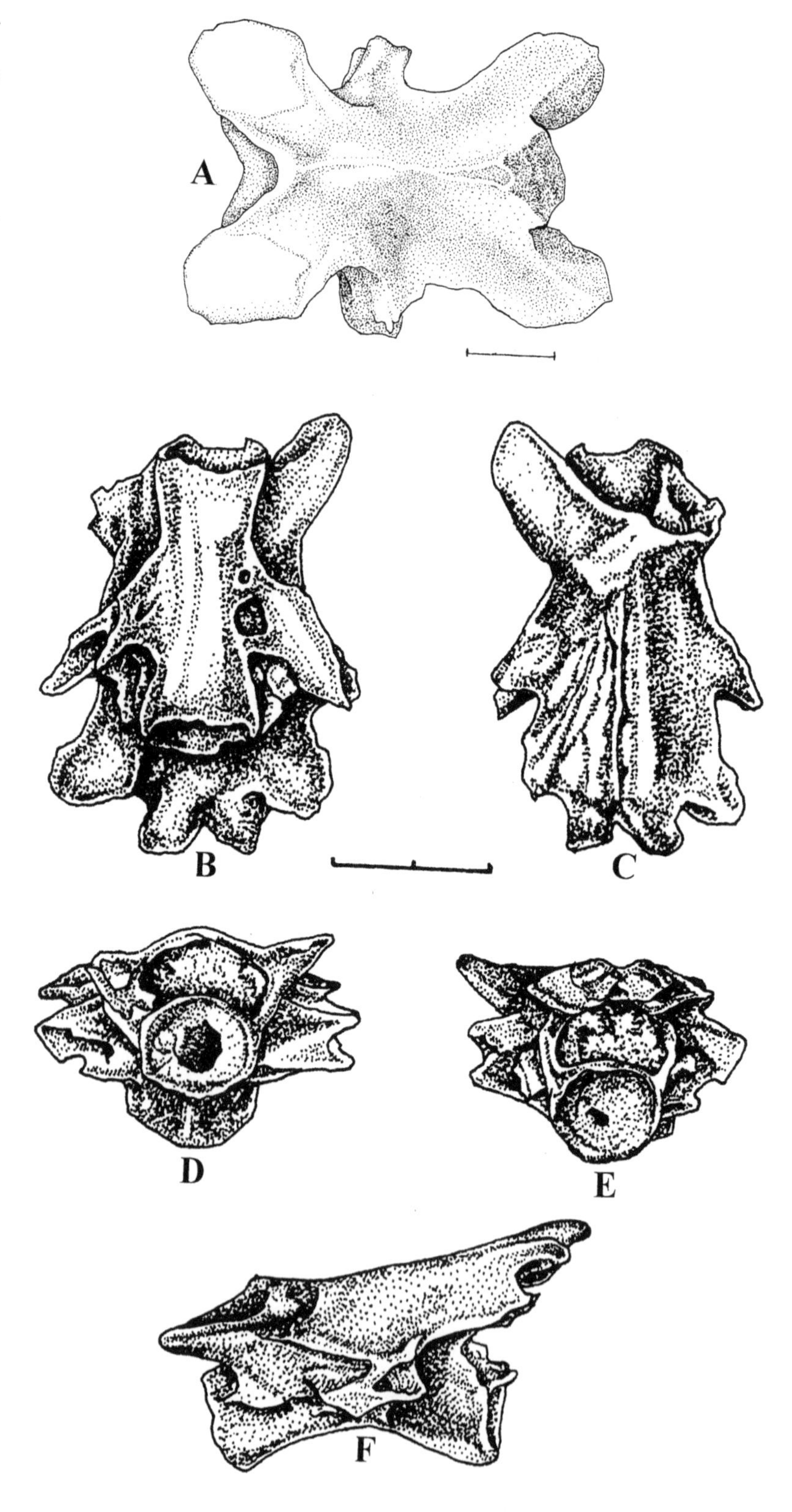

(1995a). Clark's Cave, Bath County, Virginia—Fay (1988), Holman (1995a). Kingston Saltpeter Cave, Bartow County, Georgia—Fay (1988), Holman (1995a). Natural Chimneys site, Augusta County, Virginia—Fay (1988), Holman (1995a).

The Marbled Salamander is a medium-sized (about 90–100 mm long), rather plump species, with somewhat glistening whitish bands on a blackish background color. The markings tend to be gray in females and white in males. This species occupies a variety of habitats, from wet floodplains to dry hillsides. Unlike most salamanders, the Marbled Salamander breeds in the fall. The female lays her eggs in depressions, which fill up with autumn rains. This species is restricted to the United States and ranges from southern New England to northern Florida and then west to southern Illinois, southeastern Oklahoma, and eastern Texas. Isolated colonies occur in the northwestern tip of Indiana and extreme southeastern Michigan (they have not been seen for years in Michigan), as well as in northeastern Ohio and adjacent northwestern Pennsylvania; along Lake Erie in northeastern Ohio and extreme northwestern Pennsylvania; and finally in two isolated colonies in southern Missouri.

Identification of Fossils. Rogers (1976) identified nine vertebrae from the Beck Ranch Local Fauna of Scurry County, Texas, as *Ambystoma opacum* (at present, the western extent of the geographic range of *A. opacum* is about 100 km to the east of Beck Ranch).

In Rogers's (1976) account of the identification process, reference is made to Tihen (1958), who reported that the postzygapophyses usually extend as far as or farther posteriorly than the neural arch (at least in the posterior part of the trunk vertebrae series) in the *Ambystoma maculatum* group. But Rogers's observations indicated that this situation also exists in the *Ambystoma opacum* group as well. Actually, this condition is also indicated by the figure of *A. opacum* in Tihen (1958, fig. c, p. 16). However, Rogers stated that the *A. opacum* group can be distinguished from the *A. maculatum* group on the basis of the height of the postzygapophyseal area: the neural arch is more upswept, resulting in a much higher and more triangular postzygapophyseal area than in the *A. maculatum* group. The closely related *Ambystoma talpoideum*, also of the *A. opacum* group, was eliminated because the postzygapophyses did not extend as far beyond the neural spine as in *A. opacum*. The identification of *A. opacum* from the Pliocene of Texas has stood up since 1976. Yet the fact is that a distinguishing character of the trunk vertebrae of *Ambystoma tigrinum* (a very common species in the west today) is the upswept posterior neural arch area, at least in middle trunk vertebrae.

General Comment. If the above identification is correct, this is the second record of the rather ancient occurrence of a modern species of ambystomatid salamanders that we have discussed (*Ambystoma maculatum* being known from the Late Miocene of Kansas).

*AMBYSTOMA PRISCUM HOLMAN, 1987

FIGS. 57B–57F

Holotype. A trunk vertebra (MSUVP 1077, Michigan State University Museum, Vertebrate Paleontology Collection).

Type Locality. Egelhoff Local Fauna, Keya Paha County, Nebraska.

Horizon. Middle Miocene (medial Barstovian NALMA).

Other Material. Referred topotypic specimens include four vertebrae (MSUVP 1078).

Holman (1996c) referred one vertebra (MSUVP 1382) to *Ambystoma* cf. *Ambystoma priscum.* This vertebra came from the Glad Tidings Prospect, Knox County, Nebraska (Middle Miocene: late Barstovian NALMA). Thus, all the known material is from the Miocene of Nebraska.

Emended Diagnosis. An *Ambystoma* of the *A. maculatum* group (Tihen, 1958), but differing from all of the fossil and modern species of this group in having the posterior end of the neural arch more deeply notched and more posteriorly produced.

Description and Comparison. This description of the holotype (MSUVP 1077) is modified from Holman (1987), and from time to time comparison is made with *Ambystoma minshalli.* In dorsal view, the posterior end of the neural arch is deeply notched, creating a roughly V-shaped structure that extends well posterior to the ends of the postzygapophyses. This is not the case in *A. minshalli* that I have seen (e.g., the holotype of *A. minshalli* in dorsal view [Tihen and Chantell, 1963, fig. 4, p. 509]). A weak but not faint dorsal ridge (neural spine) is present. The dorsal ridge is usually more pronounced in the middle of its extent in *A. minshalli* than in *Ambystoma priscum.* The rib-bearers are broken off, but distinct remnants are directed very sharply posteriad, in fact much more so than in most *A. minshalli* I have seen. The left prezygapophyseal articular facet is ovoid and elongated. The right prezygapophyseal articular facet is broken. The general proportions of the type are elongate as in the *Ambystoma maculatum* group of Tihen (1958).

In lateral view, the posterior part of the neural arch is not upswept as in the *Ambystoma tigrinum* group. In fact, the dorsal border of the neural arch is very straight. In anterior view, the prezygapophyses are slightly tilted upward. The neural canal is wider than it is long and is roughly ovoid. The cotyle is round. In posterior view, the postzygapophyses are moderately tilted upward, the cotyle is round, and the neural canal is wider than it is long and is ovoid. In ventral view, the posterior end of the neural arch is deeply notched and extends well posterior to the ends of the postzygapophyses. The centrum is constricted medially into an hourglass shape. The postzygapophyseal articular facets are ovoid. Measurements and ratios of the holotype are as follows: greatest length of vertebra, 5.0 mm; length through zygapophyses, 4.8 mm; length of centrum, 3.5 mm; width of centrum, 1.3; length of centrum divided by width of centrum, 2.69.

Turning now to the referred specimens, we find that although three of the four vertebrae are too fragmentary for accurate measurements, all have the deeply notched and posteriorly produced neural arches. For the vertebra that is complete enough to be measured, the measurements and ratios are comparable to those of Tihen (1958) and are as follows: greatest length of vertebra, 4.5 mm; length through zygapophyses, 4.2 mm; width through prezygapophyses, 2.5 mm; width through postzygapophyses, 2.5 mm; combined zygapophyseal widths divided by length through zygapophyses, 1.19; length through zygapophyses divided by width through zygapophyses, 2.5.

General Comments. The only other extinct Barstovian (and probably Late Miocene: Hemphillian NALMA) *Ambystoma* is *A. minshalli,* which has a much wider distribution than *Ambystoma priscum.* It certainly is not unusual to have had two extinct salamanders of the *Ambystoma maculatum* group present in the Middle Miocene of the High Plains, especially considering the widespread riparian habitats and warmer, moister climates that must have been available at that time (see Barstovian NALMA faunal lists in Voorhies [1990]).

There is a good possibility that *Ambystoma priscum* was a larger species than *Ambystoma minshalli,* at least as indicated by vertebral length (greatest vertebral length in *A. priscum* is 4.5–5.0 mm [mean 4.75], and holotype vertebra of *A. minshalli* is 3.9 mm). Moreover, the more distinct and centrally upraised dorsal ridge (neural spine) and more posteriorly directed rib-bearers of *A. priscum* may also be noteworthy characters that separate the two species.

Forty-eight *Ambystoma* vertebrae other than the *Ambystoma minshalli* vertebrae previously reported from the Egelhoff Local Fauna by Chantell (1971) were reported from this locality by Holman (1987). Seventeen identifiable vertebrae were easily differentiated from *Ambystoma priscum* vertebrae and assigned to *A. minshalli.* (The other 31 vertebrae were small and had the posterior end of the neural arch broken off; these were assigned to *Ambystoma* sp. indet.) The rarity of *A. priscum* may suggest that this form was a more upland form than *A. minshalli* and therefore less likely to be preserved as a fossil.

Ambystoma texanum (Matthes, 1855)

Small-mouthed Salamander

Figs. 58A, 58B

Fossil Localities. **Pleistocene (Rancholabrean NALMA):** Clear Creek Fauna, Denton County, Texas—Holman (1963, 1995a). Howard Ranch Local Fauna, Hardeman County, Texas—Holman (1964, 1995a). Sims Bayou Fauna, Harris County, Texas—Holman (1965a, 1995a).

The Small-mouthed Salamander is a somewhat small species, about 100–135 mm long. Both its mouth and its head are small compared with other ambystomatids. The background color of this animal is brownish, and it has grayish markings on the body; thus, it is one of the less colorful salamanders encountered in the field. In Texas, this salamander is heavily speckled with light markings, with especially large ones on the lower sides of the body (Conant and Collins, 1998). It is usually found in damp areas in a variety of local habitats. Its range covers extreme southeastern Michigan, northern Ohio adjacent Ontario, then west to southeastern Nebraska and south through eastern Texas to the Gulf. It does not occur, however, in the Ozark highland areas in Missouri and Arkansas, and it is absent from the coastal part of Louisiana.

Identification of Fossils. Holman (1963) made some comments about the identification of two *Ambystoma texanum* vertebrae from the Pleistocene of the Clear Creek Fauna of Denton County, Texas (near Dallas). The fossil locality is well within the general range of the living species. One of the fossil vertebrae was from the middle or posterior portion of

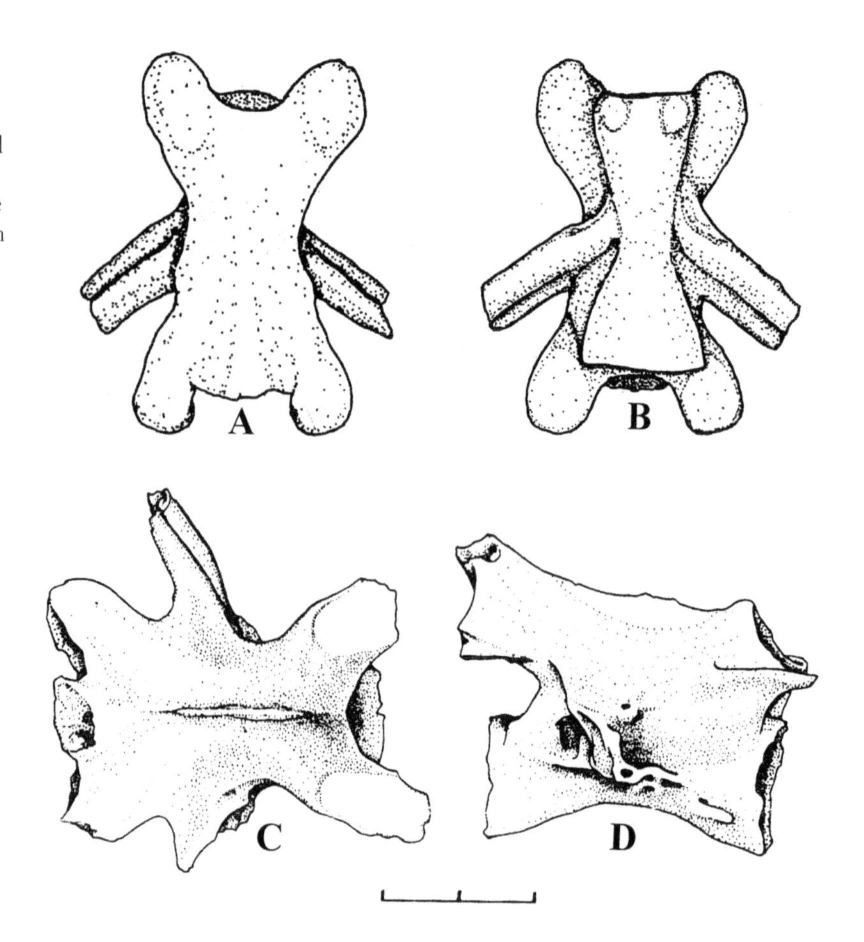

FIGURE 58. Modern *Ambystoma texanum* trunk vertebra in (A) dorsal and (B) ventral views. *Ambystoma tigrinum* from the Late Miocene of Kansas in (C) dorsal and (D) lateral views. Scale bar = 2 mm and applies only to (C) and (D).

the trunk region and had the postzygapophyses projecting well anterior to the anterior projection of the neural arch as in the *Ambystoma maculatum* and *Ambystoma texanum* groups. The only other member of the above groups that occur anywhere near the Clear Creek area today is *A. maculatum*, whose range in the area actually is somewhat east of the Dallas–Fort Worth area (see Conant and Collins, 1998, map, p. 438). The vertebra was indistinguishable from vertebrae in skeletons of modern *A. texanum* and was smaller than vertebrae of eight skeletons of *A. maculatum*. The other fossil *Ambystoma* vertebra was tentatively assigned to *A. texanum* on the basis of its similar size.

Comment. The lack of fossil records of *Ambystoma texanum*, given the number of records of other North American fossil *Ambystoma*, is somewhat surprising, because the Small-mouthed Salamander is rather common throughout its natural range today.

AMBYSTOMA TIGRINUM (GREEN, 1825)
TIGER SALAMANDER

FIGS. 58C, 58D, 59

Fossil Localities. **Late Miocene (Clarendonian NALMA):** McGinley's Stadium site, Brown County, Nebraska—Voorhies (1990). Serendipity Quarry, Keya Paha County, Nebraska—Voorhies, 1990). Wa-

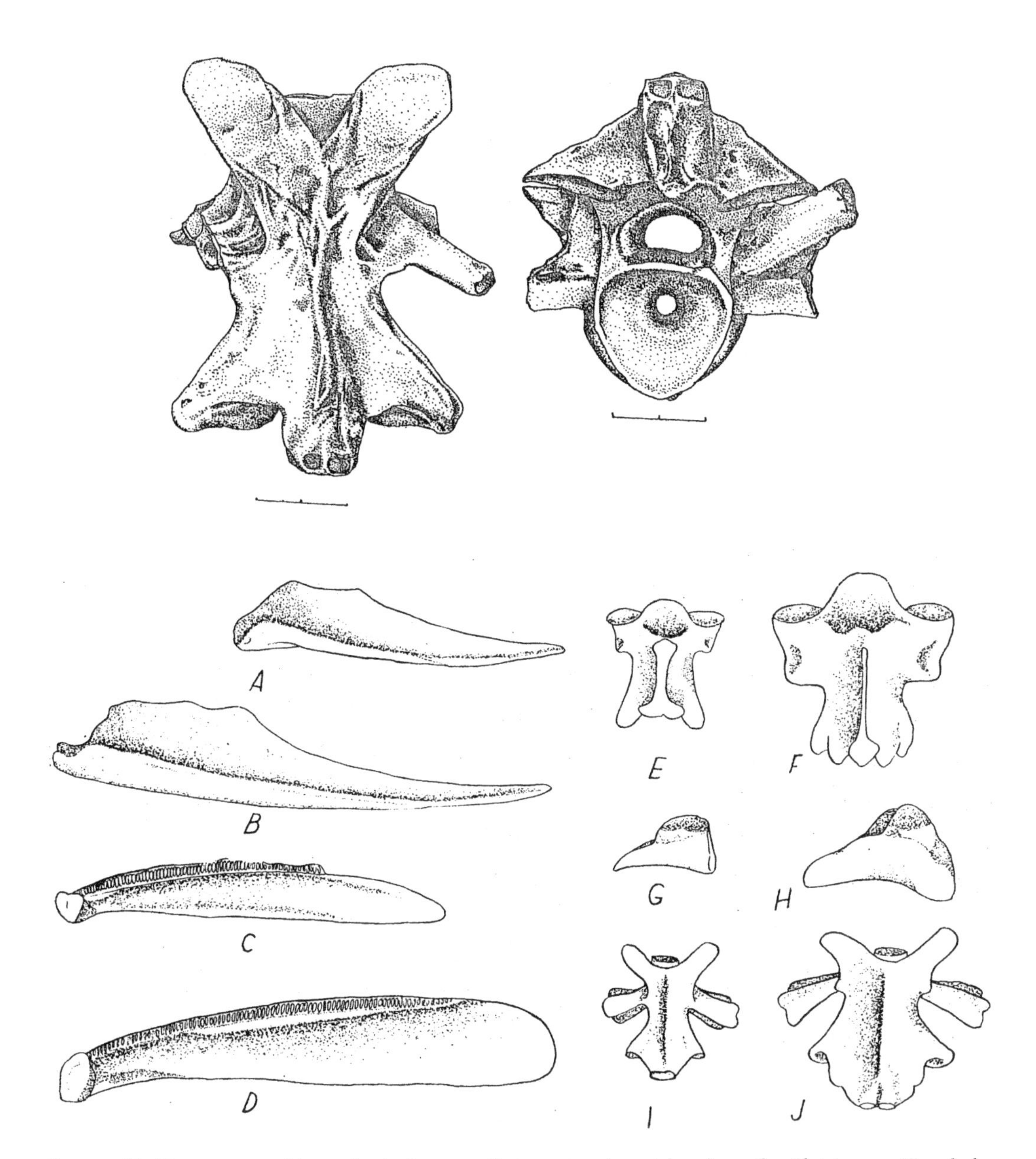

FIGURE 59. Upper group: Neotenic *Ambystoma tigrinum* trunk vertebra from the Pleistocene (Rancholabrean) of Texas in (left) dorsal and (right) posterior views. (A)–(J) Comparison of elements of modern neotenic *Ambystoma tigrinum mavortium* with the same elements of a presumed neotenic, cannibalistic *A. tigrinum* from the Pleistocene (Rancholabrean) of Kansas. (A) Angular, (C) dentary, (E) atlas, (G) quadrate, and (I) trunk vertebrae of the modern neotenic *A. t. mavortium*. (B), (D), (F), (H), (J) The same elements, respectively, in a Kansas Pleistocene cannibalistic neotene. Scale bars = 2 mm and apply only to the upper group.

Keeney Local Fauna, Trego County, Kansas—Holman (1975b). **Pliocene (Blancan NALMA):** Beck Ranch Local Fauna, Scurry County, Texas—Rogers (1976) (material assigned by Rogers [1976] to "**Ambystoma hibbardi* or *A. tigrinum*" is here assigned to *Ambystoma tigrinum*). Big Springs Quarry, Antelope County, Nebraska—Rogers (1984). Borchers Fauna, Meade County, Kansas—Tihen (1955). Crooked Creek Formation, Meade County, Kansas—Tihen (1955). Dixon Fauna, Kingman County, Kansas—Tihen (1955). Gila Conglomerate, Cochise County, Arizona—Brattstrom (1955). Glens Ferry Formation (including Hagerman, Flatiron Butte, Grand View, and Tyson Ranch local faunas), southern Idaho—Mead et al. (1998) (listed as *Ambystoma* cf. *A. tigrinum*). Hornets Nest Quarry, Knox County, Nebraska—Rogers (1984). Kentuck Assemblage, McPherson County, Kansas—Tihen (1955). Nash Local Fauna, Meade County, Kansas—Holman (1979). Sam Cave site, Rio Arriba County, New Mexico—Rogers et al. (2000). Sand Draw Fauna, Brown County, Nebraska—Holman (1972a). Sanders Fauna, Meade County, Kansas—Tihen (1955). White Rock Fauna, Republic County, Kansas—Eshelman (1975). **Pleistocene (Irvingtonian NALMA):** Alamosa Local Fauna, Alamosa County, Colorado—Rogers (1987), Holman (1995a). Albert Ahrens Local Fauna, Nuckolls County, Nebraska—Ford (1992), Holman (1995a). Cudahy Fauna, Meade County, Kansas—Tihen (1955). Cumberland Cave, Allegany County, Maryland—Holman (1977b, 1995a). Curtis Ranch Fauna, Cochise County, Arizona—Brattstrom (1955), Holman (1995a). Fyllan Cave, Travis County, Texas—Holman and Winkler (1987), Holman (1995a). Hall Ash Fauna, Seward County, Kansas—Rogers (1982), Holman (1995a). Hamilton Cave, Pendleton County, West Virginia—Holman and Grady (1989), Holman (1995a). Java Local Fauna, Walworth County, South Dakota—Holman (1977c). Seymour Formation, Knox County, Texas—Hibbard and Dalquest (1966). Vera Local Fauna, Knox County, Texas—Parmley (1988), Holman (1995a). **Pleistocene (Rancholabrean NALMA):** Arredondo site, Alachua County, Florida—Lynch (1965), Holman (1995a). Ben Franklin Local Fauna, Fannin and Delta counties, Texas—Holman (1963, 1995a). Butler Spring Local Fauna, Meade County, Kansas—Holman (1986, 1995a). Carrol Creek Local Fauna, Donley County, Texas—Kasper and Parmley (1990), Holman (1995a). Cheek Bend Cave, Maury County, Tennessee—Miller (1992), Holman (1995a). Crankshaft Pit, Jefferson County, Missouri—Parmalee et al. (1969), Holman (1995a). Devil's Den chamber 3, Levy County, Florida—Holman (1978b, 1995a). Dry Cave Fauna unit II, Eddy County, New Mexico—Harris (1987), Holman (1995a). Duck Creek Local Fauna, Ellis County, Kansas—Holman (1984, 1995a). High Terrace Sands, Meade County, Kansas—Tihen (1955). Howell's Ridge Cave, Grant County, New Mexico—Van Devender and Worthington (1977), Holman (1995a). Illusion Lake site, Tahoka Formation, Bailey County, Texas—Holman (1975a). Isle of Hope site, Chatham County, Georgia—Hulbert and Pratt (1998). Jinglebob Fauna, Meade County, Kansas—Tihen (1955). Jones Fauna, Meade County, Kansas—Tihen (1942, 1955), Holman (1995a). Kingsdown Formation, Meade County, Kansas—Tihen (1955). Kingston Saltpeter Cave, Bartow County, Georgia—Fay

(1988), Holman (1995a). Ladds Quarry site, Bartow County, Georgia—Wilson (1975), Holman (1967, 1985a, 1985b, 1995a). Locality 7, Meade County, Kansas—Tihen (1955). Lubbock Lake site, Clovis-period level, Lubbock County, Texas—Johnson (1987), Holman (1995a). Miller's Cave, Llano County, Texas—Holman (1966b, 1995a). Natural Chimneys site, Augusta County, Virginia—Fay (1988), Holman (1995a). Orange Lake site, Marion County, Florida (Holman, 1959b, 1995a). Peccary Cave, Newton County, Arkansas—Davis (1973) (referred to *Ambystoma* cf. *Ambystoma tigrinum*), Holman (1995a). Peter Bear locality, Brown County, Nebraska—Voorhies (1990). Reddick 1 site, Marion County, Florida—Holman (1995a; see list of additional reference there). Sandahl Local Fauna, McPherson County, Kansas—Holman (1971, 1995a), Preston (1979). Schulze Cave Fauna, Edwards County, Texas—Parmley (1986), Holman (1969a, 1969c, 1995a). Slaton Local Fauna, Lubbock County, Texas—Holman (1969a–1969c, 1995a). Smith Falls locality, Cherry County, Nebraska—Voorhies (1990). Smith Springs locality, Cherry County, Nebraska—Voorhies, 1990). U-Bar Cave Fauna, Hidalgo County, New Mexico—Harris (1987), Holman (1995a). Vanhem Formation, Meade County, Kansas—Tihen (1942, 1955). Williams Local Fauna, Rice County, Kansas—Preston (1979), Holman (1984, 1995a). Williston IIIA site, Levy County, Florida—Holman (1959a, 1995a, 1996a).

More accounts have been given of fossil *Ambystoma tigrinum* and its often large, aquatic neotenic forms than of any other salamander species is North America. There are probably three reasons for the abundance of fossil remains of this species: (1) metamorphosed adults tend to burrow deeply into the ground, where they are more likely to fossilize; (2) these salamanders are larger and have a more robust bone structure than other North American salamander species; and (3) large die-offs occurred in the large neotenic forms of this species during droughts in the Central Plains regions during the Pleistocene.

Modern Tiger Salamanders are large, robust caudates that tend to have dark bodies covered with lighter spots or bars. The average ones range from about 175 to 200 mm long, but the neotenic ones get much larger. These animals breed very early in the spring, even in northern areas in the United States, and they tend to do this in deeper water than other salamanders, such as the Spotted Salamander. *Ambystoma tigrinum* occurs from the southern limits of boreal forests in central Alberta and Saskatchewan, Canada; south to Florida and Puebla, Mexico; and west to Washington and Arizona and northern Sonora, Mexico. An isolated population occurs west of Sierra Nevada, between Sonoma and Santa Barbara counties, California (Frost, 1985).

In Field and Lab. As a youth I kept Tiger Salamanders as pets from time to time, as did several other buddies of mine. In the spring and sometimes in autumn these animals would fall into window wells during their travels, and everybody would take them to school to show the teacher. My pet Tiger Salamanders would eat about anything that moved and was small enough to swallow, but they also readily learned to eat inert objects, such as bits of red meat or pieces of fish, which they would poke around the terrarium to find. Adults tend to burrow deeply, and evidently they stay in their burrows a long time in dry weather. I have only seen them in spring and fall.

I was surprised to find live individuals of these salamanders in the dry prairies of western Kansas after 2 years of excavating fossil amphibians and reptiles there. We had previously seen plenty of live frogs, turtles, snakes, and lizards in the area, but no salamanders. One day our crew was excavating a bank of sandy soil to get to the fossiliferous layer below, and we started digging up live Tiger Salamanders about 3 feet (about 0.9 m) below the surface. Later that summer, after a heavy rainstorm, we saw dozens of Tiger Salamanders out on the roads at night. We knew these wandering caudates were fed upon by Great Horned Owls, as we found Tiger Salamander bones in owl pellets in a large fissure in a butte a day or two after the storm (Holman, 1976b).

Identification of Fossils. Fossil Tiger Salamander vertebrae outnumber by a large percentage other elements of their skeleton that are found as fossils. In fact, vertebrae are sometimes the only bones of this species found in fossil sites. Essential characters in the identification of *Ambystoma tigrinum* vertebrae are (1) the general large size and sometimes very large size of these ambystomatid vertebrae; and (2) the upswept posterior part of the neural arch in the trunk vertebrae (see Fig. 58D). Moreover, very large ambystomatid vertebrae with a large, round, open notochordal canal usually represent neotenic Tiger Salamanders (Fig. 59, top group). Other views of fossil neotenic Tiger Salamander vertebrae and cranial elements are shown in Figs. 59A–59J.

General Comments. One of the most spectacular fossil finds of Tiger Salamanders, or any other salamanders for that matter, occurred in 1939 and 1940 as C. W. Hibbard (1940) and H. T. U. Smith (1940) collected fossils from a very late Pleistocene (Rancholabrean NALMA) sink deposit in Meade County, Kansas. The fauna of this locality is known as the Jones Fauna (listed above). Here, remains from more than 1250 individuals were recovered and studied by Tihen (1942). These bones represented a large colony of neotenic Tiger Salamanders. The average length of mature individuals was 250 mm (about 9.84 inches). Some large specimens are said to have represented individuals 400 mm (about 15.75 inches) in length!

The large number of fossil Tiger Salamander remains of individuals allowed Tihen (1942) to trace the development of particular fossil bones from a very young stage of life in the salamanders to the condition found in the bones of old individuals. The evidence indicated that at the time the salamander remains were deposited, the neotenic condition had developed in the group to such an extent that in the majority of the population, metamorphosis was no longer possible. In other words, they were now obligate neotenes. These assumptions explain such a high concentration of individuals in a small area, as well as the relatively high percentage of presumed cannibalistic individuals present. Cannibalistic neotenic *Ambystoma tigrinum* (Powers, 1907; Tihen, 1942) appear where there is a great concentration of individuals and a consequent food shortage. Morphological characters indicating cannibalism in *A. tigrinum* neotenes are their very large size and changes in the structure of the bones of their skeletons (Figs. 59A–59J).

The Jones Fauna fossil sink was probably formed in the early part of the Pleistocene (Tihen, 1942). The sides may have been very steep at first, so any Tiger Salamanders present at the time of the down-sinking of the structure, or that accidentally entered at a slightly later date, were trapped. The inability of Tiger Salamander larvae to leave the sink and lead a normal life on land might account for the first appearance of neoteny. This actually seemed somewhat unlikely to Tihen (1942), given the very large size of the sink. He suggested that a more likely explanation was that at some period (which could have been at the time of formation of the sink or at a somewhat later time) the surrounding area was exceedingly dry and offered little shelter. Such conditions could cause the appearance of neoteny. No matter what factors caused the first appearance of neoteny in the area, such factors must have been around long enough that the salamanders adapted to neoteny as the normal condition. Crowding and lack of food, other than the salamanders themselves, would have led to the natural selection of the big cannibals.

Tihen (1942, 1955) mentioned two other Pleistocene fossil sites in Kansas that had neotenic Tiger Salamander remains. Curiously, he found that all three of these sites were associated with glacial stages. All five Kansas sites that were associated with interglacial ages, however, had normal populations of *Ambystoma tigrinum*. This seems to be backwards, but modern studies of Pleistocene climates in North America indicate that both glacial and interglacial Pleistocene sites had more equable climates than were suspected by scientists back in 1942, when they thought that glacial stages were supposed to be universally cold and wet; interglacial stages, warm and dry (see Holman, 1995a).

Holman (1975a) found four large *Ambystoma tigrinum* vertebrae (Fig. 59, top row), which he thought represented a neotenic individual or individuals from the Pleistocene Illusion Lake site in Bailey County, Texas. These vertebrae were of large size and had a large, round, persistent notochordal canal. The four fossils are larger than their counterparts in the skeletons of seven Recent normal adult A. *tigrinum*. The greatest length of the fossil vertebrae ranges from 8.4 to 10.7 mm (mean, 9.46 mm). The length of the adult Recent vertebrae ranges from 4.5 to 7.6 mm (mean, 6.09 mm). In a study of 13 Recent normal Tiger Salamanders, Holman (1975a) found that a larval form had a much enlarged, persistent notochordal canal; that a metamorphosed subadult had a moderately enlarged, persistent notochordal canal; that three of seven adults had a constricted persistent notochordal canal; and that four of seven adults had an interrupted notochordal canal.

**AMBYSTOMA TIHENI* HOLMAN, 1968B

FIG. 60

Holotype. A trunk vertebra (SMNH 1431, Royal Saskatchewan Museum of Natural History).

Type Locality. Calf Creek Local Fauna, near Eastend, Saskatchewan, Canada.

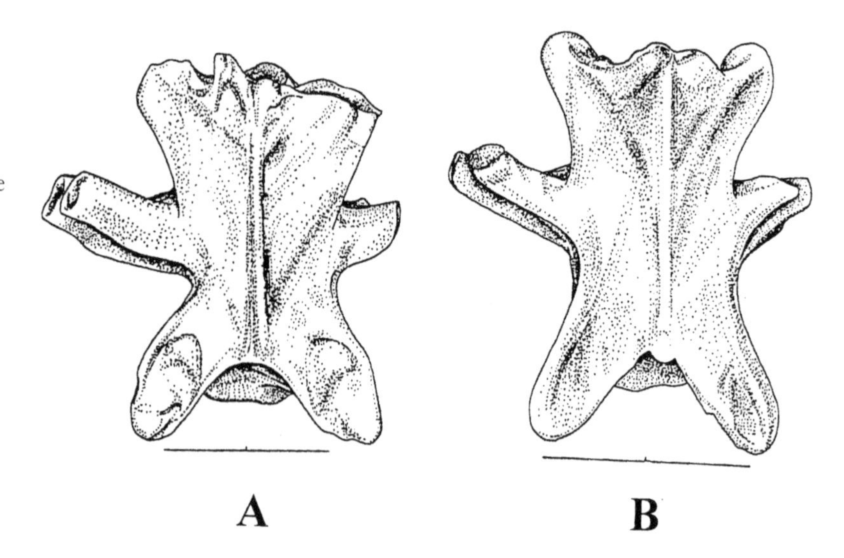

FIGURE 60. Holotype (A) and paratype (B) trunk vertebrae, both in dorsal view, of **Ambystoma tiheni* from the Late Eocene of Saskatchewan. Scale bars = 2 mm.

Horizon. Late Eocene (Chadronian NALMA)

Other material. The only other known material is a paratype trunk vertebra (SMNH 1432) from the type locality.

Diagnosis. The diagnosis is directly from Holman (1968b): "An *Ambystoma* similar in size and proportions to the *Ambystoma opacum* species group of *Tihen* (1958, p. 19, Table 1), but differs in having (1) the neural arch more depressed, (2) the foramina on the ventral part of the centrum obsolete or absent, (3) the ends of the centrum less widely flared, and (4) the transverse processes usually more robust."

Description of the Holotype. This description is modified from Holman (1968b). In dorsal view the prezygapophyseal articular facets are ovoid and about twice as long as they are wide. The neural arch is depressed. The neural spine is relatively prominent. The centrum extends anterior to the edge of the neural arch. The posterior tip of the neural arch ends slightly behind the posterior edge of the postzygapophyses. The rib-bearers are robust, slightly backswept, and sharply divided into dorsal and ventral portions.

In ventral view, the prezygapophyses are ovoid and extend well beyond the anterior end of the centrum. The centrum is constricted at its middle and has its ends only moderately flaring. Foramina on the centrum are obsolete. The ventral segments of the rib-bearers are more anteriorly directed than are the dorsal segments. The postzygapophyseal articular facets are rounded. Measurements and ratios are as follows: length through zygapophyses, 4.2 mm; width through prezygapophyses, 2.7 mm; width through postzygapophyses, 2.8 mm; combined zygapophyseal width divided by length through zygapophyses, 1.31.

Description of the Paratype. This description is also modified from Holman (1968b). The differences between the holotype and the paratype are slight and are attributable to individual or intracolumnar variations. The neural arch ends at the level of the posterior end of the postzygapophyses. The neural spine is slightly thicker than in the holotype but not as robust. The transverse processes are slightly less robust in the paratype. No foramina are observable on the ventral side of the centrum. Measurements and ratios are as follows: length through zygapophyses, 3.8 mm; width through prezygapophyses, 2.6 mm; width through postzygapophyses, 2.4 mm; combined zygapophyseal width divided by length through zygapophyses 1.50.

General Comments. This is the earliest fossil record of the genus *Ambystoma*. The fossil vertebrae fit best with the *Ambystoma opacum* group of ambystomatids (*A. opacum* and *Ambystoma talpoideum*) on the basis of size and vertebral proportions. But the fossils differ on the basis of some qualitative characters. Definite assignment of *Ambystoma tiheni* to the *A. opacum* group should await more fossil material from Late Eocene deposits. Nevertheless, the mere suggestion that species groups of *Ambystoma* were established in the Eocene is of considerable interest relative to the evolutionary history of this fascinating genus. *Ambystoma tiheni* is put in the context of its faunal associates in Holman (1972b, 1976c).

Ambystoma sp. indet.

There are many publication records of fossil "*Ambystoma* sp." or "*Ambystoma* sp. indet." These are frequently listed along with other identified *Ambystoma* and are often fragmentary remains of vertebrae. All these records are within the modern range of the genus, as far as I am aware, except one, and that occurrence is from the Vero Beach (strata 2 and 3) locality, Indian River County, Florida. This record is from Weigel (1962) and was confirmed again by Holman (1995a). At present *Ambystoma talpoideum* and *Ambystoma tigrinum tigrinum* extend farther down peninsular Florida than any other species of the genus—to about the level of Citrus County, well north of Tampa on the Gulf Coast. The Vero Beach locality on the Atlantic Coast lies roughly 250 km southwest of this. I have no explanation for the absence of ambystomatid salamanders in the southern half of peninsular Florida today, other than that winters may have become too warm for them during the Holocene warming trend.

General Evolutionary Comments on Ambystoma. The genus *Ambystoma* forms a morphologically similar group of species, yet arguably a potentially evolutionarily plastic group of taxa. Several living *Ambystoma* species freely hybridize with one another, producing morphologically distinct individuals, as well as all-female populations. Nevertheless, it is questionable whether these kinds of individuals and populations have ever developed truly stable species, or will do so in the future.

With regard to the process of neoteny, the most widespread North American species of Ambystoma, *Ambystoma tigrinum*, has produced large, obligate neotenic, cannibalistic populations in both Pleistocene and modern times in the United States. These populations have different skeletal morphologies (see Fig. 59) but have not differentiated into stable species as yet in the United States.

Finally, there is an exceedingly remote possibility that the single "saddle-backed" *Ambystoma* vertebra from the Pleistocene of Colorado, which was named *Ambystoma alamosensis* by Rogers (1987), represented a "hopeful monster" (an organism with a drastic mutation in its genome) that could have produced an "instant clade" that lasted a while. Realistically, however, the odds are very much against hopeful monsters' surviving beyond a generation or two.

Family Salamandridae Gray, 1825
Salamandrids

The family Salamandridae is a successful group, with eight European–West Asian genera and five Asian genera, but only two American genera, *Notophthalmus* and *Taricha*. About 54 species are currently recognized. Salamandrids are primarily specialized for aquatic feeding, and terrestrial feeding adaptations are found only in the Old World *Salamandra* group genera. The salamandrids are highly variable in size and appearance. Two Old World genera, *Salamandra* and *Pleurodeles*, reach total lengths of more than 200 mm (Duellman and Trueb, 1986). All the salamandrids have robust limbs, and many of the aquatic species have caudal and dorsal body fins. All salamandrids have toxic skin secretions, some of which can produce unpleasant reactions or even fatal results in other animals. Many genera have bright colors that are displayed in various warning postures. The larvae have four pairs of gill slits, as well as large external gills.

The complex distribution of salamandrids is modified from Duellman and Trueb (1986) as follows. They are distributed mainly in Europe and Asia, where they occur from the British Isles and Scandinavia east to the Ural Mountains in Russia and south into the Iberian Peninsula, Asia Minor, and some of the Greek islands, as well as Corsica and Sardinia. In central and eastern Asia, this family occurs from northern India, Burma, Thailand, and Viet Nam south to Hong Kong and east through China and the Japanese archipelago. Two mainly European genera occur in extreme northwestern Africa, and two others are endemic to North America.

Some osteological characters used to define the Salamandridae are modified from Duellman and Trueb (1986). The premaxillae are paired in primitive salamandrid genera and fused in advanced ones. Septomaxillae, lacrimals, and the ypsiloid cartilage occur in all of them. The exoccipital, prootic, and opisthotic are all fused. An internal carotid foramen is absent. A frontosquamosal arch is usually present, but it is reduced or absent in some derived genera. The columella is fused with the operculum. The teeth are pedicellate. Palatal dentition extends posteriorly on the lateral edges of the vomers. The vertebrae of salamandrids are opisthocoelous. All but the first two spinal nerves exit intravertebrally.

Salamandrid characters that Estes (1981) considered to be derived are slightly modified as follows. The palatal tooth row is extended posteriorly by the lateral growth of the vomer. The vomerine teeth are replaced from the medial side. The angular, lacrimal, septomaxilla, and second epibranchial are absent. The columella is fused. Frontosquamosal arches are often present. The vertebrae are truly opisthocoelous.

The ancestry of the two modern North American salamandrid genera, *Notophthalmus* from eastern North America and *Taricha* from the Pacific Coast region, is not yet settled.

Genus *Notophthalmus* Rafinesque, 1820

Eastern Newts

Genotype. Notophthalmus viridescens (Rafinesque, 1820).

Character States Compared with Those of Taricha. Estes (1981) pointed out derived character states in *Notophthalmus* that are more advanced than in *Taricha*. In *Notophthalmus* (1) the maxilla extends beyond the eye but falls far short of the quadrate; (2) the operculum is cartilaginous; (3) the interradial cartilage is absent; (4) the hyoglossus is single; and (5) the radioglossals are paired. *Notophthalmus* is similar to *Taricha* in other derived character states.

Identification of Fossils. Notophthalmus vertebrae may be distinguished from *Taricha* vertebrae on the basis of (1) having an extensive, pitted dermal cap; and (2) having a higher neural spine.

**Notophthalmus crassus* Tihen, 1974

Figs. 61A–61C

Holotype. A trunk vertebra (UNSM 61040, University of Nebraska State Museum).

Type Locality. Flint Hill locality equivalent (see Tihen, 1974), Batesland Formation, Bennett County, South Dakota.

Horizon. Early Miocene (Hemingfordian NALMA).

Other Material. No other material is known.

Diagnosis. The diagnosis presented here is directly from Tihen (1974, p. 214): "A species of the genus *Notophthalmus* differing from all other known species in its greater size and in the possession of relatively broad zygapophyseal proportions (see Table 1) [in Tihen, 1974], in combination with a high massive neural spine whose dorsal portion is very markedly expanded for its entire length."

Description of the Holotype. The description of the type specimen is modified from Tihen (1974). The holotype is relatively complete, but it has been subject to erosion to the extent that all the projecting parts and surfaces are somewhat rounded off. Measurements to the nearest 0.05 mm are as follows: length of centrum, 3.35 mm; width of centrum at cotyle, 1.45 mm; total width including transverse processes, 4.00 mm; total height, 4.10 mm; zygapophyseal length, 3.40 mm; anterior zygapophyseal width, 2.25 mm; height of neural spine (from level of lateral borders of posterior zygapophyses), 1.85 mm. Because of the erosion of the projecting surface, these measurements probably differ slightly from those of an unworn specimen, but they are nevertheless fairly close approximations.

The entire vertebra is massive and solid. The neural spine is strongly expanded dorsally, with the sides of the expansion diverging somewhat sharply anteriorly. The dorsal surface of the expansion was probably flat and ornamented originally, as in other *Notophthalmus*. However, in this specimen this dorsal surface appears smooth and has somewhat rounded edges; this is presumably because of wear. The rib-bearers are short and directed posteriorly at a fairly strong angle. The cotyle is slightly flattened dorsally, but it is not distinctly oval as in *Taricha*. The centrum is very

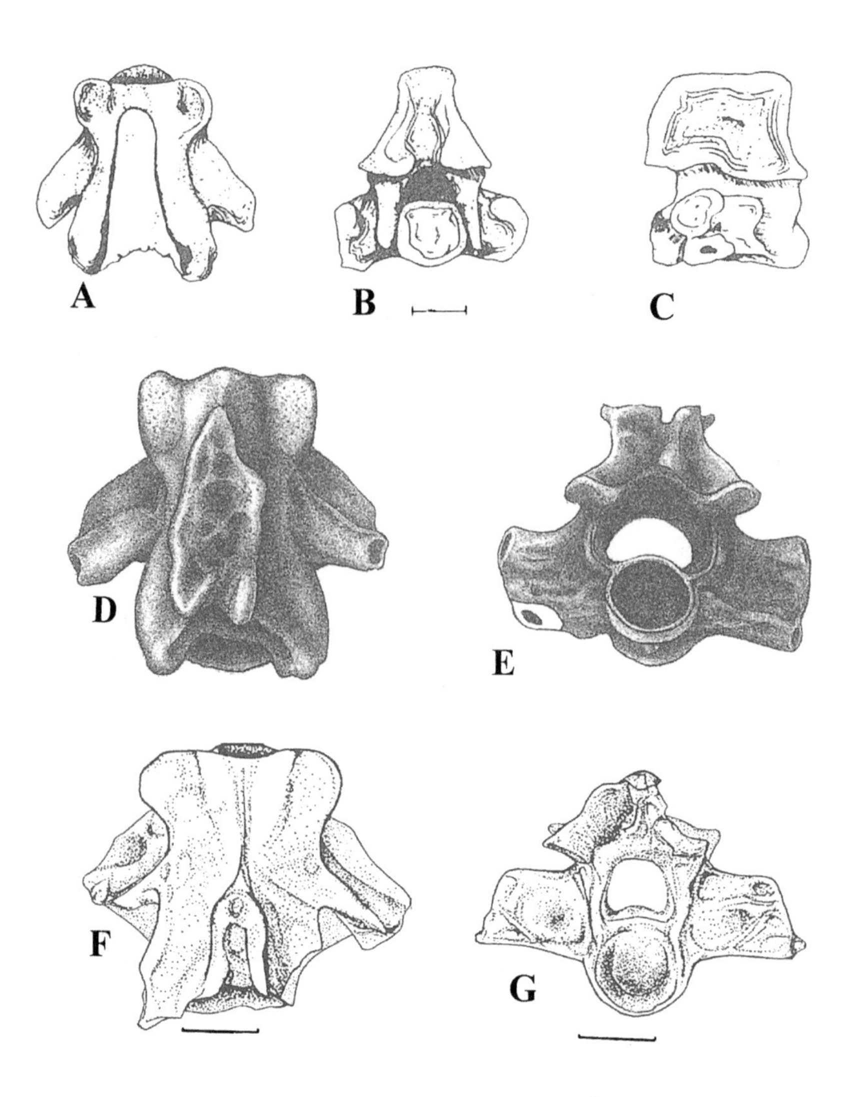

FIGURE 61. Holotype trunk vertebra of *Notophthalmus crassus from the Early Miocene of Montana in (A) dorsal, (B) posterior, and (C) lateral views. Holotype trunk vertebra of *Notophthalmus robustus from the Early Miocene of Florida in (D) dorsal and (E) posterior views. Holotype trunk vertebra of *Notophthalmus slaughteri from the Middle Miocene of Texas in (F) dorsal and (G) posterior views. Scale bars = 1 mm. The top scale bar applies to (A)–(C). No scale was provided for (D) or (E).

wide relative to its length; this feature is believed to be artifactual, because of the erosion of the condyle.

General Comments. Tihen (1974, p. 215) stated, "Zygapophyseal proportions of the new species are comparable to those of *N. robustus, N. slaughteri* [fossil forms] and *N. m. meridionalis* [a living form]; it resembles the last form also in the shape of the cotyle. But it differs markedly from all 3 of those forms in the great height of the neural spine, being fully comparable to the average Floridian *N. viridescens* in this respect." Tihen (1974) further commented that there were no clear indications of the closest relatives of *Notophthalmus crassus.* He was not able to compare *N. crassus* with any of the Old World salamandrids with similar vertebral structure. It is to be hoped that additional fossil material will shed more light on this interesting species.

*NOTOPHTHALMUS ROBUSTUS ESTES, 1963

FIGS. 61D, 61E

Holotype. A trunk vertebra (MCZ 3384, Museum of Comparative Zoology, Harvard University).

Type Locality. Thomas Farm, Gilchrist County, Florida.

Horizon. Early Miocene (Hemingfordian NALMA).

Other Material. Topotypic material consists of the following vertebrae: trunk vertebra (MCZ 3383); a first vertebra (MCZ 3385); nine other unnumbered specimens; two trunk vertebrae (UF 6503, Florida Museum of Natural History); a trunk vertebra and two unnumbered vertebrae (MCZ 3597); and one unnumbered vertebra (FGS, Florida Geological Survey).

Diagnosis. The diagnosis is taken directly from Estes (1981, p. 83): "A *Notophthalmus* differing from the Recent species in having more robust vertebrae, in which the neural spines are relatively low, and the rib-bearers relatively short and stubby."

Description. The following description is slightly modified from Estes (1981). The vertebrae are robust, and the centrum length does not exceed 2 mm. The vertebrae are opisthocoelous, with the small, projecting condyle set apart from the body of the centrum by a prominent constriction. Short and stubby rib-bearers are connected throughout their length by a sheet of bone. The ventral rib-bearers are connected anteriorly to the centrum by well-developed or weak ventral lamina, depending on the position in the vertebral column. The zygapophyses are relatively small and placed near the centrum. The neural arch is heavy and fairly low. This arch is capped by a flat-topped, sculptured, and pitted deposit of dermal bone. This pitted structure extends anteroposteriorly about two-thirds of the total length of the vertebra. The neural arch and the dermal encrusted area are deeply notched posteriorly. The cervical vertebra (atlas) has a well-developed interglenoid tubercle that projects anteriorly between small oval cotyles. The neural arch is anteriorly expanded over the neural canal.

General Comments. Estes (1981) pointed out that *Notophthalmus robustus* is a primitive species of the genus in that it has more robust proportions and a low neural spine that differs more from Holocene species than those species differ from each other.

*NOTOPHTHALMUS SLAUGHTERI HOLMAN, 1966A

FIGS. 61F, 61G

Holotype. A trunk vertebra (SMPSMU 61870, Shuler Museum of Paleontology, Southern Methodist University).

Type Locality. Trinity River Local Fauna, San Jacinto County, Texas.

Horizon. Middle Miocene (early Barstovian NALMA).

Other Material. No other material has become available.

Emended Diagnosis. A Middle Miocene (early Barstovian NALMA) species of *Notophthalmus* that resembles the Early Miocene (Hemingfordian NALMA) species *Notophthalmus crassus* Tihen (1974) and *Notophthalmus robustus* (Estes, 1963) more than modern species but differs from both species in being more lightly built. It further differs from N.

robustus in having a compressed rather than depressed outline of the neural canal in posterior view, an ovoid rather than a round cotyle, and round rather than strongly oval prezygapophyseal articular facets. It differs from *N. crassus* in having a much lower neural spine and less vaulted neural arch; it differs from *N. robustus* in having a more vaulted neural arch.

Description of the Holotype. The holotype is fairly lightly built, as in modern species of *Notophthalmus.* This vertebra is relatively complete but has part of its right postzygapophysis missing. It also has the anterior part of its neural spine and some of the dermal cap of the top of the neural spine missing. In dorsal view, the prezygapophyseal articular facets are rounded and end at about the level of the anterior end of the broken neural spine. The rib-bearers are reflected posteriorly, but not quite as much as in the holotype of *Notophthalmus robustus.* A bony web extends posteriorly from the ventral rib-bearer. The posterolateral outline of the dorsal edge of the neural arch appears to be somewhat sinuate, but it is possible that this is the result of breakage and erosion. The outline of the neural spine in dorsal view is rounded anteriorly, and it flares slightly laterally posteriorly.

In posterior view, the neural arch in *Notophthalmus slaughteri* is much less vaulted than in *Notophthalmus crassus* but more vaulted than in *Notophthalmus robustus.* The dorsal roof of the neural arch is moderately thick and slightly convex. It is more convex in the holotype of *N. crassus.* The postzygapophyseal articular facets are moderately tilted upward. The neural canal is loaf-of-bread shaped, rather than kidney-bean shaped as in the holotype of *N. robustus.* The shape of the cotyle is ovoid and slightly compressed. The upper right rib-bearer extends laterally at about a right angle to the axis of the vertebra in this view. It is less robust than in *N. crassus* and *N. robustus.* This rib-bearer is quite cone shaped as it extends laterally where it is broken off. A significant web of bone extends ventrally from the broken portion of the lower rib-bearer.

General Comments. Estes (1981) assigned *Notophthalmus slaughteri* to *Notophthalmus robustus,* although Tihen (1974) had recognized both taxa in his discussion of the relationships of *Notophthalmus crassus.* Holman (1996b) accepted the Estes (1981) change, but after a review of fossil salamanders for this book, Holman has resurrected this species from synonymy with *N. robustus.* The temporal and geographic range of the three extinct species of *Notophthalmus* is now as follows: *N. robustus* is known only from the Early Miocene (Hemingfordian NALMA) of the Thomas Farm Local Fauna of Florida. This deposit represents a time span of about 19–18 Ma BP (Woodburne, 1987). *Notophthalmus crassus* is known only from the Early Miocene (Hemingfordian NALMA) of the Flint Hill locality equivalent in the Batesland Formation of South Dakota. This deposit also represents a time span of about 19–18 Ma BP. *Notophthalmus slaughteri* is known only from the Middle Miocene (early Barstovian NALMA) of the Trinity River Local Fauna of Texas. This deposit represents a time span of about 15.2–14.8 Ma BP. Oddly, no other fossils of *Notophthalmus* are known until the Pleistocene.

It appears that the two 19–18 million year old species of *Notophthalmus* were more robust, primitive species of the genus and that the at least

3 million years younger *Notophthalmus slaughteri* had developed a more lightly built vertebral form, approaching somewhat the condition of the modern species.

Notophthalmus viridescens (Rafinesque, 1820)
Eastern Newt

Fossil Localities. **Pleistocene (Irvingtonian NALMA), Kansan glacial stage:** Cumberland Cave, Allegany County, Maryland—Holman (1977b, 1995a) (previously listed as *Notophthalmus* cf. *Notophthalmus viridescens* by Holman). Hamilton Cave, Pendleton County, West Virginia—Holman and Grady (1989), Holman (1995a). **Pleistocene (Rancholabrean NALMA):** Clark's Cave, Bath County, Virginia—Fay (1988), Holman (1995a). Frankstown Cave, Blair County, Pennsylvania—Fay (1988), Holman (1995a). Kingston Saltpeter Cave, Bartow County, Georgia—Fay (1988), Holman (1995a). Natural Chimneys site, Augusta County, Virginia—Fay (1988), Holman (1995a). New Paris 4 site, Bedford County, Pennsylvania—Fay (1988), Holman (1995a). New Trout Cave, Pendleton County, West Virginia—Holman and Grady (1987), Holman (1995a).

Eastern Newts have a roughened skin and are not as slimy as most other salamanders. They are mainly aquatic forms, but they have an eft stage that sometimes spends considerable time on land. Eastern Newts are relatively small salamanders, averaging about 60–120 mm long. The newt (aquatic) stage prefers clean, calm water and occurs in a variety of such habitats. It may occur in water 30–40 m deep, and in the northern part of North America it is active under the ice all winter long. This species has a wide range in eastern North America, occurring from the Canadian Maritime provinces through peninsular Florida, thence west to eastern Texas and north to Minnesota and southwestern Ontario.

Identification of Fossils. If you find a Pleistocene (Rancholabrean NALMA) salamander fossil vertebra with a large, conspicuous, roughened dermal cap on top of its neural spine—north of Florida and southeastern Alabama (see Conant and Collins, 1998, maps, pp. 443, 444)—you can be fairly certain that it is a vertebra of *Notophthalmus viridescens.*

Notophthalmus viridescens (Rafinesque, 1820) or *Notophthalmus perstriatus* (Bishop, 1941)
Eastern Newt or Striped Newt

Fossil Locality. **Pleistocene (Rancholabrean NALMA):** Arredondo site, Alachua County, Florida—Holman (1962, 1995a).

This vertebra was found in an area where both species of the above newts occur at present (see Conant and Collins, 1998, maps, pp. 443, 444). I have not been able to distinguish these two species of the Eastern Newt group from each other on the basis of vertebral characters. At one time these two newts were recognized as subspecies of *Notophthalmus viridescens* (see Schmidt, 1953).

Notophthalmus sp. indet.

Fossil Localities. **Pleistocene (Irvingtonian NALMA):** Fyllan Cave, Travis County, Texas—Holman and Winkler (1987), Holman (1995a).

Haile XVIA site, Alachua County, Florida—Meylan (1984), Holman (1995a).

The authors in these papers were not able to carry the identification of fossils of *Notophthalmus* any farther than the generic level, although in each case the bones represented the Eastern Newt complex of species.

Genus *Taricha* Gray, 1850

Pacific Newts

Genotype. Taricha torosa (Rathke, 1833).

Taricha is most closely related to the other North American newt genus *Notophthalmus* than to any other salamandrid. In the previous section on *Notophthalmus* it was pointed out how *Taricha* was less derived than *Notophthalmus*. The Pacific Newts occur from southern Alaska southward to San Diego, California.

Identification of Fossils. Taricha lacks the extensive pitted dermal cap on the top of the vertebrae that occurs in *Notophthalmus* and has a lower neural spine than that genus.

Estes (1981) listed shared derived character states of *Taricha*, which are slightly modified as follows. The premaxillae are fused, and the nasals are separated. With regard to the hyobranchium, the first basibranchial and the second ceratobranchial are cartilaginous and mineralized, and the second basibranchial is absent; and the anterior radii are absent, and the interradial cartilage is reduced. The rib processes do not protrude from the body wall. The neural spines of the vertebrae are relatively high, but little or no dermal cap structure is present. The cotyles of the vertebrae are horizontally oval. Caudosacral ribs are absent.

**Taricha lindoei* Naylor, 1979

Fig. 62 (top, partial skeleton)

Holotype. Impressions of a partial skeleton (UALVP 13870, University of Alberta Laboratory for Vertebrate Palaeontology).

Type Locality. John Day Formation, Wheeler County, Oregon.

Horizon. Late Oligocene (Arikareean NALMA).

Other Material. No other material is known.

Diagnosis. The diagnosis is directly from Estes (1981, p. 85): "Differs from *Taricha oligocenica* in having a narrow scapular blade and lacking dermal bone on the neural spine; differs from Holocene *Taricha* in having longer tubercular processes on the ribs."

Description. As with many frog and salamander fossils that are impressions of partial skeletons, it is difficult to identify certain parts, and once these parts are identified, they are often difficult to describe. The description here is modified from Estes (1981). The vertebrae have long, thin neural spines that lack a cap of dermal bone. The atlas has a robust, sloping neural spine. The skull is somewhat long and narrow and has an identifiable frontosquamosal arch. The dermal sculpture of the skull is weakly developed. The scapular region of the scapulocoracoid is broad. The tubercular processes of the ribs are elongate.

Comments. Compared with *Taricha oligocenica* (Van Frank, 1955) (to follow later) of the same horizon (Late Oligocene: Arikareean

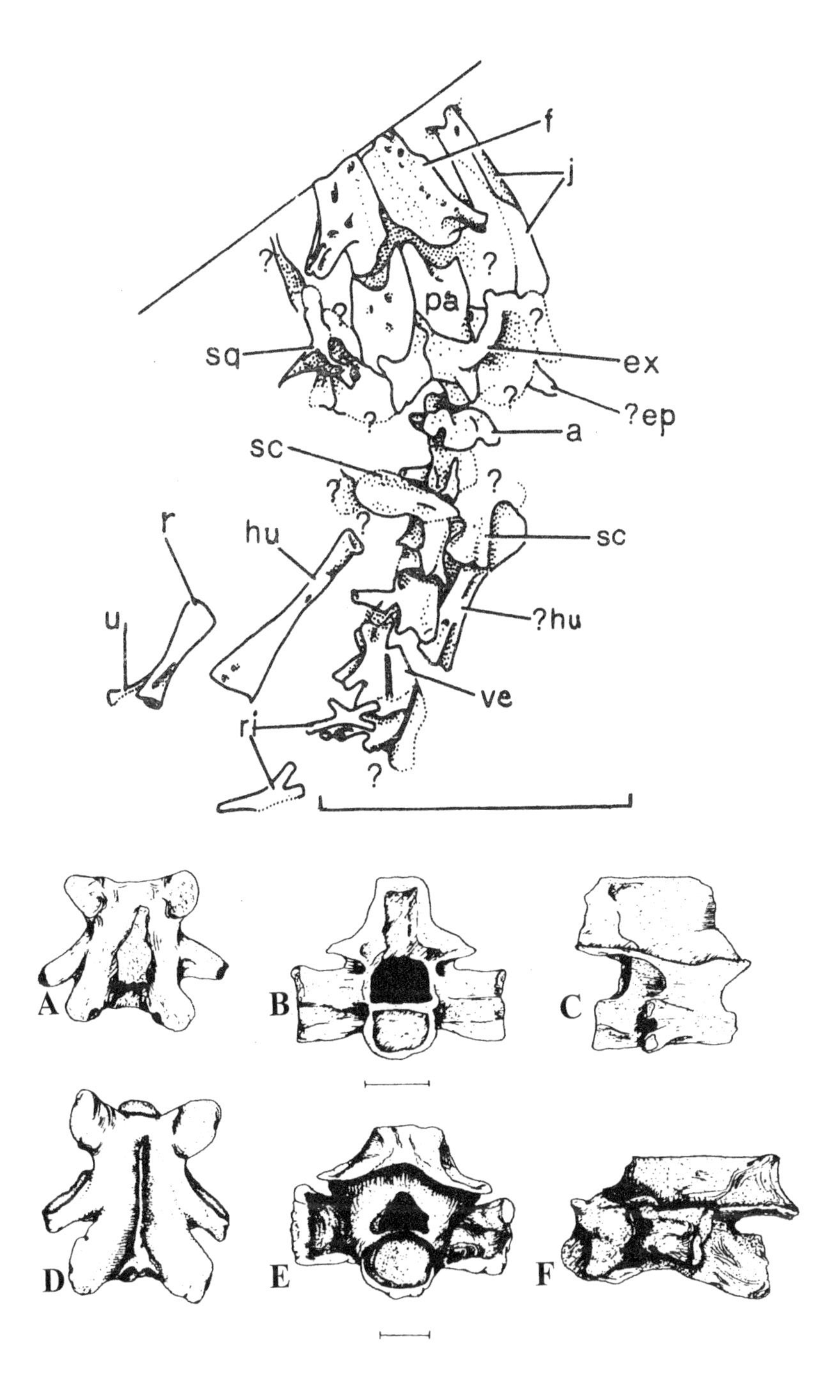

FIGURE 62. Top figure: Dorsal view of partial skeleton of the holotype of **Taricha lindoei* from the Late Oligocene of Oregon. Holotype of **Taricha miocenica* from the Early Miocene of South Dakota in (A) dorsal, (B) posterior, and (C) lateral views. Modern *Taricha granulosa* in (D) dorsal, (E) posterior, and (F) lateral views. Abbreviations: a, atlas; ep, epibranchial; ex, exoccipital; f, frontal; hu, humerus; j, jugal; pa, parietal; r, radius; ri, rib; sc, scapula; sq, squamosal; u, ulna; ve, vertebra. Scale bar = 10 mm in the top figure. Other scale bars = 1 mm.

NALMA), *Taricha lindoei* has a broader scapular region, but it also lacks the dermal cap on the neural spine that *T. oligocenica* and the modern species of the genus have. The specimen of *T. lindoei*, however, is only about half as large as the type specimen of *T. oligocenica*. Naylor (1979) did not think *T. lindoei* was a larval or juvenile specimen but instead thought it might be a young individual because of its narrow skull and the relatively anterior position of the quadrate condyles. There is still a chance, however, that the absence of dermal bone atop the neural spines is an ontogenetic feature, rather than a true similarity to modern species of *Taricha*.

**Taricha miocenica* Tihen, 1974

Figs. 62A–62C

Holotype. An essentially intact vertebra (MPUM 1479, Museum of Paleontology, University of Montana).

Type Locality. Cabbage Patch Beds, Granite County, Montana.

Horizon. Late Oligocene to Early Miocene (Arikareean NALMA).

Other Material. Referred topotypic material consists of a fragment of a right dentary containing approximately the posterior one-half of the tooth row, with spaces for about 24 teeth (MPUM 1918); right and left scapulocoracoids (MPUM 1928, 1929, respectively); a left femur (MPUM 1946); and several vertebrae (MPUM 1961, 1969–1984, 1986–2001, 2018).

Diagnosis. The diagnosis is directly from Tihen (1974, p. 216): "A *Taricha* of the subgenus *Palaeotaricha* [this subgenus not recognized by Estes, 1981] differing from *T.* (*P.*) *oligocenica* in its somewhat smaller size and in the less extensive expansion and capping of the neural spine. The expanded dorsal surface of the spine is V-shaped, as contrasted with an approximate rectangle in *T.* (*P.*) *oligocenica*, and the tops of successive neural spines would not be in contact with each other anteriorly and posteriorly in intact specimens."

Description of the Holotype. The description of the holotype is modified from Tihen (1974). The measurements of the holotype to the nearest 0.05 mm are as follows: length of centrum, 2.45 mm; width of centrum at cotyle, 1.05 mm; total width including transverse processes, 3.30 mm; total height, 2.55 mm; zygapophyseal length, 2.90 mm; anterior zygapophyseal width, 2.20 mm; posterior zygapophyseal width, 2.40 mm; height of neural spine (from level of lateral borders of the posterior zygapophyses), 0.95 mm.

The anterior end of the condyle is somewhat flattened, probably as a result of erosion. Moreover, a small portion of one lateral border of the condyle is broken away; otherwise, the specimen is complete. Except for the difference in the neural spine and the greater apparent width of the centrum, all the vertebral characters and proportions fully agree with those of modern *Taricha*. The cotyle of the type vertebra is distinctly oval. The top of the neural spine is distinctly expanded, with the lateral borders of the expansion diverging fairly sharply to a point slightly anterior to the posterior end of the spine, from which they slightly converge. The surface is pitted in the manner of *Notophthalmus*.

Description of the Referred Material. The description of the referred vertebrae is also modified from Tihen (1974). In all, 34 other topotypic vertebrae are present. Eleven of these were not preserved well enough to measure. The size and proportional measurements of the holotype are very close to the average for the 25 vertebrae left. The shape and degree of the expansion of the top of the neural spine vary somewhat. This expansion is usually somewhat less than in the holotype, which has its lateral borders diverging less sharply and continuing to diverge to the posterior edge of the spine. In general, the neural spine appears to be more comparable in height and in form and extent of the cap to that of non-Floridian *Notophthalmus viridescens* than to that of any other taxon Tihen (1974) had seen. Tihen also stated that the scapulocoracoids, femur, and dentary fragment were all fully comparable to corresponding parts in modern *Taricha* and showed no noteworthy characters.

General Comments. The size of the specimens of *Taricha miocenica* indicates that they represent individuals about 50 mm in SVL. The range of size and the average size are similar to those of immediately pre- and post-metamorphic individuals of living *Taricha*. Tihen (1974) believed that they represented aquatic adults for the most part; thus, *T. miocenica* is smaller than any other known member of the genus.

**Taricha oligocenica* (Van Frank, 1955)
(#*Palaeotaricha oligocenica* Van Frank, 1955)

Holotype. A nearly complete skeleton (UOMNH F5405, University of Oregon Museum of Natural History).

Type Locality. Willamette Formation, Lane County, Oregon.

Horizon. Late Oligocene (Arikareean NALMA).

Other Material. No other material is known.

Diagnosis. The diagnosis is directly from Estes (1981, p. 85): "Differs from *Taricha miocenica* in larger size and in having greater expansion and capping of neural spine with dermal bone."

Description. The description is modified from Estes (1981). The premaxillae are fused. The frontosquamosal arches are complete. The paroccipital processes are extruded. The vertebrae have laterally expanded, broad, flat, rectangular dermal caps that are sculptured and lie against each other, both anteriorly and posteriorly. The neural spines of the vertebrae are high. The ribs have prominent uncinate processes.

Comments. Originally, Van Frank (1955) proposed that *Taricha oligocenica* be placed in a new genus—namely, *Palaeotaricha*—on the basis of several differences in the modern genus. Later, after additional preparation of the specimen was made, Tihen (1974) was able to show that the species differs from modern species of the genus only by having a much more extensive dermal cap on the vertebrae.

Taricha sp. indet.

Fossil Locality. **Late Miocene (late Hemphillian NALMA):** Buchanan Tunnel locality, Tuolumne County, California—Peabody (1959).

General Comments. This identification was confidently made on the basis of trackways by Peabody (1959), who made detailed comparisons with modern salamander trackways.

Comments on Fossil Salamandrids in North America. The three named fossil *Taricha*—two from the Late Oligocene of Oregon and one from the Late Oligocene of Montana—are remarkably similar to one another and to modern *Taricha* in almost every detail. The same situation obtains with the three named fossil *Notophthalmus*, two of which are known from the Early Miocene of Florida and South Dakota, respectively, and the third of which is known from the Middle Miocene of Texas.

The most obvious difference in the osteology of these two is that by modern times all species of *Taricha* (Figs. 62D–62F) had lost the prominent dermal cap on top of the vertebra, whereas *Notophthalmus* has retained this cap throughout its history. However, the dermal cap character itself is somewhat of an interpretive "red herring," as it can separate from North American fossil salamandrid vertebrae during the fossilization process. This separation is even more likely in fossil neotenic individuals!

Family #Batrachosauroididae Auffenberg, 1958

Batrachosauroidids

Several strange groups of fossil salamanders cannot be confidently placed within an evolutionary framework that includes modern salamander families. In North America, two of these odd families are the Batrachosauroididae and the Scapherpetontidae.

The most definable group of genera within the Batrachosauroididae has the following characteristics. The atlas (cervical vertebra) lacks an interglenoid tubercle (odontoid) and has a pair of deeply concave anterior condyles. The vertebrae may be amphicoelous or opisthocoelous and have slender, posteriorly directed neural spines. The skulls are of the larval type, and when well preserved, have prominent, ossified ceratohyal and ceratobranchial elements of the hyobranchium. It appears that all of them lack spinal foramina on both the trunk and the caudal vertebrae.

Genus #*Batrachosauroides* Taylor and Hesse, 1943

Genotype. Batrachosauroides dissimulans Taylor and Hesse, 1943.

Diagnosis. These animals were probably somewhat elongated, and the limbs may have been reduced. The vertebrae are massive and lack basapophyses. The skull is neotenic, but the maxilla is present. Both the prefrontal and the frontal are invested with dermal sculpture. The premaxillae appear to be fused. The occipital condyles are stalked. Pterygoid teeth are absent. All the members of the Batrachosauroididae are extinct. They occurred from the Cretaceous to the Late Miocene in North America and in the mid-Cretaceous of central Asia. An unnamed vertebra of this family is known from the Late Cretaceous of France.

Genus #*Batrachosauroides dissimulans* Taylor and Hesse, 1943

Fig. 63

Holotype. A skull (see Fig. 63) that is now lost (MAMCT 2344, Museum, Texas A&M University, College Station).

Type Locality. Fleming Formation, San Jacinto County, Texas.

Horizon. Middle Miocene (early Barstovian NALMA).

Other Material. Other material of this species from the Early Miocene (Hemingfordian NALMA) includes vertebrae (MAMCT 40067-38, 40067-100) from the lower Oakville Formation, Hidalgo Bluff, Texas; and an atlas (UF 7802, Florida Museum of Natural History) from the Thomas Farm, Gilchrist County, Florida.

Material from the Middle Miocene (early Barstovian NALMA) includes a vertebra (UF 2013) from the Fleming Formation, San Jacinto County, Texas; vertebrae (MAMCT 31057-57, 31057-94, 31057-120) from

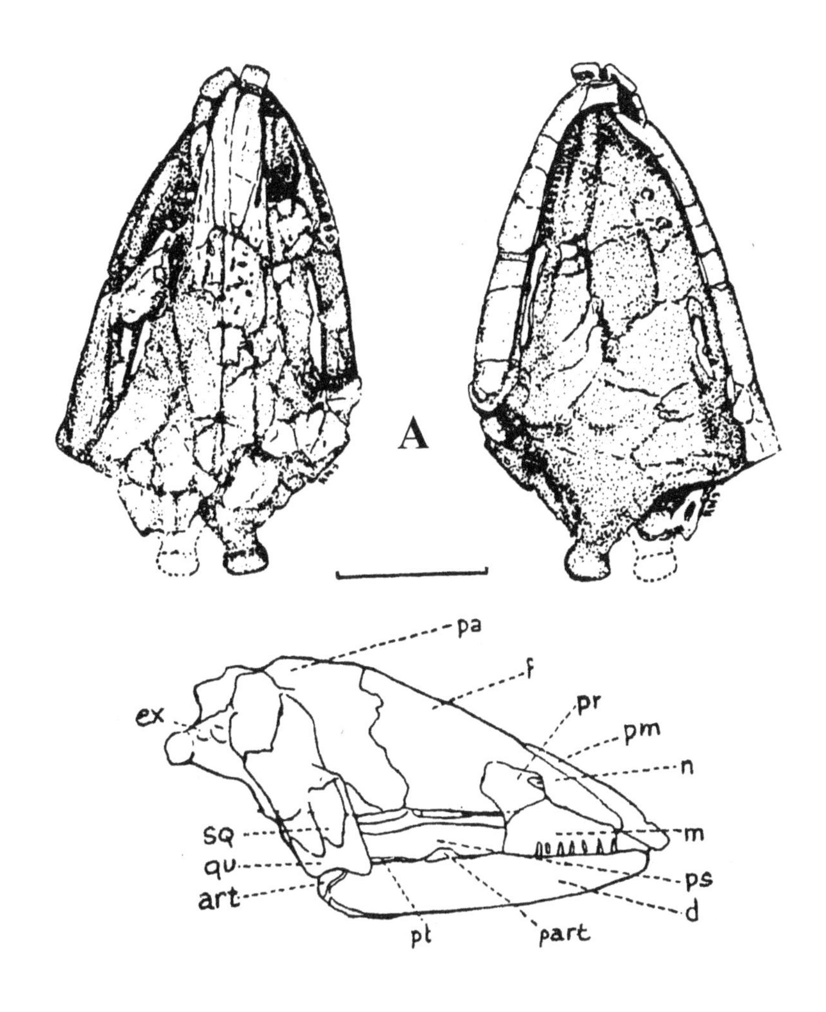

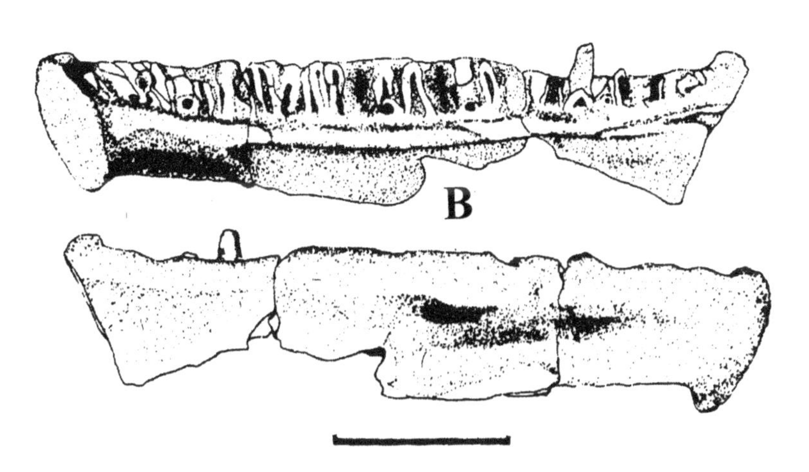

FIGURE 63. (A) Holotype skull of #*Batrachosauroides dissimulans* from the Miocene of Texas in (left) dorsal, (right) ventral, and (bottom) lateral views. After Taylor and Hesse (1943). (B) Right dentary of *B. dissimulans* from the Miocene of Texas (upper) in lingual and (lower) labial views. Abbreviations: art, articular; d, dentary; ex, exoccipital; f, frontal; m, maxilla; n, nasal; pa, parietal; part, prearticular; pm, premaxilla; pr, prefrontal; ps, parasphenoid; pt, pterygoid; qu, quadrate; sq, squamosal. Scale bar = 10 mm in group (A) and applies to all figures. Scale bar = 5 mm in group (B) and applies to both figures.

the Fleming Formation, Polk County, Texas; a premaxilla, a maxilla, skull fragments, and vertebrae (USNM 18241, United States National Museum) from the Oakville Formation, San Jacinto County, Texas; a dentary (UT 31057-11a, University of Texas), from the Fleming Formation, Polk County, Texas; a vertebra (SMPSMU 63672-13, Shuler Museum of Paleontology, Southern Methodist University) from the Moscow site, Polk County, Texas; a complete vertebra (UF 111741, Florida Museum of Natural History), exceptionally well preserved (Bryant, 1991, figs. 5a, 5b, p. 476), from the upper Torreya Formation, Willacoochee Creek Fauna, Gadsden County, Florida.

Diagnosis. This diagnosis is quoted directly from Estes (1981, p. 30): "A species of *Batrachosauroides* differing from *B. gotoi* in having more robust centra and with rib-bearers more closely appressed."

Description of the Holotype. This description is modified from Estes (1981). The holotype, now lost, provides most of the information about the genus *Batrachosauroides*. The premaxilla was probably paired. An estimated nine teeth were present on this bone in life. The surface of the premaxillary spine is smooth except for a very few pits and grooves that extend posteriorly about one-third the length of the skull. The surface of the maxilla is mainly smooth except for a little pitting. Parts of 18–19 pleurodont (growing from the inside part of the jaw) teeth are visible. The prefrontal is well developed and roughly sculptured. This bone lies between the maxilla and the frontal. It is thin, posteriorly thickened, and bent downward, and it has an extensive dorsal process. The nasals apparently make contact with the prefrontals. The paired frontals extend anteriorly about as far as the anterior ends of the maxillae and terminate in sharp points that do not separate the nasals from the premaxillae. The frontals extend back to the level of the posterior end of the lower jaw. The dorsal surfaces of the anterior and medial portions of the frontals are deeply sculptured as far back as the distinct temporal ridge, beyond which they descend into the temporal fossa.

The parietals join at the midline and form a very narrow, low crest. The exoccipitals have prominent, stalked condyles that are ovate and have their posterior articular surfaces somewhat flattened. The quadrate is in an oblique position on the skull. This bone is anteroventrally oriented and is thickened, especially in its ventral portion. In lateral view, the quadrate is mainly hidden by the squamosal. The parasphenoid is extensive posteriorly and apparently separates the vomers anteriorly. The pterygoids are wide and extend forward from the quadrate, contacting the parasphenoid for part of its length. The anterior end of the pterygoids contacts the vomers. At least 25 non-pedicellate teeth are present on the dentary. The posterior teeth are somewhat circular or somewhat oval in cross section. The anterior teeth are much larger and are oblong or triangular in cross section.

The outer face of the dentary is traversed by a broad, fairly deep longitudinal channel. This channel is narrower and deeper anteriorly and becomes shallower and finally disappears toward its posterior portion. A foramen can be seen below about the level of the 11th dentary tooth. Little or no pitting occurs on the bone. In lateral view, the suture between the dentary and the angular may be seen only in the extreme posterior part of the jaw. From the lower part of the jaw the angular can be traced nearly its entire length until it ends as a splint-like point. The angular is concave posteriorly, with a deep groove between the angular and dentary present on the upper surface of the jaw.

Turning to the vertebral column, we find that the atlas is a robust, well-ossified bone that has two deeply concave, suboval cotyles, which are separated by a small, ridged interglenoid tubercle (odontoid). This tubercle is weakly produced. The atlas is short and bears several foramina on its concave ventral surface. The neural arch of the atlas is short and robust, and the posterior zygapophyses are well developed. The single posterior cotyle of the atlas is deeply concave.

The trunk vertebrae are large and opisthocoelous. The centra are longer than high, and the cotyle is rounded. The notochordal canal is well defined in the center of the cotyle. The condyle is well developed and has a deep, rounded depression in its center. The centrum has a ridge-like median keel ventrally that is nearly straight in lateral view. On either side of the median keel, a single subcentral foramen may be found. The neural arch is massive and is broken in most known specimens. Its width at the narrowest part of the zygapophyseal ridges is slightly greater than the width of the centrum. Spinal nerve foramina are not present. The neural canal is slightly flattened in anterior view and has a well-developed median epipophyseal ridge present on its floor. The articulating surfaces of the prezygapophyses are oval and longer than wide and directed slightly laterally in dorsal view. The zygapophyseal ridges are rounded and not well developed. Transverse processes are well developed, having two closely appressed rib-bearers that are connected by a vertical septum. The rib-bearers' articular surfaces extend almost to the posterior edge of the centrum. In the more anterior vertebrae the centra are shorter in proportion to their width, their transverse processes are directed more dorsally, and their hypapophyseal keels are more ridgelike and more highly developed. The first few vertebrae that follow the atlas and second vertebra differ from the other anterior vertebrae in having thickened and flattened keels, thicker neural spines, and larger neural canals.

General Comments. The skull of *Batrachosauroides dissimulans* is generally of the larval type, and the presence of large maxillary bones indicates that it had about the same degree of neoteny that *Amphiuma* has today. The presence of stalked occipital condyles and a solid skull in this taxon indicates it was probably an active, voracious predator, as in *Amphiuma*. All the known material of *B. dissimulans* comes from low-energy aquatic environments in Miocene deposits in Florida and Texas, indicating that this taxon must have lived in situations similar to the ones *Amphiuma* and *Siren* live in today.

#*BATRACHOSAUROIDES GOTOI* ESTES, 1969B

FIG. 64

Holotype. A relatively complete posterior trunk vertebra (PU 18012, Princeton University).

Type Locality. White Butte site, Golden Valley Formation, Stark County, North Dakota.

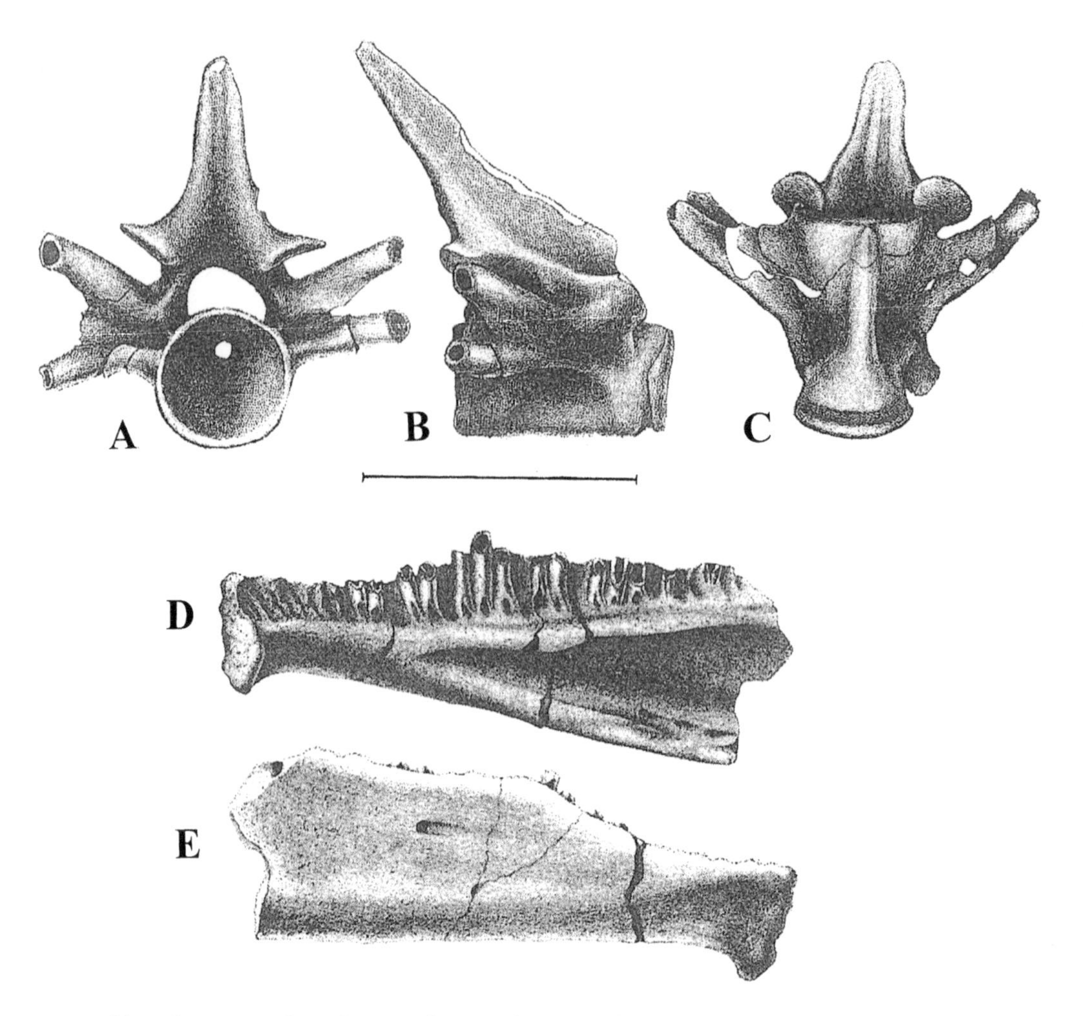

FIGURE 64. Holotype vertebra of #*Batrachosauroides gotoi* of the Early Eocene of North Dakota in (A) posterior, (B) lateral, and (C) ventral views. Right dentary of *B. gotoi* from the Early Eocene of South Dakota in (D) lingual and (E) labial views. Scale bar = 10 mm and applies to all figures.

Horizon. Early Eocene (Wasatchian NALMA)

Other Material. Referred topotypic material includes an almost complete (probably posterior) trunk vertebra (PU 17275). Other referred topotypic specimens are fragmentary and include unnumbered fragments of four trunk vertebrae, one of which is probably from the anterior end of the vertebral column; and right dentaries broken posteriorly (PU 17400a, 17400b).

Diagnosis. This diagnosis is directly from Estes (1981, p. 32): "A *Batrachosauroides* similar in size to the Miocene *B. dissimulans* but differing in having rib-bearers separated rather than closely appressed and in having a slightly less robust build, especially of the neural arch."

Description. The description is modified from Estes (1969b, 1981). The dentary symphysis is strong, ventroposteriorly expanded, and deep posteriorly. The Meckelian groove is deep and reaches anteriorly to the level of the 10th dentary tooth. A very shallow dental sulcus is present. The outer surface of the dentary is smooth and has a single foramen and a deep anteroposterior depression present on this surface. The teeth number at least 32 and are apparently non-pedicellate.

Turning to the vertebrae, we find that the range in length of the centra of the measurable vertebrae is 6.0–11.9 mm. The vertebrae are robust and opisthocoelous and have a prominent notochordal pit that is continuous with the chordal foramen in the two measurable specimens and is closed in the other vertebrae. A distinct subcentral keel occurs on the type specimen (PU 18012). On the unnumbered vertebrae the keel is straight in lateral view. On PU 17275 the keel is broadly flattened, and this situation also obtains in the unnumbered vertebrae. The subcentral foramina are tiny and paired, except in PU 17275.

The rib-bearers are subcylindrical and are widely separated distally. They are webbed with bone nearly to their tips. Rib articulation facets are well developed on the rib-bearers, and their tips project slightly beyond the posterior border of the centrum. The neural canal is oval and slightly compressed. The neural arch is flattened dorsally on either side of a prominent neural spine. The neural spine has a strong anterior keel that begins near the anterior border of the neural arch. The neural spine itself is long and robust and terminates in finished bone. Ventrally, the neural spine is ridged and has strong intervertebral muscle scars. The postzygapophyses are large and tilted strongly upward. The zygapophyseal ridges are distinct, low, and rounded. On a single rib fragment, the ventral head as well as part of the shaft is preserved. A low ridge is present on the anterior face of the shaft, and the head is subcylindrical, with a circular, hollow articular facet.

General Comments. Estes (1969b, 1981) postulated that *Batrachosauroides gotoi* was probably a large and strong-limbed species without any special aquatic adaptations. He suggested that it was more primitive than *Batrachosauroides dissimulans* (Estes, 1988). He believed that the Miocene *B. dissimulans*, with its narrow, *Amphiuma*-like skull and more closely appressed rib-bearers, may have been more aquatically adapted. A questionable "?*Batrachosauroides*" vertebra (Carnegie Museum of Natural History) from the Oligocene of Wyoming assigned to this genus by Setoguchi (1978) is very small and amphicoelous and should not be assigned to this genus.

Batrachosauroides sp. indet.

Albright (1994) assigned a posterior trunk vertebra from the Toledo Bend Local Fauna (Fleming Formation, Early Miocene, Texas) to *Batrachosauroides* and stated that it was very similar to *Batrachosauroides dissimulans*.

Genus #*Opisthotriton* Auffenberg, 1961

Genotype. Opisthotriton kayi Auffenberg, 1961.

#*Opisthotriton gidleyi* Sullivan, 1991

Fig. 65

Holotype. An incomplete skull (AMNH 2667, American Museum of Natural History).

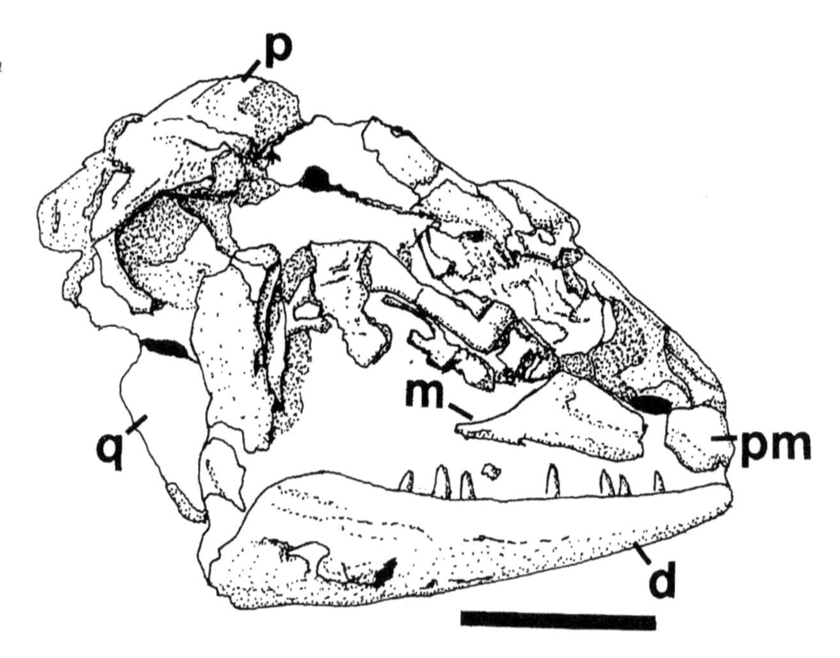

FIGURE 65. Holotype skull of #*Opisthotriton gidleyi* from the Paleocene of Wyoming in lateral view. Abbreviations: d, dentary; m, maxilla; p, parietal; pm, premaxilla; q, quadrate. Scale bar = 5 mm.

Type Locality. Gidley Quarry, Lebo Formation (= Fort Union No. 2 of previous workers), Sweet Grass County, Wyoming.

Horizon. "Late early" (= middle) Paleocene after Sullivan (1991).

Diagnosis. The diagnosis is directly from Sullivan (1991). "A batrachosauroidid that differs from *O. kayi* in having a vertical quadrate and teeth extending posteriorly on the dentary."

Description of the Holotype. This description is modified from Sullivan (1991). The holotype partial skull is only partially distorted. It is smaller than those specimens previously referred to *Opisthotriton kayi*, measuring about 17 mm in length from below the occipital region to the symphyseal end of the right dentary. Its height is about 12 mm as measured from the posterior ventral surface of the mandible to the skull roof. The skull width is estimated at 7 mm on the basis of the width of the otoccipital.

Only the right dentary is preserved, but it is nearly complete, with perhaps a few teeth missing. This dentary forms the entire mandible, with the exception of a small postdentary component that includes the articular, angular, and prearticular. In posterior view these elements strongly articulate with the quadrate. Four complete dentary teeth are preserved, and fragments of 4 others and 13–15 spaces for additional teeth indicate a total tooth count of 19 or 20 teeth.

Turning to the other bones of the skull, we find that the premaxillae each bear four teeth. The ascending processes of the premaxillae are nearly complete; they are flattened and extend upward to the parietal. The left nasal lies ventral to the posterior ascending process of the maxillary border; only part of the right nasal is visible. A bone in the general vicinity of a left prefrontal is partially exposed posterior to the left nasal and was tentatively interpreted as that element. The paired frontals form the dorsomedial part of the skull, but details of their structure cannot be observed. The parietals show a moderately developed sagittal crest. Paired temporal–occipital crests extend obliquely from the parietals and extend forward to articulate medially with the frontals. The otoccipital has a prominent right occipital condyle, but the left condyle is missing.

The squamosal forms a prominent bar with an expanded proximal surface that articulates with both the parietal and the otoccipital. The left squamosal is complete and is situated lateral to the left quadrate. The right squamosal is incomplete and lies lateral to the right quadrate. Both quadrates are oriented perpendicularly. They are robust and lie medially to the squamosals. The paired vomers are about 4 mm in length and articulate with each another anteriorly. A tooth-bearing ridge parallels the tooth row of the premaxillae and maxillae. A vomerine tooth count cannot be made. There is no evidence of a vomerine articulation with the parasphenoid medially as in *Batrachosauroides*. The pterygoid has a broad shelf that is strongly arched and is reflected posteriorly.

Both ceratohyals are present. The right one is robust and complete, but the left ceratohyal is incomplete posteriorly, and its anterior end is obscured by matrix. Anteriorly, the right ceratohyal is cone shaped in lateral view; when viewed posteriorly it may be seen to bear an ascending process that is tapered and has a prominent median flange.

General Comments. Sullivan (1991) did not state whether the dentary teeth in *Opisthotriton gidleyi* were pedicellate or non-pedicellate, although I assume the latter. James B. Gardner is currently studying the validity of this species.

#*Opisthotriton kayi*, Auffenberg, 1961

Fig. 66

Holotype. A middle trunk vertebra (CM 6468, Carnegie Museum of Natural History).

Type Locality. Lance Formation, Niobrara County, Wyoming.

Horizon. Late Cretaceous.

Other Material. There is very abundant material of *Opisthotriton kayi* in the CM, UCMP, MCZ, and PU collections. Late Cretaceous localities are in the Lance Formation of Niobrara County, Wyoming; the Hell Creek Formation of McCone County, Montana; the Laramie Formation of Weld County, Colorado; and the Frenchman Formation of Saskatchewan, Canada (see, for instance, Estes, 1964; Estes et al. 1969; Carpenter, 1979).

The main Paleocene localities are the Early Paleocene Tullock Formation, McCone County, Montana (Van Valen and Sloan, 1965); the "late early" Paleocene Lebo Formation of Sweet Grass County, Montana (Sullivan, 1991); the Middle Paleocene Tongue River Formation, Carter County, Montana (Estes, 1976); and the Late Paleocene Fort Union Formation, Park County, Wyoming (Estes, 1975), the latter locality yielding the only articulated material that has been described, including a skull, vertebral column, and parts of the hyobranchial skeleton.

Emended Diagnosis. This diagnosis should also serve as a diagnosis for the genus itself. Vertebrae opisthocoelous, with a prominent uncalcified depression in the center of the condyle. (*Opisthotriton gidleyi*, unlike

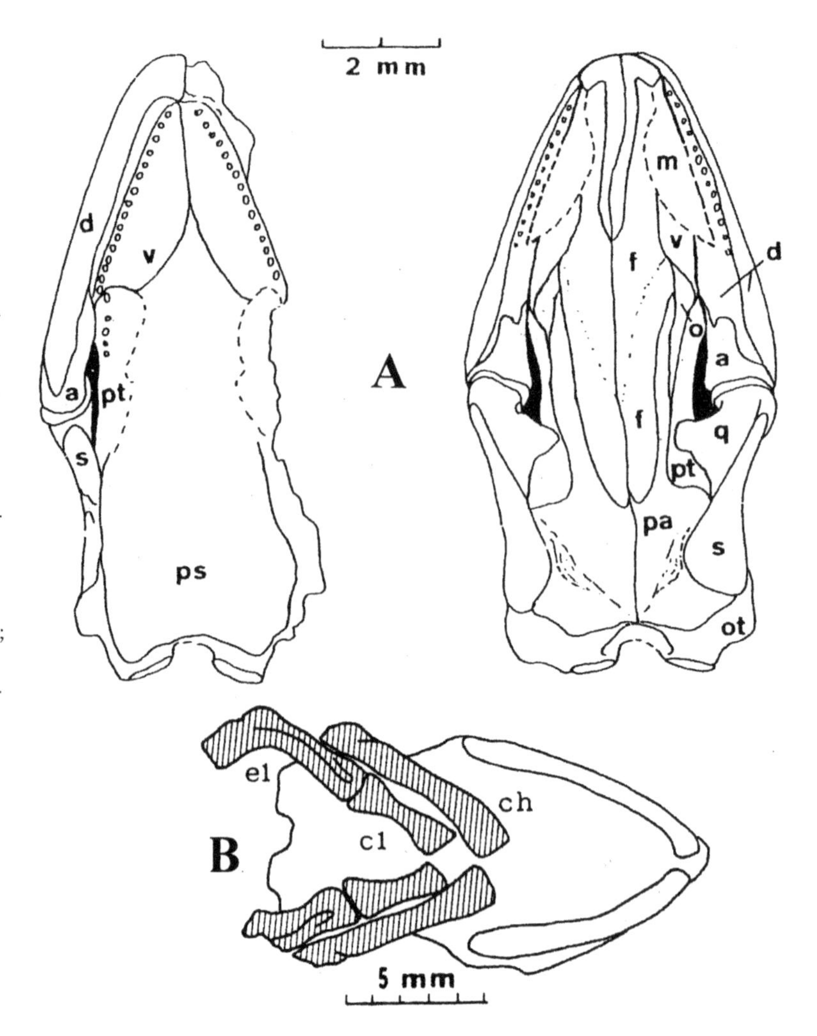

FIGURE 66. Outline drawings of skull and hyobranchial skeleton of #*Opisthotriton kayi* from the Late Cretaceous of Wyoming. (A) Skull in ventral (left) and dorsal (right) views. (B) Hyobranchial skeleton as a shaded area relative to the skull outline. Abbreviations for (A): a, articular; d, dentary; f, frontal; m, maxilla; o, orbitosphenoid; ot, otoccipital; pa, parietal; ps, parasphenoid; pt, palatopterygoid; q, quadrate; s, squamosal; v, vomer. Abbreviations for (B): ch, ceratohyal; cl, first ceratobranchial; el, first epibranchial. Scale bar in (A) applies to both views.

Opisthotriton kayi, has a vertically directed quadrate and teeth that do not extend to the posterior end of the dentary.) Differs from *Prodesmodon copei* in having generally larger vertebrae, with basapophyses extending forward past the midpoint of the centrum, and presence of uncalcified depression in condyle. Differs from *Palaeoproteus*, an Old World genus, in opisthocoely. Differs from *Batrachosauroides* in having basapophyses, generally smaller and less massive vertebrae, more flattened neural arch and neural spine, and more closely appressed rib-bearers.

Description. This description is modified from Estes (1981). The skull is flattened and tapers anteriorly. It is of the generally neotenic type. In dorsal view, the premaxillae are paired, and their elongated dorsal processes lie in grooves on the frontal. The frontals themselves are flattened and expanded posteriorly; then they narrow and reach the level of the optic area as a midline wedge. Anterolateral projections of the frontal occur at the level of the posterior tips of the premaxillae. These projections are capped with a transverse ridge. The orbitosphenoids are notched for the passage of the olfactory tracts, the bones themselves extending laterally beyond the frontals and terminating abruptly at about the midpoint of the skull length. Most of the occipital area of the skull is covered by the parietals. These bones extend far forward, covering the orbitosphenoids laterally. Posteriorly, the parietals enclose the tips of the frontals on the midline.

The otoccipital has prominent condyles, and the jugular foramina are large. Small facets for the interglenoid tubercle of the atlas are present on the midline. Anteriorly, the otoccipital encloses the posterior border of the fenestra ovalis. The prootic is a separate ossification capped by the squamosal, which is poorly preserved and hardly visible. The fenestra ovalis is small.

In ventral view, the parasphenoid is broad and flat. It reaches posteriorly almost to the occipital condyles and extends anteriorly to the vomers and palatopterygoids. The palatopterygoid is toothed anteriorly and meets the vomer. The vomer is strongly curved, with a prominent tooth-bearing process, occurring laterally, that has about 20 teeth. Flat palatal processes of the vomer meet along their entire length along the midline. Anterodorsally, a prominent, flat process to the premaxilla and a lateral expansion to the maxilla are visible. The maxillae are flat, and their ascending processes are extensive and toothed throughout their length. The number of these teeth ranges from 20 to 24.

The mandible is composed only of a dentary and a fused "postdentary" bony element. The dentary itself has about 25 teeth. The dentary symphysis is expanded, with a muscular crest present anteroventrally. The posterior end of the mandible is expanded, with a prominent "coronoid" process present. This process extends posteriorly almost to the end of the condyle.

Pedicellate or "subpedicellate" teeth with a zone of weakness are present on the premaxillae maxillae, vomers, and palatopterygoids. The teeth are long and conical and slightly recurved. This is an important difference when this genus is compared with *Batrachosauroides*, which has non-pedicellate teeth.

In the hyobranchial skeleton, the ceratohyal is a robust tapering element that is rounded anteriorly and strongly curved posteriorly. The ceratohyal has an angular process. The posterior end of the element is flattened. The ceratobranchial is short and robust and is expanded at each end. The epibranchial is strongly curved, and ventrally it has a deep groove for the branchial artery.

Turning to the vertebral column and ribs, we find that the atlas has prominent double cotyles and a small, prominent, knobbed intercotylar process (odontoid). All the other vertebrae have a ring of calcified material on the anterior cotyle. A prominent notochordal pit is present in the center of the condyle. The vertebrae have prominent neural arches that are dorsally flattened and have prominent neural spines. The rib-bearers are well developed and are closely appressed and two-headed. Well-developed horizontal laminae (alar processes) of the ventral rib-bearers are present. Hollow posterior cotyles are present on all the vertebrae. Strong ventroposterior basapophyses are present and extend beyond the midpoint of the centra. Very well developed but thin ventral hypapophyseal keels are present. The ribs are small and two-headed, with the heads close together. These ribs occur only on the anterior trunk vertebrae. The humerus is very reduced. A trochanter is connected to the head of the humerus, and both the head and the distal end are unossified.

General Comments on Opisthotriton. Opisthotriton is a fairly well known fossil salamander, because articulated and disarticulated material has been found from both Cretaceous and Paleocene rocks. Obviously, it survived the great extinction event at the end of the Cretaceous that wiped out the last of the dinosaurs. But *Opishthotriton* itself did not extend beyond the Paleocene, as far as I am aware.

Opisthotriton is a larval form that is specialized in having reduced limbs and an elongate vertebral column composed of at least 40 presacral vertebrae. But modern *Amphiuma* and *Siren* (forms that are thought to live in habitats similar to those of *Opisthotriton*) have about 55 and 45 presacrals, respectively (Estes, 1981). So perhaps *Opisthotriton* was outcompeted or was more prone to predators than amphiumids and sirenids, which were longer and probably squirmier.

Genus #*Parrisia* Denton and O'Neill, 1998

Genotype. Parrisia neocesariensis Denton and O'Neill, 1998.

Diagnosis. The diagnosis is the same as for the genotype and only known species, *Parrisia neocesariensis.*

#*Parrisia neocesariensis* Denton and O'Neill, 1998

Fig. 67A

Holotype. A complete atlas (NJSM 16609, New Jersey State Museum).

Type Locality. Ellisdale site, Monmouth County, New Jersey.

Horizon. Late Cretaceous (Campanian).

Other Material. Topotypic material consists of dentaries (NJSM 14691, 15280, 15651, 15828, 15875, 16752); atlantes (NJSM 14690B, 16613); second cervical vertebra (NJSM 14207); anterior thoracic vertebra (NJSM 16756); anterior caudal vertebra (NJSM 15042A); caudal vertebrae (NJSM 15042a, 16555); three posterior caudal vertebrae (NJSM 15042).

Diagnosis. The diagnosis is directly from Denton and O'Neill (1998, p. 485): "A paedomorphic [neotenic] salamander, with functional anatomy reflecting an aquatic habitus, which is referred to the family Batrachosauroididae on the basis of the following characters: atlas with prominent, deeply concave anterior cotyles; lack of a well developed atlantal intercotylar process (odontoid); a ring or dome of calcified cartilage on the opisthocoelous vertebral condyles, which bear a persistent notochordal pit.

"Distinguishable from all other members of the family by the following characters: posterior cotyle of the centrum displaced ventrally relative to the anterior condyle/cotyle when viewed from the lateral aspect; atlantal cotyles surmounted by dorsomedial spinous protuberances; neural spine tips finished in cartilage; caudal vertebrae with paired inter-vertebral nerve openings; shallow dentary with a smooth lateral surface and an enlarged medial projection of the ventral edge; and teeth mesiodistally

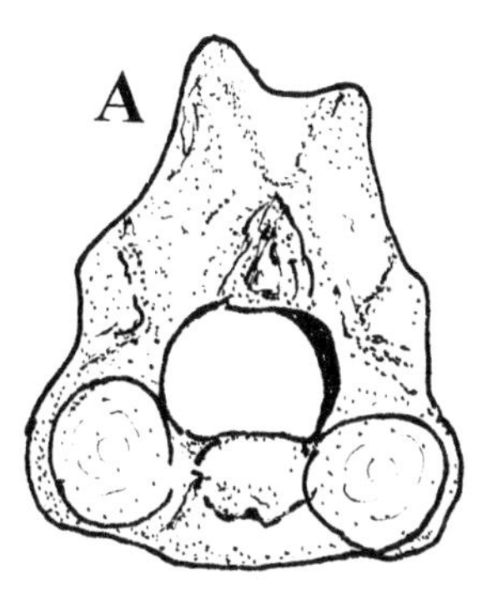

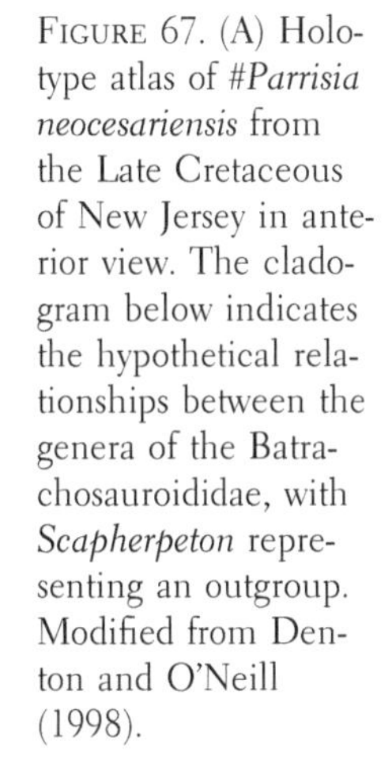

FIGURE 67. (A) Holotype atlas of #*Parrisia neocesariensis* from the Late Cretaceous of New Jersey in anterior view. The cladogram below indicates the hypothetical relationships between the genera of the Batrachosauroididae, with *Scapherpeton* representing an outgroup. Modified from Denton and O'Neill (1998).

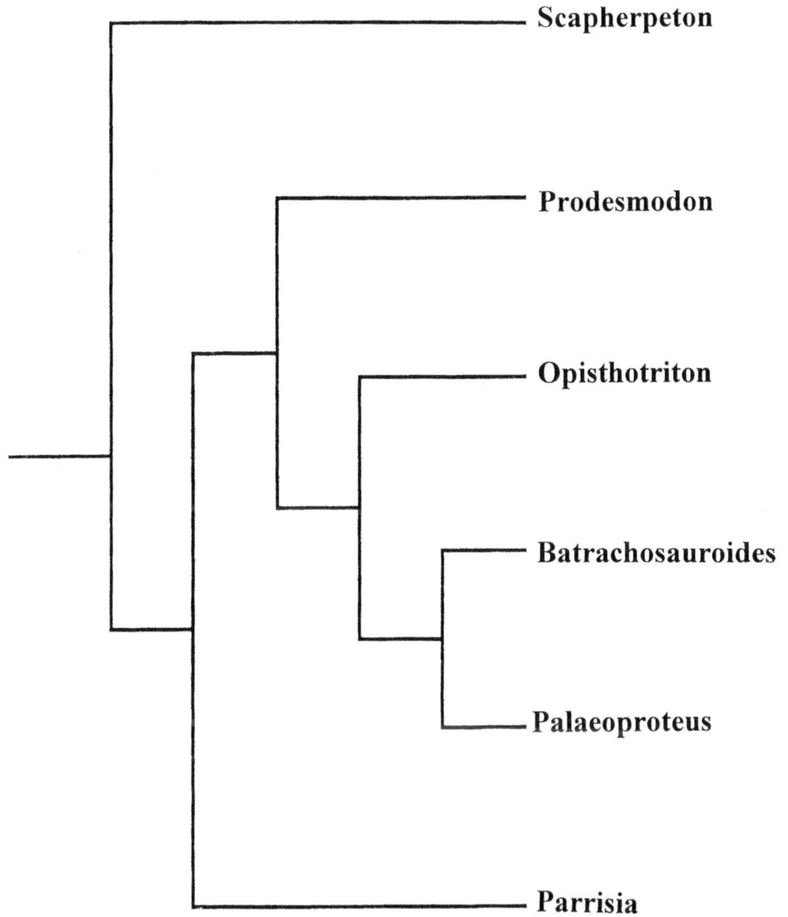

compressed. Differs from all other members of the family except *Prodesmodon* in having convex, fully ossified condyles on the cervical and anterior thoracic vertebrae. Differs from all other members of the family except *Paleoproteus* [Old World genus] in having a deeply forked atlantal neural spine."

Description. This description is modified from Denton and O'Neill

(1998). All the atlantes from the type locality are identifiable as batrachosauroidids on the basis of having large anterior cotyles and poorly developed odontoid processes. In all the topotypic specimens, the odontoid protrudes slightly as a small ridge that is located at one-third of the height of the cotylar surfaces on either side and is similar to *Opisthotriton* in this character. The atlantes, however, differ from those of *Opisthotriton* in that in all three growth stages the neural canal is widely open and is not laterally compressed, although this is a variable character in the latter.

In atlas NJSM 16613 the anterior cotylar surfaces are ovoid (compressed rather than depressed), whereas these surfaces in *Opisthotriton* are circular in outline. On the other hand, there is considerable variation between individual atlantes in *Parrisia,* the cotyles becoming more circular and deeply cupped in the growth series. Denton and O'Neill (1998) interpreted the circular cotyles as a derived condition, with compressed ovoid cotyles as the primitive condition, as present in *Scapherpeton, Habrosaurus, Prosiren,* most non-batrachosauroidid salamanders, and some individuals of *Opisthotriton. Parrisia* lacks the blade-like central ridge of the neural spine that occurs in *Opisthotriton;* rather, it has a broad, flat protuberance that begins as a paired process lying dorsal to the neural canal and broadens posteriorly to cover the entire surface of the forked neural spine, apparently a true derived character in *Parrisia.* In lateral view, some of the most important differences between the atlas of *Parrisia* and that of *Opisthotriton* can be observed. In the holotype atlas (NJSM 16609) the anterior cotyles are mounted above by paired spinous protuberances, which form the medial edge of the neural canal. These structures are not present in *Opisthotriton.* In *Parrisia,* the neural arch extends posteriorly at about a 45-degree angle to the plain of the cotyles. This leaves an opening on the dorsal side of the neural canal that extends one-third the length of the centrum. But the anterior edge of the neural arch of *Opisthotriton* extends dorsally in the same plane as the anterior edge of the cotyles. In *Opisthotriton,* the neural spine makes a sharp 90-degree angle at its dorsal edge and extends posteriorly and parallel with the axial plane of the centrum. The entire neural spine and neural arch in *Parrisia* are nearly twice the length of the centrum and extend posteriorly to the edge of the posterior cotyle. This is not true in *Opisthotriton,* where these structures end just above the cotylar rim.

The posterior cotyle of NJSM 16609 is ventrally directed such that it is at an angle greater than 20 degrees off the axial plane of the centrum. The only other batrachosauroidid that has this character is the poorly known *Peratosauroides problematica* (Naylor, 1981b) of the Late Miocene of California. Nevertheless, *Peratosauroides* has a robust neural crest, unlike that in *Parrisia.* In *Opisthotriton* the posterior cotyle is on the same plane as the anterior cotyles. Also, the articular surfaces of the postzygapophyses are parallel with the axial plane of the centrum. But in *Opisthotriton* these same processes extend posteriorly at a 45-degree angle. This results in the zygapophyses of the atlas being more prominent in *Parrisia* than in *Opisthotriton.*

The second cervical, anterior trunk, posterior trunk, anterior caudal, and posterior caudal vertebrae were also compared with *Opisthotriton* by Denton and O'Neill (1998).

Of several dentaries that are probably referable to *Parrisia*, a partial left dentary (NJSM 14691) is best preserved. This dentary is broken at the anterior and posterior ends, but it suggests that the intact bone was not as deep as is usual in batrachosauroidids. The mandibular symphysis is missing, but the referred dentaries show that this structure was not as enlarged as in *Batrachosauroides* and is actually more similar to that in *Opisthotriton*. NJSM 14691 is 6 mm long and has three complete teeth. The Meckelian groove is widely open, starting anteriorly at the level of the seventh tooth. The subdental ridge is poorly developed and appears as a thin shelf beneath the teeth. In most other members of the family this structure tends to be deep and well developed. The lower halves of nine teeth are said to be present, although this number cannot be seen on fig. 6b in Denton and O'Neill (1998), where NJSM 14691 is depicted. The authors stated that all nine teeth were broken at an equivalent point and that this suggests that an eroded zone in the three complete teeth represents the limit of the pedicel. The tooth bases that are intact all show enlarged resorption pits. The teeth are rather widely spaced, and in this respect they are similar to those of both *Opisthotriton* and *Batrachosauroides*. All the *Parrisia* teeth show striking anteroposterior compression. This is unlike the situation in *Opisthotriton* and *Batrachosauroides*, whose teeth are cone shaped throughout their length. The crowns of the intact teeth of *Parrisia* are gently recurved. In general, the teeth of *Parrisia* resemble those of *Prodesmodon*.

The lateral wall of the dentary is deflected medially, where it forms a large ventromedial shelf that represents the floor of the Meckelian groove. A similar shelf is seen only in the Old World genus *Palaeoproteus* among the Batrachosauroididae, but it is also seen in *Necturus* and *Prodesmodon*. In lateral view, there is no evidence of mental foramina as in *Opisthotriton* and *Prodesmodon*. In external view, the dentary is concave and has a moderately convex area that starts just below the ninth tooth and extends posteriorly.

General Comments. In their discussion section, the authors (Denton and O'Neill, 1998) concluded that *Parrisia* is most similar to *Prodesmodon* of the Cretaceous of North America and *Opisthotriton kayi* of the Cretaceous and Paleocene of North America. They also presented a cladogram (modified in Fig. 67 of this volume) that depicts their proposed phylogeny for some genera of Batrachosauroididae, with *Scapherpeton* as an outgroup. Denton and O'Neill suggested that the ecological role of *Parrisia* may have been the equivalent of that of extant genera of elongated, neotenic salamanders, such as *Amphiuma*, *Siren*, and *Pseudobranchus*, that are still found in coastal environments like those where *Parrisia* formerly lived.

For the general readership, cladograms (trees) are based on characters that are interpreted by their authors as derived ones ("a character or character state not present in ancestral stock," Lincoln et al., 1982, p. 66). Such cladograms are especially tenuous in vertebrate paleontology, as (among other problems) (1) numbers of individual fossils are usually so few that individual variation in chosen characters cannot be determined; (2) fossil taxa are normally identified and described on the basis of a few fragmentary pieces (such as is the case with most of the salamander taxa in this book), with the result that there are few characters to evaluate; and (3) only a limited number of series of modern salamander skeletons are available for comparative study.

GENUS #*PERATOSAUROIDES* NAYLOR, 1981B

Genotype. Peratosauroides problematica Naylor, 1981b.

Diagnosis. The diagnosis is the same as for the genotype and only known species, *Peratosauroides problematica.*

#*PERATOSAUROIDES PROBLEMATICA* NAYLOR, 1981B

FIGS. 68A, 68B

Holotype. An atlas (UCMP 75465, University of California Museum of Paleontology).

Type Locality. San Pablo Formation, Stanislaus County, California.

Horizon. Late Miocene (Hemphillian NALMA).

Other Material. A topotypic trunk vertebra is a referred specimen (UCMP 123433).

Diagnosis. The diagnosis is modified from Estes (1981) and (Naylor (1981b). An atlas of a Late Miocene batrachosauroidid salamander from

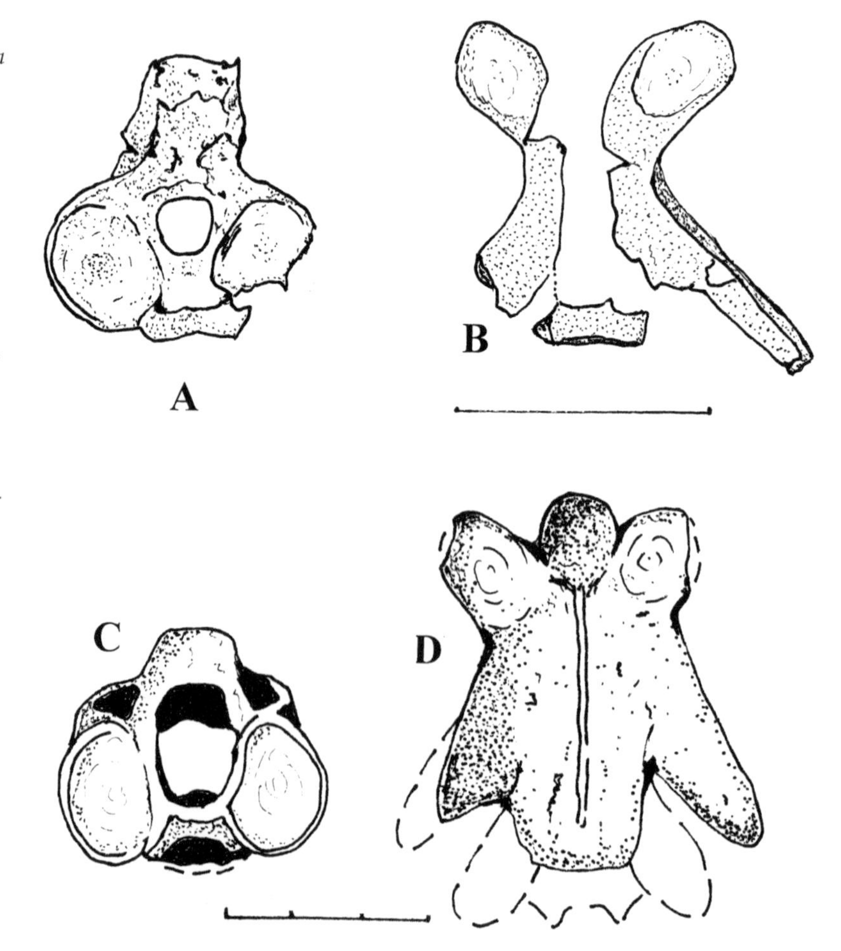

FIGURE 68. #*Peratosauroides problematica* holotype atlas (A) in anterior view and referred vertebra (B) in dorsal view; both are from the Late Miocene of California. #*Prodesmodon copei* holotype atlas (C) in anterior view and referred vertebra (D) in dorsal view; both are from the Late Cretaceous of Wyoming. Scale bars = 3 mm and apply to both figures in (A), (B) and (C), (D).

California. The greatest width of the atlas in anterior view is about 12.5 mm. It differs from other batrachosauroidids in having the posterior cotyle of the atlas below the level of the anterior cotyles; a strong ventrolateral ridge on the centrum; a produced but non-articular interglenoid tubercle on the atlas; and a well-developed neural spine that is anteroposteriorly expanded.

Description. The holotype atlas is robust in form and bears large anterior cotyles that are oval and somewhat deflected laterally. The right cotyle is slightly broken ventrally. The neural canal is ovoid in anterior view, whereas the posterior cotyle is circular. The foramen for the spinal nerve is somewhat anteriorly placed. The trunk vertebra is not well preserved but shows a few features. The rib-bearers are elongate, anteromedially projected, and connected to each other by a bony webbing. The postzygapophyseal articular facets are large, posterolaterally directed, and ovoid.

General Comments. This monotypic genus, described on the basis of a minimal amount of vertebral material, is significant because of its late appearance in the Late Miocene of California. Estes (1981) made the following comments. The holotype atlas closely resembles that of other taxa of batrachosauroidids in having deeply concave anterior cotyles and an obsolete interglenoid tubercle. The closest resemblance is to the atlantes of *Prodesmodon* and *Opisthotriton*. The trunk vertebra not only resembles other batrachosauroidids but also shows some similarity to the modern Mudpuppy genus *Necturus*. *Peratosauroides* is significant in being the latest fossil record of the family Batrachosauroididae.

GENUS #*PRODESMODON* ESTES, 1964

Genotype. Prodesmodon copei Estes, 1964.

(*CUTTYSARKUS MCNALLYI* ESTES, 1964)

Diagnosis. The diagnosis is the same as for the genotype and only known species, *Prodesmodon copei.*

#*PRODESMODON COPEI* ESTES, 1964

FIGS. 68C, 68D

Holotype. A trunk vertebra (UCMP 55783, University of California Museum of Paleontology).

Type Locality. Lance Formation, Niobrara County, Wyoming.

Horizon. Late Cretaceous.

Other Material. Abundant material consisting of vertebrae and dentaries reside in the AMNH, UALVP, UCMP, and UM collections. Late Cretaceous referred material comes from the Mesaverde Formation of Wyoming (Campanian); the San Juan Basin, New Mexico; the Lance Formation, Wyoming. Paleocene specimens are from the Hell Creek Formation, Montana (Maastrichtian), and the Tullock Formation, Montana (Puercan NALMA).

Diagnosis. The diagnosis is taken directly from Estes (1981, p. 39): "Differs from *Opisthotriton kayi* in usually having the basapophyseal crests not extending beyond midpoint of centrum and in having completely ossified condyle; differs from *Batrachosauroides* ssp. in possession of basapophyses, complete ossification of condyle and smaller size. Differs from both of the above and *Palaeoproteus* in dentary structure."

Description. The description is modified from Estes (1964, 1981). The vertebrae are opisthocoelous and have the condyle fully formed, except occasionally there is a small, medial notochordal pit present. A medial subcentral keel is present. It is usually low and concave, but it may be deep and straight edged. The narrow anterior end of the keel begins just posterior to the condyle and extends posteriorly, meeting the edge of the cotyle between the prominent posteriorly keeled basapophyses. The basapophyses themselves originate posterior to the midpoint of the centrum and extend posteriorly, close to the median keel. Their tips may extend beyond the edge of the cotyle. The transverse processes of the vertebrae have broad, wing-like ventral lamina, the anterior border of which may be either straight or concave. These transverse processes point posterolaterally and narrow to a blunt tip, rather than having a rib facet. The neural arch is flattened dorsally. A slightly raised neural spine that extends the length of the neural arch may be present. The neural arch bifurcates posteriorly. The prezygapophyseal articular facets are in the form of elongate ovals.

The second cervical vertebra of *Prodesmodon* is shorter than the trunk vertebrae and lacks the expanded ventral laminae of the latter. The two-headed rib-bearers are distinct, but they are closely appressed. The ventral rib-bearer is larger and bears a rib facet; this facet is lacking in the dorsal rib-bearer. The neural arch has small but distinct processes that somewhat resemble pterypophyseal processes.

The atlas is a compact element that has two vertically oriented, concave cotyles that are separated by the neural canal. These cotyles are of an elongate ovoid shape and are very slightly concave on their inner side. The lower part of the bony base of the neural canal that lies between the cotyles bears a very small, ridgelike interglenoid tubercle. Small posterior basapophyses are present on the atlas. The neural arch has a swollen neural spine, which is somewhat loaf-of-bread shaped in anterior view. This spine separates flattened lateral areas that extend posteriorly into the forked posterior border of the neural arch. The posterior cotyle of the atlas is hemispherical, and the posterior zygapophyses are prominent.

The dentary is short and subtriangular. External foramina are lacking in the bone. Twelve to 14 non-pedicellate teeth, which are slightly recurved and pointed, occur on the dentary.

General Comments. This is another batrachosauroidid genus that is based on a minimum amount of material. On the other hand, the vertebrae are very distinctive. Actually, one complete, complex salamander vertebra that can be turned around and looked at from many angles can be much more taxonomically useful than a distorted, two-dimensional impression of a fossil caudate.

Genus #*Prosiren* Goin and Auffenberg, 1958

Genotype. Prosiren elinorae Goin and Auffenberg, 1958.

Diagnosis. The diagnosis is the same as for the genotype and only known species, *Prosiren elinorae.*

#*Prosiren elinorae* Goin and Auffenberg, 1958

Fig. 69

Holotype. A trunk vertebra (FMNH PR-391, Field Museum of Natural History).

Type Locality. Antlers Formation, Montague County, Texas.

Horizon. Early Cretaceous (Albian).

Other Material. Other material of *Prosiren elinorae* from the Antlers Formation, Montague County, Texas, includes a trunk vertebra (FMNH PR-390; an atlas (FMNH PR-801); trunk vertebrae (FMNH PR-392, PR-802–PR-804); a right premaxilla (FMNH PR-805); and right dentaries (FMNH PR-806, PR-807). Material of *P. elinorae* from the Antlers Formation, Wise County, Texas, includes an anterior trunk vertebra (SMPSMU 61042, Shuler Museum of Paleontology, Southern Methodist University); a trunk vertebra (SMPSMU 61047); and a right humerus (SMPSMU 61041). Winkler et al. (1990) identified *P. elinorae* from six Early Cretaceous (Comanchean) localities in north-central Texas.

Emended Diagnosis. Because of the uncertain placement of *Prosiren* in the classification system, I am substituting a more detailed diagnosis of *Prosiren elinorae* than usually occurs for fossil salamanders. Maxillae and dentaries with very weakly tricusped, non-pedicellate teeth; premaxillae with roughened, obsoletely pitted surface; Meckelian groove closed; mandibular symphysis interlocking; dentary teeth somewhat heterodont; atlas having well-developed interglenoid tubercle (odontoid); vertebrae

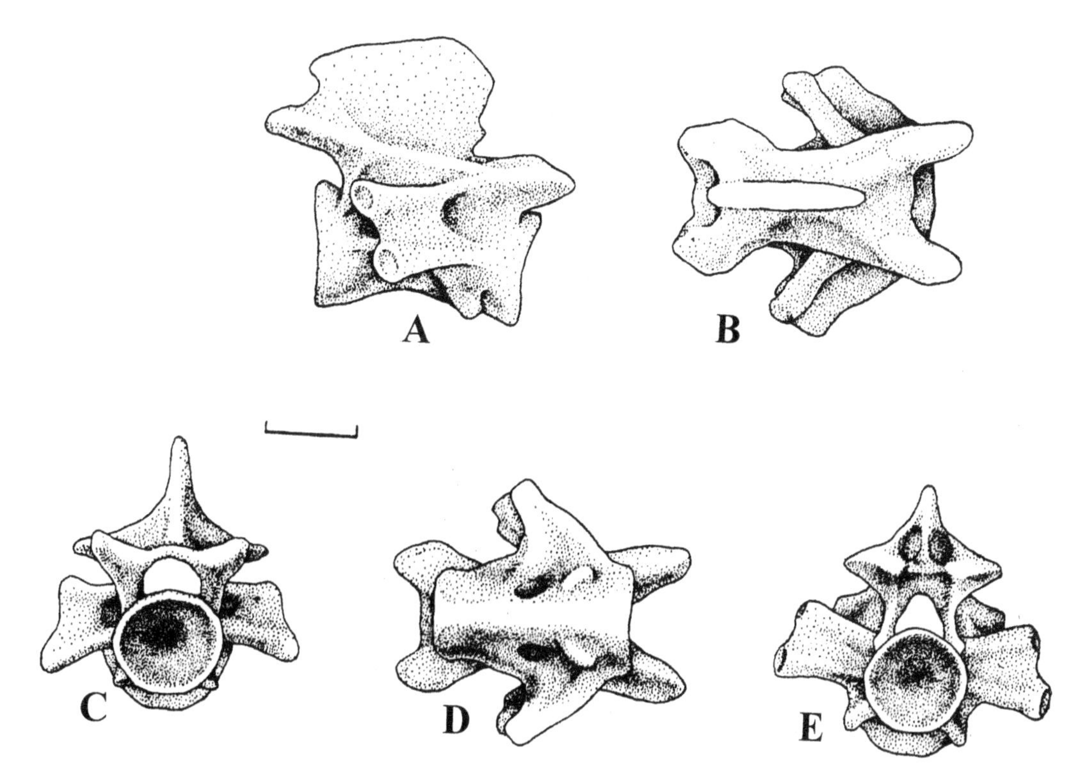

Figure 69. Trunk vertebra of #*Prosiren elinorae* from the Early Cretaceous of Texas in (A) lateral, (B) dorsal, (C) anterior, (D) ventral, and (E) posterior views. Scale bar = 1 mm and applies to all figures.

amphicoelous with robust neural arches and rather large, thin, expanded neural spines; zygapophyses long and narrow; rib-bearers bicapitate, with heads well separated and connected by bony septum; no alar processes; prominent anterior basapophyses on anterior trunk vertebrae.

General Comments. I have tentatively grouped *Prosiren* with the Batrachosauroididae, following the suggestion of Milner (2000) that this genus may be an early member of this family. Estes (1981) was aware of the ongoing study of Fox and Naylor (not published until 1982) in which dentaries of *Prosiren* were reassigned to *Albanerpeton arthridion* (a non-caudate lissamphibian). However, Estes still retained these jaws as part of the *Prosiren* hypodigm in his 1981 work. James D. Gardner is currently re-describing *Prosiren* on the basis of Estes's original material, as well as additional vertebrae.

General Comments on the Batrachosauroididae. When the late Alfred S. Romer used to deliver his many popular lectures on vertebrate paleontology, he often referred to the placoderms (an odd assortment of early fishlike vertebrates) as the "funny fishes." I have been tempted to refer to the batrachosauroidids as the "silly salamanders." Actually, one definitely has the right to pose the question, are some batrachosauroidids really salamanders?

#Family Scapherpetontidae Auffenberg and Goin, 1959

Scapherpetontids

The family Scapherpetontidae, like the Batrachosauroididae, is an enigmatic caudate group that is of uncertain position (Milner, 2000). The group is composed of three genera: *Lisserpeton*, *Piceoerpeton*, and *Scapherpeton*. All these genera occur in the Late Cretaceous and Paleocene of North America, with *Piceoerpeton* extending into the Eocene on this continent. The family may also include *Eoscapherpeton* from the Cretaceous of central Asia (see Nessov, 1988; Nessov et al., 1996). All these genera are considered to have been large, neotenic creatures that probably were at least superficially similar to cryptobranchids.

Edwards (1976) relegated this family to the Dicamptodontidae on the basis of the presence of spinal nerve foramina in the caudal vertebrae and the lack of these foramina in the precaudal series of vertebrae. Milner (2000), however, rejected this idea and tentatively retained the family. Estes (1981) pointed out that it is difficult to diagnosis the family Scapherpetontidae because many of its distinctive characters are the results of neoteny, but he did present a tentative diagnosis that I have modified as follows.

Tentative Diagnosis. Scapherpetontids are large, extinct salamanders that have vertebral centrum lengths up to 20 mm long. The neural spine is long and slender. A subcentral keel is usually present, and it is often very prominent. The atlas may or may not have an interglenoid tubercle (odontoid). The nasals appear to be present. The premaxillae have short, blunt dorsal processes. The maxilla has a reduced facial portion (pars facialis). Limbs may be normal or reduced. The vomer has an anterior row of teeth that are parallel to those of the maxilla and premaxilla, or these vomerine teeth may tend toward a transverse arrangement. The exoccipital bears a hypoglossal foramen. The angular is separate from the prearticular, as it is in the Cryptobranchidae.

#Genus *Lisserpeton*

Genotype. Lisserpeton bairdi Estes, 1965b.

Diagnosis. The diagnosis is the same as for the genotype and only known species, *Lisserpeton bairdi.*

#*Lisserpeton bairdi* Estes, 1965b

Fig. 70

Holotype. A nearly complete trunk vertebra (AMNH 8123, American Museum of Natural History).

Type Locality. Hell Creek Formation, McCone County, Montana.

Horizon. Paleocene (Maastrichtian).

Other Material. Many vertebrae, as well as dentaries and parietals, reside in the UCMP, AMNH, and PU collections. Collection localities

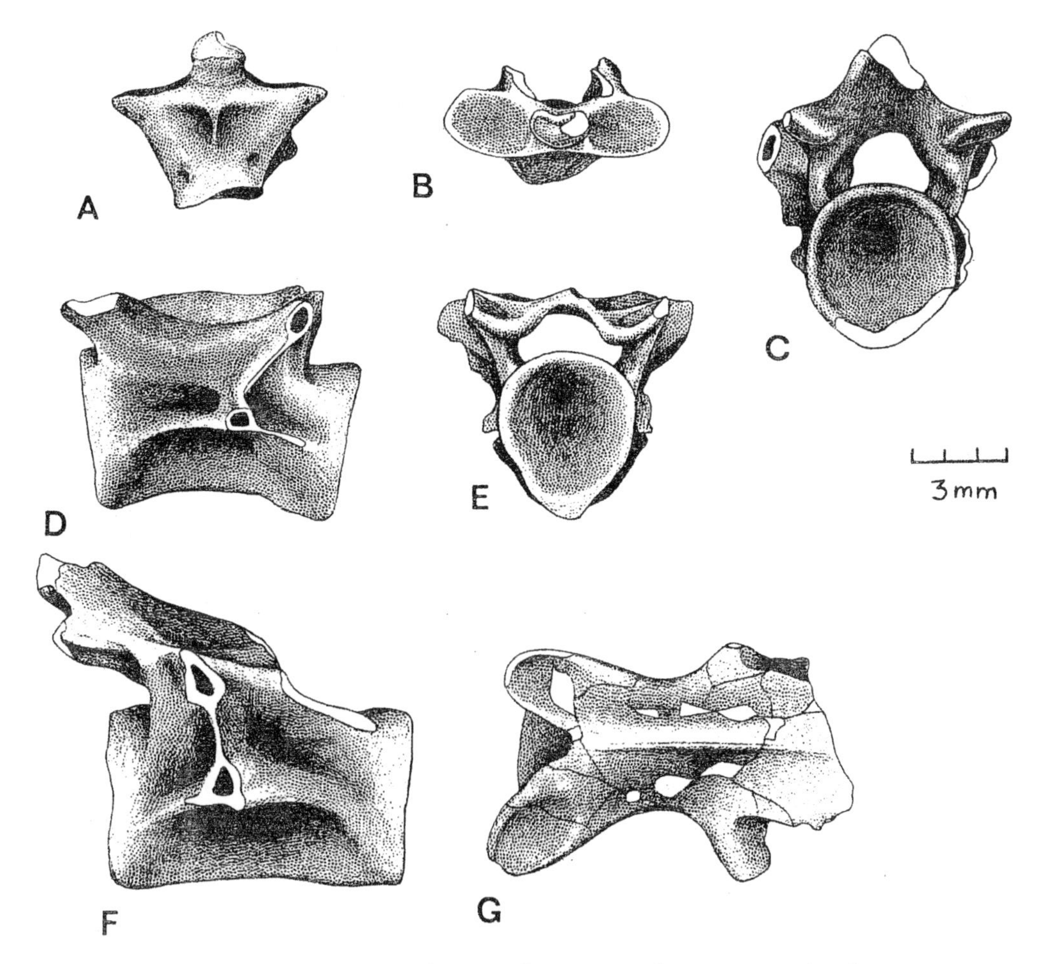

Figure 70. #*Lisserpeton bairdi* from the Paleocene of Montana. Atlas in (A) ventral and (B) anterior views. Various trunk vertebrae in (C) posterior, (D) lateral, (E) anterior, (F) lateral, and (G) dorsal views. Scale bar applies to all figures.

include the following. Late Cretaceous: Laramie Formation, Colorado; Lance Formation, Wyoming. Early Paleocene: Hell Creek Formation, Montana; Tullock Formation, Montana. Middle Paleocene: Tongue River Formation, Montana. Late Paleocene: Fort Union Formation, Wyoming. Refer to Estes (1965b, 1975, 1976), Carpenter (1979), and Krause (1980).

Diagnosis: The diagnosis is directly from Estes (1981, p. 52): "Differs from the related *Scapherpeton tectum* in lacking any cotylar ossification, cotyles relatively more rounded; deep fossae present on sides of a thin subcentral keel; dorsal rib-bearer emerging at level of dorsal surface of neural arch; presence of well-developed zygapophyseal ridge; teeth relatively fewer and larger; parietal with deep, well-defined fossa for muscle attachment. Differs from dicamptodontids in larger size, presence of a subcentral keel on vertebrae and maxilla with much reduced pars facialis."

Description. This description is modified from Estes (1965b, 1981). The vertebrae of *Lisserpeton* are deeply amphicoelous. Ossification does not occur in the cotyles, which are rounded and slightly oval. The subcentral keel is thin and has deep fossae on either side. This keel may be straight or concave in lateral view. The neural spine is prominent and is finished in cartilage. The zygapophyses are moderately large, and their medial edges occur near the midline. A strongly developed zygapophyseal ridge forms a sharp edge between the side and the top of the neural arch. The rib-bearers have two heads; they are webbed with bone and diverge from one another. The dorsal rib-bearers emerge at the level of the dorsal surface of the neural arch. The atlas is robust and has its anterior cotyles dorsoventrally compressed. A prominent interglenoid tubercle that is present on the atlas is not markedly constricted at its base. The centrum of the atlas is relatively long and has deep, paired fossae on its ventral surface.

Turning to the cranial skeleton, we find that the maxilla is long and slender, but it is also strongly built. The facial portion of the maxilla occurs well back and is very narrow. The number of maxillary teeth is about 17. The dentaries are deep bones with a prominent dorsal symphysis. The teeth on the dentary are subpedicellate. A prominent lateral groove occurs on the medial side of the dentary. The premaxillary bones are paired and have short, blunt posterior processes that diverge laterally. Strong, flattened articular surfaces for the vomer are present behind the tooth row. The parietals are paired and broadly L-shaped. A sagittal crest is present, and its overlap with the frontal is extensive. The parietal articulation with the squamosal is thickened. A sharp-edged depression for muscle attachment occurs anterolaterally on the parietal.

General Comments. Lisserpeton is closely related to *Scapherpeton*, from which it mainly differs in its more delicate skeletal construction and ossification and a few other fairly minor details. *Lisserpeton* was probably a permanently aquatic form and may have been fairly elongate, with reduced limbs. James B. Gardner is currently studying additional differences between *Lisserpeton* and *Scapherpeton*.

Genus #*Piceoerpeton* Meszoely, 1967

Genotype. Piceoerpeton willwoodense Meszoely, 1967.

Diagnosis. The diagnosis is the same as for the genotype and only known species, *Piceoerpeton willwoodense*.

#*Piceoerpeton willwoodense* Meszoely, 1967
(*Piceoerpeton willwoodensis*, 1967)
Fig. 71

Holotype. A trunk vertebra with the transverse processes and zygapophyses missing (PU 18021, Princeton University).

Type Locality. Willwood Formation, Park County, Wyoming.

Horizon. Early Eocene.

Other Material. Referred material consists of trunk vertebrae, an atlas, a skull, premaxillae, a dentary, various skull elements, and limb bones; these are collectively from the National Museum of Canada (NMC), PU, UALVP, UM, UMMPV. Collection localities include these from the Late Paleocene: Paskapoo Formation, Alberta; Ravenscrag Formation, Roche Percee, Saskatchewan; Tongue River or Sentinel Butte Formation, Montana; Tongue River Formation, Fallon County, Montana; and Polecat Bench Formation, Park County, Wyoming. Early Eocene localities include the following: Willwood Formation, Park County, Wyoming; and Eureka Sound Formation, Canadian Arctic Archipelago.

Diagnosis. This diagnosis is modified mainly from Estes (1969b,1981) and Naylor and Krause (1981). A robust, large-limbed scapherpetontid from the Late Paleocene of Saskatchewan, Montana, and Wyoming and the Early Eocene of Wyoming and Arctic Ellesmere Island. Differs from *Scapherpeton* and *Lisserpeton* in having a more robust skeleton and larger limbs, interglenoid tubercle (odontoid process) obsolete, anterior atlantal cotyles more rounded, interdigitating premaxillary and maxillary sutures, and more strongly built mandibular symphysis. Differs from *Scapherpeton* in having rounded cotyles lacking calcified tissues, dorsal rib-bearers originating from dorsolateral margin of neural arch, and palatine process of premaxillae expanded medially.

Description. This description is modified from Estes (1969b, 1981) and Naylor and Krause (1981). The vertebrae are amphicoelous and lack calcified tissue on the cotyles. The centrum length of the vertebrae ranges up to 20 mm. The subcentral keel is either obsolete or absent. Basapophyses are missing. Large and well-defined fossae occur anterior to the transverse processes. The rib-bearers have two heads and are widely divergent; nevertheless, they are at least partly connected by bone. Weak ventral laminae occur on the transverse processes. The neural spine is massive, prominent, and extensive. No spinal nerve foramina are present on the vertebrae.

The atlas has an obsolete interglenoid tubercle. The atlantal cotyles are oval and

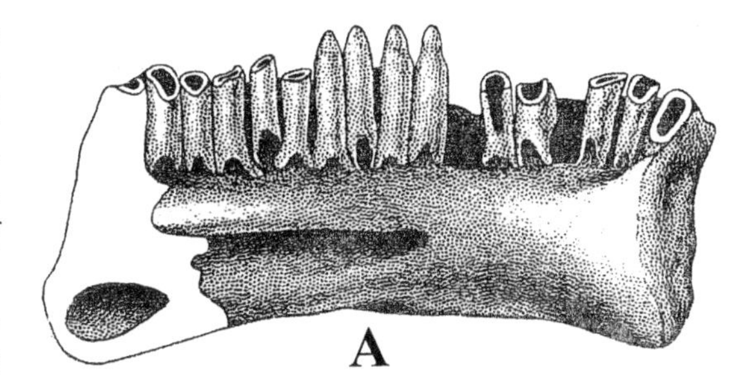

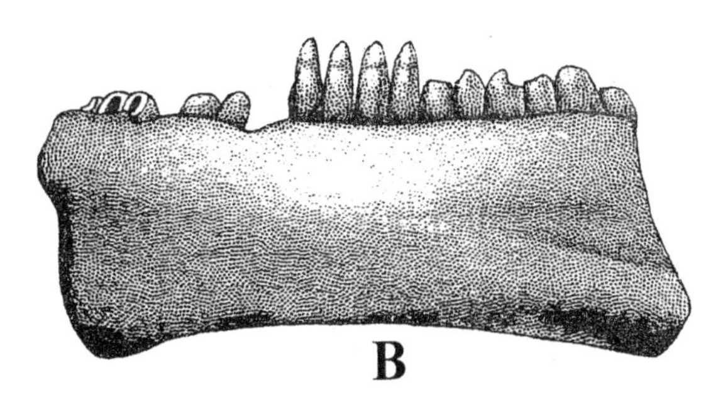

Figure 71. Dentary of #*Piceoerpeton willwoodense* from the Paleocene of Wyoming in (A) lingual and (B) labial views. No scale is given, but the preserved specimen is about 43 mm long.

deeply concave. Spinal nerve foramina are present just behind the cotyles. The neural spine is robust. Transverse processes are absent. The greatest width of the atlas is 33 mm, and its greatest length is 21 mm; the greatest width of one atlantal anterior cotyle is 14.5 mm.

The dentary is robust, and its symphysis is strongly built. The Meckelian groove is narrow and extends anteriorly to about the 10th dentary tooth. A prominent longitudinal labial depression is present. The dental sulcus is obsolete or absent. The teeth are robust and non-pedicellate. The maxilla has a short, blunt facial portion and has an interdigitating suture with the premaxilla. The premaxillae meet on a complex midline suture. The nasal processes indicate a very flat snout.

The limb bones are massive. The distal articulating portion of the femur is kidney-bean shaped. The greatest distal width of the femur is 18 mm.

General Comments. This is a giant salamander—if it is indeed a true salamander. Naylor and Krause (1981) pointed out that *Piceoerpeton* is closely related to or actually descended from *Lisserpeton*. *Piceoerpeton* lived in lowland environments. Its distribution in time and space indicates it lived in a climate that had a fairly cool mean annual temperature. Moreover, the presence of growth zones in various parts of the skeleton of this species indicate pronounced seasonality in the environment where it lived (Naylor and Krause, 1981).

GENUS #*SCAPHERPETON* COPE, 1877

Genotype. Scapherpeton tectum Cope, 1877.

Diagnosis. The diagnosis is the same as for the genotype and only known species, *Scapherpeton tectum.*

#*SCAPHERPETON TECTUM* COPE, 1877

(*HEDRONCHUS STERNBERGI* COPE, 1877; *HEMITRYPUS JORDANIANUS* COPE, 1877; *SCAPHERPETON EXCISUM* COPE, 1877; *SCAPHERPETON FAVOSUM* COPE 1877; *SCAPHERPETON LATICOLLE* COPE, 1877)

FIG. 72

Holotype. An anterior trunk vertebra (AMNH 5682, American Museum of Natural History).

Type Locality. Judith River Formation, Montana.

Horizon. Late Cretaceous.

Other Material. This species is represented by abundant material that resides in the UCMP, AMNH, MCZ, USNM, and CM collections. Collectively, these museums house vertebrae, limb and limb girdle elements, dentaries, fused prearticulars, maxillae, a frontal, parietals, and exoccipitals. All these elements are disarticulated. The main localities where the Late Cretaceous (Campanian) material was collected are as follows: the Mesaverde Formation, Wyoming; the Judith River Formation, Montana; and the Oldman Formation, Saskatchewan and Alberta. Localities where Late Cretaceous (Maastrichtian) elements were collected include the Lance Formation, Wyoming; the Laramie Formation, Colorado; and the Frenchman Formation, Saskatchewan. The Early Paleocene material came from the the Hell Creek and Tullock Formations, Montana. The Middle Paleocene material was found in the Tongue River Formation, Montana. The Late Paleocene elements were collected from the Fort Union Formation Wyoming; the Paskapoo Formation, Alberta. See Estes (1964, 1975, 1976), Van Valen and Sloan (1965), Estes et al. (1969), Sahni (1972), Carpenter (1979), and Krause (1980).

Diagnosis. The diagnosis is directly from Estes (1981, p. 50): "Differs from *Lisserpeton*band *Piceoerpeton* in having tear-drop shaped amphicoelous cotyles, small oval zygapophyses set close to midline, rib-bearers at least partially connected by a web of bone; variable but usually prominent and often convex subcentral keel; dentary slender anteriorly with oval symphysis, little expanded."

Description. The description is mainly modified from Estes (1981). The vertebrae of *Scapherpeton* are very robust. They are amphicoelous,

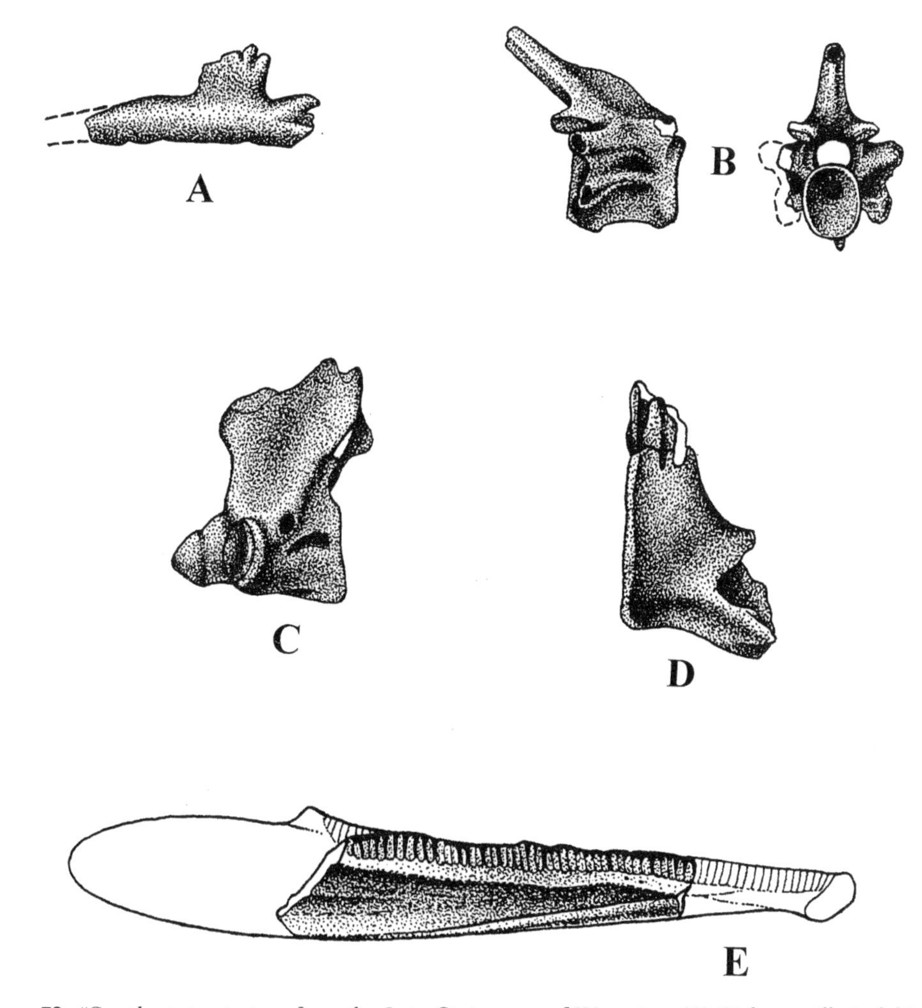

FIGURE 72. #*Scapherpeton tectum* from the Late Cretaceous of Wyoming. (A) Right maxilla in labial view. (B) Trunk vertebra in lateral (left) and posterior (right) views. (C) Atlas in left lateral view. (D) Right parietal in dorsal view. (E) Left dentary (restored from other specimens) in lingual view. All figures ×3.

and their cotyles are usually shallow and have a definitive calcified layer. The cotyles are teardrop shaped. The basapophyses and crests are absent, and the surface of the bone is usually smooth but sometimes pitted. A subcentral keel is present, and it may be concave, straight, or convex in lateral view. Subcentral foramina are often numerous, and they sometimes form channels in the subcentral keel. The transverse processes are laterally directed, and the rib-bearers are two headed and at least partially connected by a web of bone. The rib-bearers do not extend beyond the posterior border of the cotyle. A vertebral canal lies in deep pits anterior and posterior to the transverse processes. There are no foramina present for spinal nerves in the trunk vertebrae. The neural arch is low, and the neural canal is relatively small. The zygapophyses are also small and are positioned near the midline. The neural spine is elongate and thin and finished in cartilage.

The atlas has either a pitted or a smooth ventral surface. Its anterior cotyles are oval or lenticular and separated by a prominent interglenoid tubercle, which is constricted at its base. A foramen for the spinal nerve is present. The caudal vertebrae are perforated by foramina for spinal nerves, and these vertebrae also have simple hemal arches.

The dentaries are elongate and are very slender anteriorly. The dental symphysis is strong, but it is only moderately expanded. The tooth-bearing border of the dentaries is almost horizontal. Posteriorly, the jaw is also little expanded. The teeth are pedicellate, small, and very closely spaced. The Meckelian canal is open, and its dental sulcus is shallow or absent. The premaxillae are paired and only slightly curved; its tooth number is about 15. The maxillae are robust, and the dental portion is posteriorly slender. The facial portion of the maxilla is narrow but may be expanded and rugose dorsally. The tooth number of the maxilla varies from 25 to 50, depending on the size of the fossil. The vomer has an anterior marginal row of tooth bases and a thin, flat posterior process. The parietals are broadly L-shaped and have smooth surfaces. The squamosal attachment of this bone is rugose, a fossa sometimes being present for the squamosal. The frontals are long and slender and close to the midline, and their posterior tips are separated. The exoccipitals have prominent, dorsoventrally compressed condyles. A notch is present for the interglenoid tubercle. Laterally, a large vagus foramen and a very small hypoglossal foramen are present.

The ilium of *Scapherpeton* is gracile and cylindrical, and its acetabulum is projected upward from the surface of the bone. The femur is slender and bears a small, somewhat "stick-like" trochanter proximally. The head of the femur is somewhat wedge shaped in cross section, and the distal end of the bone is slightly expanded. The humerus bears an expanded, fan-shaped proximal end, and the distal end is only slightly expanded.

General Comments. In general, laterally compressed vertebrae with a prominent subcentral keel coupled with a high, slender neural spine, such as occurs in *Scapherpeton,* indicate some degree of adaptation for aquatic locomotion (Estes, 1981). The limb bones are well preserved and relatively abundant in *Scapherpeton,* but they are small compared with the size of the known vertebrae of this species.

Family Albanerpetontidae Fox and Naylor, 1982

Albanerpetontids

Fig. 73

The last two families of salamanders we discussed, the Batrachosauroididae and the Scapherpetontidae, were labeled "enigmatic caudate groups of uncertain position" by Milner (2000, p. 1435). The albanerpetontids, on the other hand, were termed "doubtful Caudata" by (Milner, 2000, p. 1437). Moreover, this group was recently characterized by Gardner (2002, p. 12) as "a clade of Middle Jurassic to Miocene salamander-like lissamphibians." However, albanerpetontids are probably a much longer step away from being salamanders (Caudata) than the batrachosauroidids or scapherpetontids are, and Gardner's (2001) evidence against the albanerpetontids being salamanders is convincing.

Here we shall consider albanerpetontids as a natural group (clade) of salamander-like amphibians (possibly an unnamed lissamphibian order) that evolved in ecological parallel with primitive caudate groups. Since

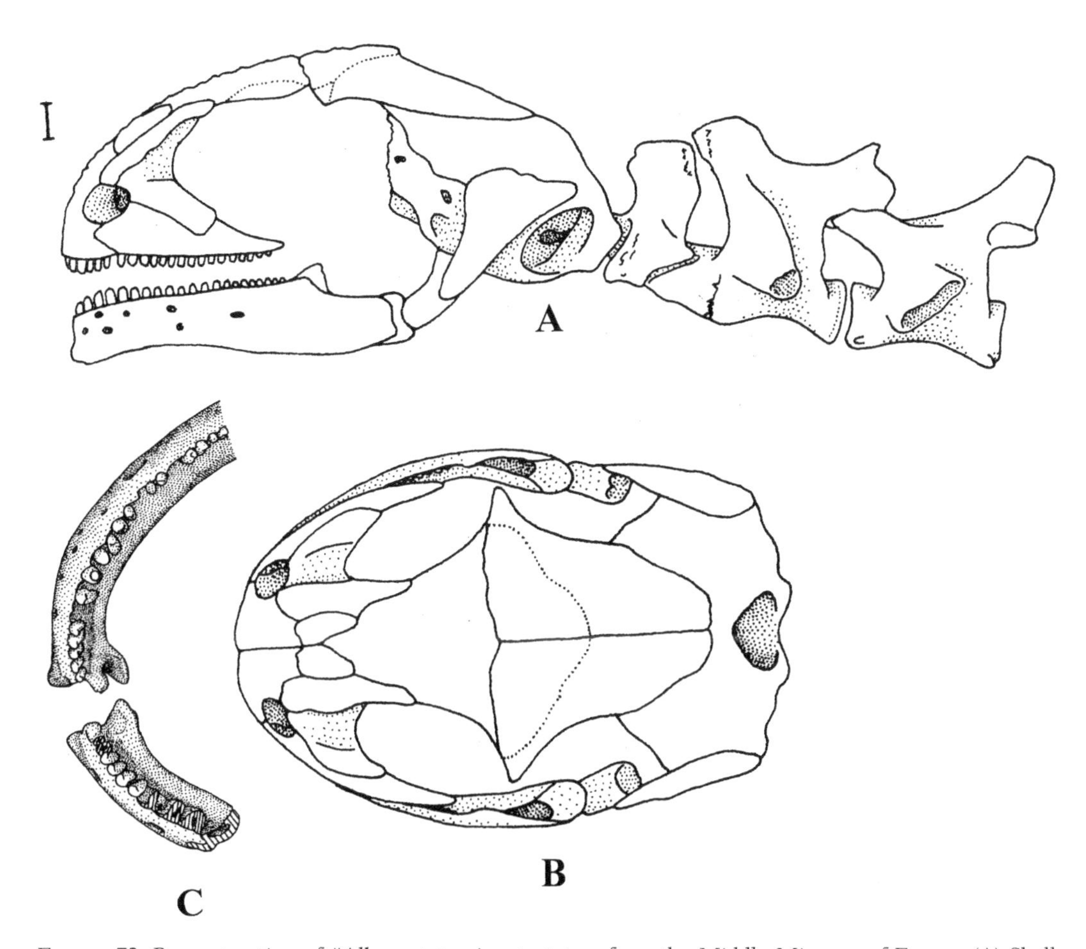

Figure 73. Reconstruction of #*Albanerpeton inexpectatum* from the Middle Miocene of France. (A) Skull and anterior part of the vertebral column in lateral view. (B) Same skull in dorsal view. (C) Dentaries of #*Albanerpeton galaktion* of the Late Cretaceous of Alberta, showing the interlocking structure of the mandibular symphysis.

in all probability they are not salamanders, we shall not present albanerpetontid salamander species here in a detailed format as we did for members of other families.

Recently, Milner (2000) accepted the family Albanerpetontidae described by Fox and Naylor (1982). Milner (1988, 2000), McGowen and Evans (1995), Gardner (1999a–1999c, 2000a–2000c, 2001, 2002), and Gardner and Averianov (1998) have published detailed accounts of this group. Among these authors, Gardner has presented the most comprehensive studies of the North American members of this family.

The Albanerpetontidae are known from Europe, northern Africa, middle Asia, and North America. They existed from Middle Jurassic to Early Pliocene times. In North America they are known from the Early Cretaceous to the Paleocene of Canada and the United States. Three genera of albanerpetontids are known, but only a single genus, *Albanerpeton,* is known from North America. This genus also occurred in Europe from the Late Cretaceous to the Early Pliocene. Some definitive characters for the group are modified from Milner (2000). The skin of the head and body is covered with small osteoderms, and some have characteristic structures on the hind legs that could be glandular masses associated with mating. The skull of the Albanerpetontidae is composed of premaxillae, maxillaries, nasals, large lacrimals, a median frontal, large paired parietals, squamosals, jugals, and quadrates. The quadrate has a very convex articular area. The anterior part of the skull roof is heavily sculptured. The parasphenoid is gracile and spike-like and differs from the parasphenoid of lissamphibians, which consists of broad structures that form the floor of the braincase. The lower jaw comprises a dentary, an articular, a prearticular, and an angular. The dentaries have an asymmetrical symphyseal connection that consists of interdigitating lobes that allow for a variety of movements between the mandibles. The articular bears a very concave articular surface.

The upper and lower teeth are non-pedicellate and pleurodont and bear three cusps. The premaxillae bears 6–7 teeth; the maxilla, 15–23; and the dentary, 23–33.

Turning to the vertebral column, we find that there are 22 presacral vertebrae, 1 sacral vertebra, and 24 or more caudal vertebrae. The atlas–axis complex is characteristic of the Albanerpetontidae in that it has a large bicondylar atlas and a small axis centrum that lacks a neural arch. The trunk vertebrae are amphicoelous and have anterior basapophyses. The rib-bearers have only a single head and are thought to be endochondral in their derivation; the ribs of the trunk region are short and straight.

In the pectoral girdle, the scapula and the coracoid are separate ossifications, which is not the case for modern caudates. The scapula bears an anterior extension. The humerus has an ossified distal head (ball) that is associated with the radius, a situation that occurs in frogs. The carpus is ossified, and the manus has four digits and a phalangeal formula of 2.3.3.3 (inner to outer).

In the pelvic girdle, the ilium, the ischium, the pubis, and the tarsus are ossified. The pes has five digits and a phalangeal formula of 2.3.4/5.3.3 (inner to outer).

James D. Gardner (personal communication, 2005) has informed me that skull bones, vertebrae, and long bones are the albanerpetontid remains most commonly recovered as fossils and that these elements are all diagnostic for albanerpetontids. The three genera are differentiated on the basis of features of the frontals and jaws. Species are differentiated on the basis of frontals and jaws as well as inferred body size. North American albanerpetontid species are as follows.

#*ALBANERPETON ARTHRIDION* FOX AND NAYLOR, 1982

Holotype. A nearly complete right premaxillary (FMNH PR-805, Field Museum of Natural History).

Type Locality. Forestburg site, Antlers Formation, Texas.

Horizon. Early Cretaceous (latest Aptian to middle Albian).

An up-to-date treatment is provided by Gardner (1999b).

#*ALBANERPETON CIFELLII* GARDNER, 1999C

Holotype. Incomplete right maxilla, the only element of this species recovered (OMNH 25400, Oklahoma Museum of Natural History).

Type Locality. Smoky Hollow Member, Straight Cliffs Formation, Garfield County, Utah.

Horizon. Late Cretaceous (late Turonian).

The only detailed treatment is by Gardner (1999c).

#*ALBANERPETON GALAKTION* FOX AND NAYLOR, 1982

FIG. 73C

Holotype. A left premaxilla (UALVP 16203, University of Alberta Laboratory for Vertebrate Palaeontology).

Type Locality. Site MR-6, upper Milk River Formation, Alberta.

Horizon. Late Cretaceous (late Santonian).

A revised diagnosis was given by Gardner (2000b).

#*ALBANERPETON GRACILIS* GARDNER, 2000B

Holotype. A nearly complete left premaxilla (RTMP 95.181.70, Royal Tyrrell Museum of Palaeontology).

Type Locality. RTMP locality 410, Dinosaur Park Formation, Dinosaur Provincial Park, Alberta.

Horizon. Late Cretaceous (middle Campanian).

This material was described in detail by Gardner (2000b).

#*ALBANERPETON NEXUOSUS* ESTES, 1981

Holotype. An almost complete left dentary (UCMP 49547, University of California Museum of Paleontology).

Type Locality. Bug Creek, Lance Formation, Niobrara County, Wyoming.

Horizon. Late Cretaceous or Early Paleocene.

A revised description and diagnosis were given by Gardner (2000b).

General Comments. Gardner (2002) suggested that the center of the history of *Albanerpeton* was in North America.

NOTE

1. Woodfordian (full glacial) date (I-4163): 19.7 ± 0.6 ka BP (L. P. Fay, personal communication, May 12, 1993).

CHAPTER

3

Chronological Accounts

Salamanders are possibly known from as early as the Late Triassic, in the form of a tiny, fragmentary skeleton of *Triassurus sixtelae*, from Uzbekistan (see Fig. 19). Two unnamed salamander atlantes (atlases) from the Early Jurassic Kayenta Formation of northeastern Arizona represent the earliest salamanders in North America, and if *Triassurus* is proven not to be a true salamander, the world. By Middle and Late Jurassic times, true salamanders were in England and Kazakhstan, respectively. The two enigmatic salamander families, Batrachosauroididae and Scapherpetontidae, appeared in North America in the Cretaceous. The Albanerpetontidae, a salamander-like group that is known from the Cretaceous and Paleocene of North America, is not considered here to be a true salamander. Salamanders of modern families did not appear in the North American fossil record until the Late Cretaceous.

Mesozoic Era

Jurassic Salamanders

On a worldwide scale, the Jurassic period is formally divided into three epochs: the Early Jurassic (206–180 Ma BP), the Middle Jurassic (180–159 Ma BP), and the Late Jurassic (159–144 Ma BP). The dates for the ages (e.g., Hettangian age, Tithonian age) within these three Jurassic epochs may be found in Fig. 23. Two tiny salamander atlases (MCZ 9017, 9018 Museum of Comparative Zoology, Harvard University) from the Early Jurassic Harvard Gold Spring Quarry of the Kayenta Formation of northeastern Arizona (Curtis and Padian, 1999) represent the most ancient salamanders in North America and possibly the world. The description of these vertebrae given here is directly from Curtis and Padian (1999, p. 23): "MCZ 9017 and 9018 are two urodele atlas vertebrae. The anterior end of each vertebra has two oval cotyles separated from each other by a ventral groove and a wedge-shaped dorsal intercotylar process.

The posterior end has a single, deeply concave (seemingly notochordal), smoothly conical cotyle with an oval opening. Only the two lateral bases of the neural arch remain and each of these is pierced by an oval foramen that is directed posteroventrally. The ventral surface of the centrum is somewhat saddle-shaped and bears up to six small foramina. No sutures are visible on the centrum. Each atlas is 1.9 mm wide across the anterior cotyles, 0.7 mm wide across the posterior cotyle, and 1.4 mm long."

An articulated salamander from the Late Jurassic Morrison Formation of North America was reported by Evans et al. (2005). This fossil, named *Iridotriton hechti*, shows a combination of primitive and derived characters. Some of its characters indicate that *Iridotriton* may be related to the crown clade salamandriforms, a group that includes several living families.

Cretaceous Salamanders

On a worldwide scale, the Cretaceous period is formally divided into only two epochs: the Early Cretaceous (144–99 Ma BP) and the Late Cretaceous (99–65 Ma BP). The enigmatic salamander family Batrachosauroididae, represented only by #*Prosiren elinorae* Goin and Auffenberg, 1958, is the only named salamander species from the Early Cretaceous of North America. It was described from the Antlers Formation of Texas, which represents the late part of the Early Cretaceous, somewhere between 112 and 99 Ma BP. This species was not eel-like but had a rather short body form. However, the expanded neural arches of the vertebrae indicate that this species may have been somewhat adapted to an aquatic life. On the other hand, *Prosiren* had strong, two-headed rib-bearers unlike the two-headed rib-bearers that are typical of aquatic or elongate salamanders. As mentioned earlier, in the *Prosiren* account, this genus is currently being restudied.

Turning to the Late Cretaceous of North America, we find the first appearance of the living family *Sirenidae* and perhaps the *Amphiumidae*. The Early Cretaceous enigmatic family Batrachosauroididae reappears; and a new enigmatic salamander group, the Scapherpetontidae, appears for the first time in North America.

The family Sirenidae is first represented in the fossil record by a single genus and two species, #*Habrosaurus dilatus* Gilmore, 1928 and #*Habrosaurus prodilatus* Gardner, 2003a. *Habrosaurus dilatus* was first described (Fig. 74) from the Late Cretaceous Lance Formation of Niobrara County, Wyoming. It was thought to be a fossil lizard when it first appeared in a book about North American fossil lizards by Gilmore (1928). *Habrosaurus prodilatus* is actually known earlier in the Late Cretaceous than is *H. dilatus:* the former species was first known from the middle Campanian; the latter, from the late Maastrichtian (see Fig. 23). *Habrosaurus dilatus* was very large for a salamander, reckoned to be longer than 1600 mm (about 63 inches) in length in the largest specimen. *Habrosaurus* must have been a swamp dweller, as is modern *Siren*, feeding on small vertebrate and invertebrate prey. Invertebrate prey may have consisted largely of mollusks and hard-bodied arthropods (Estes, 1964; Gardner, 2003a). But *Habrosaurus* itself must have had to dodge the very large, active fishes of the Late Cretaceous (Fig. 75).

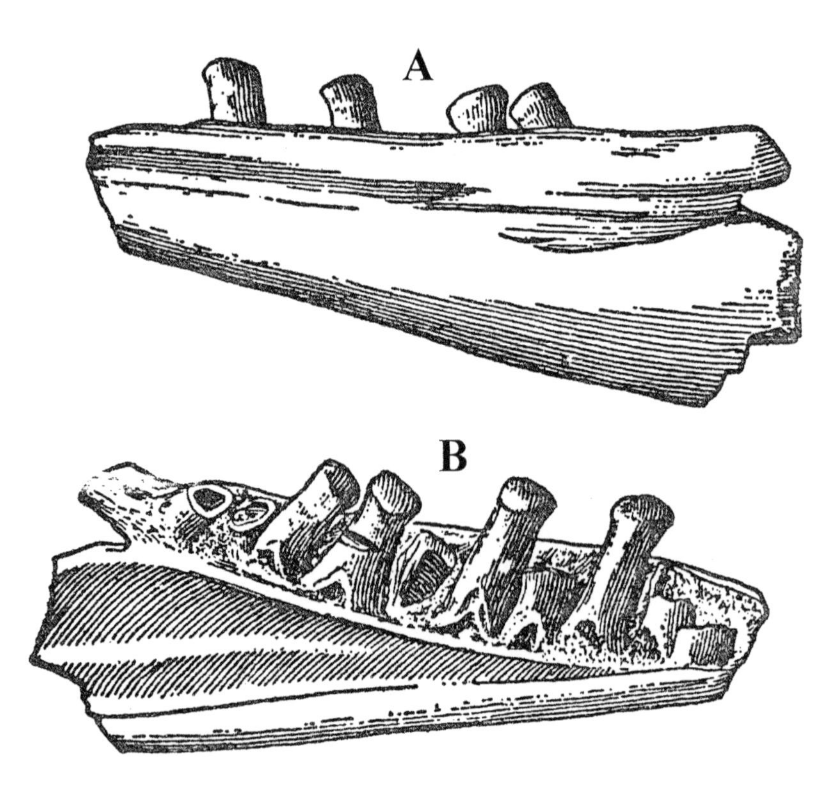

FIGURE 74. Holotype left dentary of the sirenid #*Habrosaurus dilatus* in (A) labial and (B) lingual views. From Gilmore (1928).

FIGURE 75. *Xiphactinus*, a Late Cretaceous predatory fish, adds #*Habrosaurus dilatus* to its diet in a shallow waterway in Montana.

The Amphiumidae, another eel-like family of relatively legless swamp dwellers that lives in the southeastern United States today, may be first represented in the Late Cretaceous by a single genus and species, *#Proamphiuma cretacea* Estes, 1969. But it is known from the Bug Creek Anthills site in Montana, which contains a mixture of Late Cretaceous and Early Paleocene fossils. Thus, one must be content in saying *Proamphiuma* must have lived either in the Late Cretaceous or the Late Paleocene or perhaps both. The *Proamphiuma* have vertebrae with posterior zygapophyseal crests that indicate the presence of specialized intervertebral muscles found only in *Amphiuma* (Auffenberg, 1959). The major differences between *Proamphiuma* and *Amphiuma* are the less-developed crests, somewhat narrower vertebral centra, and less prominent zygapophyses in the former.

Because of the similarities between *Proamphiuma* and *Amphiuma*, it is probable that *Proamphiuma* had similar habits, probably living in swamps or other still, shallow-water habitats. Like the modern genus, *Proamphiuma* was probably an active predator on animals such as small fish and crayfish. It is possible that this ancient salamander, in concentrated numbers, stalked small fishes at the surface of the water at night, like its modern relatives.

By Late Cretaceous times, both families of enigmatic salamanders were present. The Batrachosauroididae reappeared from the Early Cretaceous; and the family Scapherpetontidae, the other enigmatic group, appeared for the first time.

The Batrachosauroididae were represented by three species in the Late Cretaceous. *#Opisthotriton kayi* Auffenberg, 1961 is one of the best-known ancient salamanders. It existed in the form of a "giant larva" that had reduced limbs and a long body composed of 40 or more body vertebrae; its skull also had a larval aspect. More than one paleontologist (e.g., Estes 1975; Naylor 1978a, 1978c) has suggested a relationship between the Batrachosauroididae and the living family Proteidae (European Olms and the American Mudpuppy).

We get used to the fact that most early fossil frogs (see Holman, 2003) and salamanders come from the fossil beds of western Canada and the United States, but surprisingly, the batrachosauroidid *#Parrisia neocesariensis* Denton and O'Neill, 1998 is known only from the Late Cretaceous of New Jersey. The Ellisdale, New Jersey, site where *Parrisia* was collected has been interpreted as a sequence of sediments deposited in a back bay or tidal channel along a continental margin (Gallagher et al., 1986; Tashjian, 1990). The fossil fauna associated with *Parrisia* is typical of tidal faunas. It consisted of sharks teeth, mixed saltwater and freshwater bony fish remains, and bones of turtles, crocodiles, and dinosaurs.

The third batrachosauroidid genus known from the Late Cretaceous of North America is *#Prodesmodon copei* Estes, 1964 from Wyoming, New Mexico, and Montana. This genus is rather common in the Late Cretaceous Lance Formation of Wyoming but is rare in other deposits. This taxon has distinctive vertebrae that were interpreted by Estes (1964) as resembling those of the desmognathine plethodontids (living salamanders) in some ways and those of the batrachosauroidid salamander *Op-*

isthotriton kayi in others. The affinities of *Prodesmodon* with modern salamanders are rather murky at present.

The enigmatic salamander family Scapherpetontidae first appears in North America in the Late Cretaceous and is represented by two genera: #*Lisserpeton bairdi* Estes, 1965b and #*Scapherpeton tectum* Cope, 1887.

Scapherpeton tectum is one of the most well known genera in the family and is represented by relatively abundant material from Montana, Wyoming, Colorado, Alberta, and Saskatchewan. The vertebrae of this taxon are laterally compressed and have a well-developed subcentral keel. These features, as well as a well-developed, thin neural spine, probably indicate an elongate body, as well as aquatic tendencies, in *S. tectum*. Estes (1981) perceived *Scapherpeton* as being somewhat more primitive than the following genus, *Lisserpeton*, in its heavier ossification but remarked that the former genus was more specialized in its vertebral modifications for aquatic locomotion.

Lisserpeton bairdi is represented in the Late Cretaceous by fossil material from Montana, Wyoming, and Colorado. This genus is rarer than the preceding one, and in fact, no undoubted limb elements are known in this taxon. However, its skeletal anatomy indicates that there is a high probability it was a permanently aquatic form with some degree of body elongation and limb reduction.

Cenozoic Era

Paleocene Salamanders

The Paleocene is the first epoch of the Cenozoic era, and it was a very short one, lasting only about 10 Ma. This relatively short epoch (at least in geological time) was in many ways a lag time when both marine and terrestrial communities reorganized in the wake of the extinction of the dinosaurs and great marine reptiles by the end of the Cretaceous. The Paleocene epoch, like the Cretaceous period, is formally divided into only Early and Late. However, writers on fossil salamanders have referred to middle (informal division) Paleocene occurrences.

In discussing the Cenozoic era, I will often refer to North American land-mammal ages (NALMAs; see Fig. 24), especially those of the Neogene period (Miocene–Pleistocene). NALMA terms reflect recent North American stratigraphic studies well (e.g., Woodburne, 1987; Prothero and Emry, 1996) and are widely used on this continent.

Four families of Early Paleocene salamanders managed to survive the holocaust that marked the end of the Cretaceous. This holocaust saw the complete extinction of the dinosaurs and great sea reptiles, as well as other forms of life, a revolution that completely changed the biosphere as it was previously known. These hardy salamander species were *Opisthotriton kayi* and *Prodesmodon copei* of the Batrachosauroididae, from the Early Paleocene of Montana; and *Scapherpeton tectum* and *Lisserpeton bairdi* of the Scapherpetontidae, also known from the Early Paleocene of Montana. A new addition to the Batrachosauroididae in the Early Paleocene is *Opisthotriton gidleyi* Sullivan, 1991 from Montana. *Habrosaurus dilatus* of the living family Sirenidae also survived the Cretaceous holocaust and is known from Early Paleocene (middle Paleocene of Gardner 2003a) of the western North American interior. Interestingly, all of these animals were at least partially aquatic, as were the turtles and crocodilians that survived into the Cenozoic. Also, it is possible or even probable that *Proamphiuma cretacea,* the amphiumid from the mixed Late Cretaceous–Early Paleocene Bug Creek Anthills site in Montana, occurred in the Early Paleocene of North America.

Turning to the Late Paleocene, we find a striking change in the composition of the salamander fauna, as four currently living salamander families made their first appearance during this sequence of time. The Cryptobranchidae made its first appearance as the living genus *Andrias* in the form of the extinct species **Andrias saskatchewanensis* Naylor, 1981a from the Late Paleocene of Saskatchewan, Canada. The Proteidae first appeared as the living genus *Necturus* in the form of the extinct species **Necturus krausei* Naylor, 1978c, also from the Late Paleocene of Saskatchewan.

Turning to the western United States, we find that the living genus *Amphiuma* of the Amphiumidae was first represented in the Late Paleocene by the extinct species **Amphiuma jepseni* Estes, 1969c. In contrast, the earliest known sirenid, *Habrosaurus dilatus,* made its last appearance in the Late Paleocene (middle Paleocene of Gardner, 2003a), although modern sirenid genera were to follow. Finally, moving back to Canada, we find that the earliest remains of the modern family Dicamptodontidae were found in the Late Paleocene of Alberta in the form of the modern genus *Dicamptodon* and the extinct species **Dicamptodon antiquus* Naylor and Fox, 1993. Back in Sweet Grass County, Montana, trackways (Fig. 76) of dicamptodontids (probably *Dicamptodon*) were found exquisitely preserved.

The enigmatic salamander family Scapherpetontidae continued to do well in the Late Paleocene after surviving the mass extinction in the

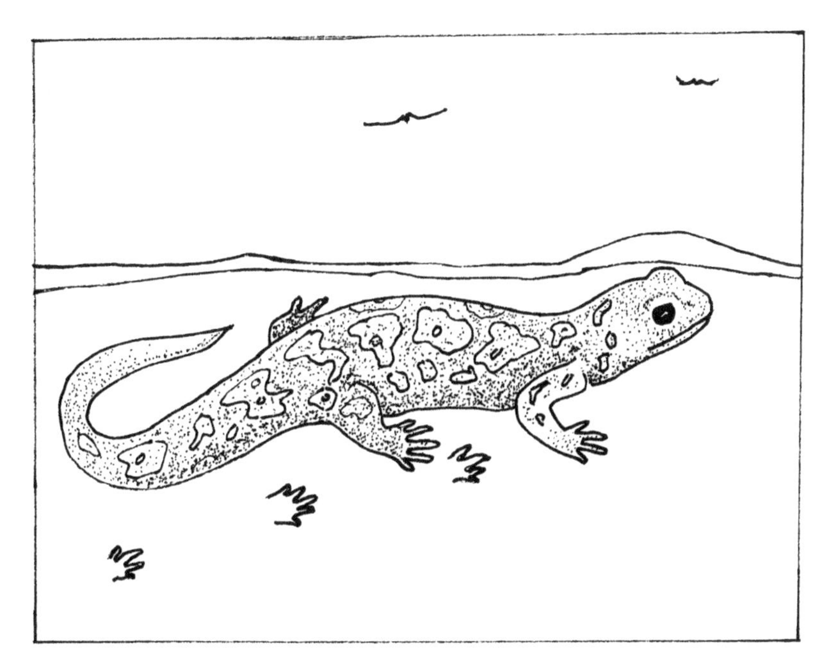

FIGURE 76. A giant salamander (*Dicamptodon*) "making tracks" in the Paleocene of Sweet Grass County, Montana.

Cretaceous. *Scapherpeton tectum* and *Lisserpeton bairdi* were carryovers from the Early Paleocene, and a new scapherpetontid, #*Piceoerpeton willwoodense* Meszoely, 1967, was recognized from the Late Paleocene of Alberta, Saskatchewan, and Wyoming. *Piceoerpeton* was a salamander of giant proportions.

It is remarkable that the genera and species of all six families of salamanders in the Late Paleocene appear to represent either fully or at least partially aquatic taxa. The Cryptobranchidae and Proteidae and their genera *Andrias* and *Necturus,* respectively, are today obligate neotenic, fully aquatic species with functional limbs, being essentially bottom crawlers. The Sirenidae of today are eel-like, obligate neotenic, fully aquatic species with tiny, nonfunctional limbs. They swim through the water like elongate fishes. Even *Dicamptodon* of the Dicamptodontidae frequently exists as a neotenic larval form with fully aquatic tendencies. The three scapherpetontid genera of the Late Paleocene, *Scapherpeton, Lisserpeton,* and *Piceoerpeton,* are undoubtedly aquatic or have aquatic tendencies.

Thus, in summary, at the family level, five of the six families in the Late Paleocene of North America are living today. All of the six known families and their respective genera and species in the Late Paleocene are fully or at least partially aquatic. According to their ecological niches in the Late Paleocene, they appear to be of three kinds: (1) large, obligate neotenic, flattened bottom crawlers (Cryptobranchidae, Proteidae); (2) large, obligate neotenic, eel-like, fast and agile swimmers (Sirenidae, Amphiumidae); and (3) large, neotenic species, probably partially to fully aquatic, with some possibly foraging on land from time to time (Dicamptodontidae, Scapherpetontidae). Perhaps this is the result of our North American fossil digs being in formerly swampy areas in the Late Paleocene of western North America. Moreover, in eastern North America, the few Cretaceous and Paleocene sites that do exist are in Atlantic coastal areas and mainly represent coastal mudflats or backwaters.

Eocene Salamanders

Author's note: #*Paleoamphiuma tetradactylum* Rieppel and Grande, 1998 is an elongate, aquatic salamander from the Early Eocene of Wyoming that was reported to be a member the family Amphiumidae. It is a sirenid that needs to be restudied so that its relationships within the family can be determined.

The Eocene epoch lasted from 54.8 to 33.7 Ma BP; thus, this epoch was about twice as long as the Paleocene. The Eocene is formally divided into Early, Middle, and Late. From what we know from the fossil record, the Eocene was a time of reorganization of terrestrial vertebrate communities that were dominated by large mammalian species. The world climate was fairly equable until near the end of the epoch, when it began to deteriorate. Many families of modern mammals originated in Eocene times. Frogs of the Eocene are much better known those of the Paleocene but were not greatly diverse. Given the length of the epoch, salamander fossils are surprisingly rare. This rarity of caudate fossils is possibly due to the drier environmental conditions in which Eocene fossil beds were deposited.

The Early Eocene has surprisingly yielded the remains of only four families (taking into account *Paleoamphiuma tetradactylum*, which is a sirenid noted above). The family Dicamptodontidae is a carryover from the Late Paleocene in the form of the dicamptodontid species #*Chrysotriton tiheni* Estes, 1981, which appears for the first time in the fossil record. It was described from the Early Eocene of South Dakota. Reappearing from the Early Paleocene we find the family Batrachosauroididae, represented by a new species, #*Batrachosauroides gotoi* Estes, 1969b, from the Early Eocene of South Dakota. Continuing on from the Late Paleocene is the scapherpetontid species *Piceoerpeton willwoodense* Meszoely, 1967. This species has been reported from the Early Eocene of Wyoming, as well as from the Canadian Arctic Archipelago (Eureka Sound Formation). Its occurrence at the latter site indicates that this area, which is currently arctic, was warmer in the Early Eocene. This Early Eocene salamander assemblage shows no striking taxonomic or ecological differences from those of earlier Cretaceous or Paleocene times.

Middle and Late Eocene North American salamander fossils are even rarer than Early Eocene ones, as only two species, representing two genera and families, are known! The family Sirenidae is represented by the first appearance of the modern genus *Siren*, in the form of **Siren dunni* Goin and Auffenberg, 1957 from the Middle Eocene Uinta Formation (Uintan NALMA) of Wyoming. The family Ambystomatidae and the genus *Ambystoma* are first represented by the extinct species **Ambystoma tiheni* Holman, 1968b from the Late Eocene Cypress Hills Formation (Chadronian NALMA) of Saskatchewan.

Siren dunni differs from *Habrosaurus dilatus*, the sirenid from the Cretaceous and Paleocene, in having obsolete basapophyses. By having this character state *S. dunni* resembles sirens of the Miocene and later times. Moreover, the straight centrum and zygapophyseal ridge indicate that *S. dunni* represents the living sirenid genus *Siren* rather than *Pseudobranchus*, also a living sirenid genus. This is also a noteworthy range extension for the genus *Siren*, which today occurs in eastern and coastal regions of the United States and southward into northeastern Mexico, then north to Lake Michigan in the Mississippi River system. It seems probable that the withdrawal of *Siren* to its present range had to do with the profound climatic deterioration that began in the Late Eocene and lasted through the Oligocene (e.g., Prothero and Emry, 1996; Holman, 2000a).

**Ambystoma tiheni* is similar in size and vertebral proportions to the *Ambystoma opacum* species group of Tihen (1958). However, it differs from this group in a few details. Estes (1981) stated that although these *Ambystoma* fossils resemble the *A. opacum* group, the small differences indicate that the definite assignment to this group should await more material. It is of interest, however, that *Ambystoma*, represented today by mainly terrestrial, burrowing species, first appears at this time and place and that indications of the development of modern species groups of *Ambystoma* are apparent in Late Eocene times.

Oligocene Salamanders

Like the Paleocene, the Oligocene Epoch of North America represents a relatively short interval of about 10 Ma (33.7–23.8 Ma BP). Also, like the Paleocene, the Oligocene is formally divided into Early and Late only. The climate began to deteriorate in earnest in the Late Eocene (e.g., Prothero and Emry, 1996; Holman, 2000a), and this continued throughout the Oligocene. All the fossil salamanders of this epoch are known from the Late Oligocene or from Late Oligocene to Early Miocene transitional sites. The latter sites will be considered here rather than in the Miocene section of the book.

The presence of two extinct species of the modern genus *Taricha* in the Late Oligocene represents the first record of the family Salamandridae in North America. **Taricha oligocenica* (Van Frank, 1955) was described from the Late Oligocene (Arikareean NALMA) of Oregon. This species was once named *Palaeotaricha oligocenica*, but Tihen (1974) demonstrated that this taxon differed from modern *Taricha* only in having a more extensive dermal cap on the top of the neural spine. **Taricha lindoei* Naylor, 1979 from the Late Oligocene of Oregon shows some resemblance to *Taricha oligocenica*, but it is only about half as large and could possibly be a larval form.

Now, turning to beds that are transitional between the Late Oligocene and Early Miocene, we find **Taricha miocenica* Tihen, 1974 from Montana. All the fossils of this species appear to be from metamorphosed aquatic adults. This species differs from *Taricha oligocenica* in being smaller and in having a less extensive dermal cap on the neural spine. Tihen (1974) believed that modern *Taricha* species could have been derived from *T. oligocenica* by way of *T. miocenica*.

Also from beds that are transitional between the Late Oligocene and Early Miocene we find the first evidence of the family Plethodontidae in North America in the presence of two modern genera, *Aneides* and *Plethodon*, from the Cabbage Patch Beds of Wyoming. Tihen and Wake (1981) went through an excruciating process to correctly identify these genera but chose not to give them specific names. The ancestry of the family Plethodontidae (which comprises most species of modern North American salamanders) is still obscure because of the lack of plethodontid fossils older than Late Oligocene–Early Miocene.

Miocene Salamanders

The Miocene saw not only the return of a more equable climate but also the rapid spread of grasslands throughout much of the world. Most of the families and genera of salamanders and even some species from the Late Miocene are living today. Noteworthy also is that the last of the enigmatic families, the Batrachosauroididae, made it through to the end of the Miocene.

Next to the Eocene, the Miocene is the longest epoch of the Cenozoic era, lasting for 18.5 MA. The Miocene is formally divided into Early, Middle, and Late. Five NALMAs occur in the Miocene. The first one, the Arikareean NALMA, overlaps the Late Oligocene; and the last one, the Hemphillian NALMA, overlaps the Early Pliocene. The Miocene has been so well studied in North America that stratigraphic intervals (e.g., early, medial, and late) within the various NALMA units are often used in faunal reports.

The Early Miocene is composed of roughly about the last one-half of the Arikareean NALMA and about the first two-thirds of the Hemingfordian NALMA (see Fig. 24). One must remember that as stratigraphic data accumulate, these boundaries change from time to time. All salamanders known from the Early Miocene belong to families and genera that are currently living, except for the odd family Batrachosauroididae, but all five species represented are extinct. The family Sirenidae is represented in the Early Miocene of Florida (Hemingfordian NALMA) by **Siren hesterna* Goin and Auffenberg, 1955. This is an interesting species that differs from all other known fossil sirenids by the wide angle (120 degrees) made by the aliform processes, as well as by the anterolateral direction of the transverse processes. Estes (1981) commented that if more material of this taxon were available it might call for the recognition of a new sirenid genus.

Turning to the Early Miocene (Arikareean NALMA) of Nebraska and Colorado, we find the giant salamander **Andrias matthewi* (Cook, 1917) (see Figs. 34, 35) represented in the North American fossil record. This genus, represented by truly giant salamanders still living in Japan and China, somehow made its way from the Old World (probably from Japan) to North America. We have previously discussed, in the systematic accounts, how very closely related the living North American salamander *Cryptobranchus* is to *Andrias.*

Two salamandrid species recorded from the Early Miocene are **Notophthalmus robustus* Estes, 1963 from the Early Miocene (Hemingfordian NALMA) of Florida and **Notophthalmus crassus* Tihen, 1974 from the Early Miocene (Hemingfordian NALMA) of South Dakota. Both of these species come from almost exactly the same horizon in the Hemingfordian (see Woodburne, fig. 6.2, 1987). *Notophthalmus robustus* is a robust, primitive species with a fairly low neural spine. It was suggested by Estes (1981) that *N. robustus* differs more from the living species than the living species differ from one another. *Notophthalmus crassus,* on the other hand, resembles living forms more closely because of its relatively high neural spine.

The enigmatic salamander family Batrachosauroididae is known on the basis of *#Batrachosauroides dissimulans* Taylor and Hesse, 1943 from the Early Miocene (Hemingfordian NALMA) of Texas and Florida.

Most of the Middle Miocene is made up of the Barstovian NALMA, which itself has been divided into early, medial, and late portions as follows (Voorhies, 1990) (oldest at the bottom):

> Late Barstovian 13.0–11.5 Ma BP
> Medial Barstovian 14.5–13.0 Ma BP
> Early Barstovian 16.0–14.5 Ma BP

Six Middle Miocene salamander families, Sirenidae, Cryptobranchidae, Amphiumidae, Ambystomatidae, Salamandridae, and #Batrachosauroididae, are known from North America. Of these, all the genera, with the exception of *Batrachosauroides,* are extant, but all the species are extinct.

Beginning with the Sirenidae, **Siren miotexana* Holman, 1977a is known from the Middle Miocene (early Barstovian NALMA) of Texas. This species is much smaller than the living species *Siren lacertina* and is similar in size to such forms as the living *Siren intermedia* and the fossil* *Siren simpsoni* of the Early Miocene of Florida. The family Cryptobranchidae occurs in the Middle Miocene (medial Barstovian NALMA) of Saskatchewan, Canada, and Nebraska in the form of the giant salamander *Andrias matthewi*. The Saskatchewan specimen consisted of two massive vertebrae with centra of 30 and 40 mm, respectively. Naylor (1981a) extrapolated a total length of 2300 mm for the individual represented by the largest vertebra!

Turning to the Amphiumidae, we find that **Amphiuma antica* Holman, 1977a is known from the Middle Miocene (early Barstovian NALMA) of Texas. This form is of rather small size and does not have particularly noteworthy features. This is the only fossil record of this family and genus in Texas (other than in the Pleistocene) that I am aware of.

The family Ambystomatidae is represented in the Middle Miocene by two species. **Ambystoma minshalli* Tihen and Chantell, 1963 occurs in the medial Barstovian NALMA in Nebraska and in the late Barstovian NALMA in both Nebraska and South Dakota. This is the common salamander of this region and time, and it is especially found in abundance at fossil sites when the screen washing techniques of Hibbard (1949) have been used. *Ambystoma minshalli* is a small salamander of the *Ambystoma maculatum* group of Tihen (1958), a group of salamanders that does not occur in the area at present. A second fossil salamander in the *A. maculatum* group, **Ambystoma priscum* Holman, 1987, is known only from the medial to late Barstovian NALMA of Nebraska. It is relatively uncommon, but it is easily distinguished from *A. minshalli* on the basis of vertebral characters.

Moving to the family Salamandridae, we note that **Notophthalmus slaughteri* Holman, 1966a is known only from the Middle Miocene (early Barstovian NALMA) of Texas. This species is 3 million years younger than *Notophthalmus robustus* and *Notophthalmus crassus* of the Early Miocene and has a more lightly built vertebral form than those more "primitive" species. Finally, we come to the last gasp of *Batrachosauroides dissimulans* in the Middle Miocene (early Barstovian NALMA) of several localities in east Texas. The family Batrachosauroididae, however, hangs on until the Late Miocene, as we shall soon find out.

Five families of salamanders are recorded in the fossil record of the Late Miocene. All are living families, with the exception of the Batrachosauroididae, which makes its last appearance. Six out of seven genera of Late Miocene salamanders are living, the exception being *#Peratosauroides* of the Batrachosauroididae. Three living species are known: *Aneides lugubris*, *Ambystoma maculatum*, and *Ambystoma tigrinum*.

The Sirenidae are represented by two modern genera, each represented by an extinct species, namely, **Pseudobranchus vetustus* Goin and Auffenberg, 1955 and **Siren simpsoni* Goin and Auffenberg, 1955, both from the Late Miocene (Hemphillian NALMA) of Florida. This is the first appearance in the fossil record of *Pseudobranchus*. Both of these sirenid species are similar to modern species within the genera represented (Estes, 1981). Both also occur within the range of these genera today.

Turning to the lungless salamanders, Plethodontidae, we find there are only two records of Late Miocene species, both from California. These include *Aneides lugubris*, a modern species, and *Batrachoseps relictus* Brame and Murray, 1968, a modern species that left intricate Late Miocene fossil trackways. Fossils assigned to *Batrachoseps* sp. are also known from the Late Miocene of California.

Four species of the genus *Ambystoma*, two of them extinct, are known from the Late Miocene. **Ambystoma kansense* (Adams and Martin, 1929) is known from the Late Miocene (Hemphillian NALMA) of Kansas. This extinct species is a neotenic species of the *Ambystoma mexicanum* group of Tihen (1958). The *A. mexicanum* group differs from the *Ambystoma tigrinum* group of Tihen mainly in its obligatory neotenic mode of existence and larger size. The living *A. mexicanum* group occupies the ancient lakes of the southern Mexican plateau. Estes (1981) pointed out that *A. kansense* may link the two species groups on the basis of its occurrence in Kansas. An abundance of fossil material was available for the study of this interesting species.

Ambystoma maculatum, a widespread species today, was identified on the basis of well-preserved material from the Late Miocene (Clarendonian NALMA) of Kansas. The WaKeeney Local Fauna in northwestern Kansas, where these fossils were excavated, is well to the west of the present range of this species in the eastern United States. **Ambystoma minshalli*, a common species from the Middle Miocene of Nebraska and South Dakota, was tentatively identified from the Late Miocene (Hemphillian NALMA) of Texas. Finally, the very widespread modern species *Ambystoma tigrinum* was identified from the Late Miocene (Clarendonian NALMA) of Nebraska and Kansas. This species still exists in both areas today.

Turning to the Salamandridae, we note that *Taricha* sp. has been identified on the basis of trackways from the Late Miocene (Hemphillian NALMA) of Kansas by the intrepid Peabody (1959). The last gasp of the Batrachosauroididae (the last of the enigmatic salamander families) finally took place in the Late Miocene (Hemphillian NALMA) of California. This taxon was named *#Peratosauroides problematica* by Naylor (1981b). This salamander was named on the basis of an atlas and a trunk vertebra.

Pliocene Salamanders

Other than the Pleistocene, the Pliocene is the shortest epoch of the Cenozoic, occurring from 5.3 to 1.8 Ma BP, a duration of 3.5 million years. The Pliocene is formally divided into Early and Late. Most of the Pliocene is occupied by the Blancan NALMA, but the Hemphillian NALMA overlaps the beginning of the Pliocene somewhat, and a very small part of the Irvingtonian NALMA of the Pleistocene overlaps the end of the Pliocene (see top right of Fig. 24). Compared with the rest of the world, few vertebrate sites of Pliocene age are known in North America. In fact, only one genus, *Ambystoma*, and four of its species (one extinct) are known in the Pliocene of this continent.

The extinct species is **Ambystoma hibbardi* Tihen, 1955 from the Pliocene (Blancan NALMA) of Kansas. This extinct species is very similar to modern *Ambystoma tigrinum* but is distinguished from it on the basis of a few, perhaps variable, characters. The main difference between *A. hibbardi* and the Late Miocene *Ambystoma kansense* is that the latter species is known only on the basis of neotenic individuals, whereas the former one consistently underwent metamorphosis. *Ambystoma opacum* (Gravenhorst, 1807), a living species, is known from the Pliocene (Blancan NALMA) of Texas, much to the west of the woodland habitats where this species exists at present. Along this same line, *Ambystoma maculatum* is known from the Late Pliocene (Blancan NALMA) of Nebraska, also much farther east than it occurs at present in North America. Finally, *A. tigrinum* is the most widespread salamander in the Pliocene of North America, occurring in the Pliocene (Blancan NALMA) of Arizona, Idaho, Kansas, Nebraska, New Mexico, and Texas.

Pleistocene Salamanders

The Pleistocene is much the shortest epoch of the Cenozoic (1.8–0.01 Ma BP), but it has many more records of fossil salamanders than any other Cenozoic epoch because there are so many Pleistocene localities. The Pleistocene has two NALMA units: (1) most of the Irvingtonian NALMA (see Fig. 24), which overlaps slightly with the end of the Pliocene and extends in the Pleistocene from 1.8 to 0.15 Ma BP; and (2) the Rancholabrean, which extends from 0.15 to 0.01 Ma BP and marks the end of the Pleistocene and the beginning of the Holocene.

The Pleistocene, which ended only about 140 human lifetimes ago is, from a human standpoint, by far the most important geologic interval. The Ice Age, as it is popularly known, is characterized by gigantic moving ice sheets that changed the face of the world in many areas, including North America, and saw a catastrophic worldwide extinction of large mammalian taxa. This was followed by the great rise of humans, whose livestock eventually replaced the large, extinct mammalian herbivores; whose cereal-producing grass species replaced native grassland communities; and whose agricultural practices and factories are now polluting the environment.

Irvingtonian North American Land-mammal Age. The Irvingtonian NALMA lasted much longer than the succeeding Rancholabrean NALMA but has only about one-third of the fossil records of salamanders that are known from the Rancholabrean. This is because there are far fewer Irvingtonian than Rancholabrean sites. Four families, all living today, are found in the Irvingtonian: Cryptobranchidae, Plethodontidae, Ambystomatidae, and Salamandridae. Four genera are known in the Plethodontidae (*Desmognathus, Gyrinophilus, Plethodon, Pseudotriton*); and one genus each is known in the Cryptobranchidae (*Cryptobranchus*), the Ambystomatidae (*Ambystoma*), and the Salamandridae (*Notophthalmus*). The family Cryptobranchidae has one species; the Plethodontidae, six; the Ambystomatidae, at least five; and the Salamandridae, at least one. Two of these 13 species (15.4%) are questionably considered extinct—one in the Cryptobranchidae and one in the Ambystomatidae.

The only Irvingtonian NALMA species of *Cryptobranchus* recorded, **Cryptobranchus guildayi* Holman, 1977b, is from Maryland and West Virginia. It is cautiously thought to represent an extinct species by its author, but it needs to be restudied.

Turning to the Irvingtonian NALMA Plethodontidae, we find that *Desmognathus fuscus, Desmognathus monticola,* and *Desmognathus och-*

rophaeus are known from West Virginia. *Gyrinophilus porphyriticus* is known from Maryland and Tennessee, and a very probable neotenic unnamed cave species of this genus is known from West Virginia. The genus *Plethodon* is represented in the Irvingtonian NALMA of Maryland on the basis of fossils that represent the very diverse *Plethodon glutinosus* complex of species. *Pseudotriton ruber*, another plethodontid species, is also known from the Irvingtonian of Maryland.

Among the Irvingtonian NALMA *Ambystoma*, **Ambystoma alamosensis* Rogers, 1987 is known from a high-altitude site in Colorado. It probably represents a neotenic individual. This species has a bizarre, saddle-shaped vertebra, not found in any other salamanders. It is almost certainly an aberrant form of *Ambystoma tigrinum*. *Ambystoma maculatum* is found in Kansas, Maryland, Texas, and West Virginia. *Ambystoma opacum* has been recorded in Maryland and West Virginia. *Ambystoma tigrinum* has the widest occurrence of any other North American Irvingtonian NALMA salamander being reported from Arizona, Colorado, Kansas, Maryland, Nebraska, South Dakota, Texas, and West Virginia. In Pendleton County, West Virginia, *A. tigrinum* is out of its present range in this portion of Appalachia.

The family Salamandridae is represented in the Irvingtonian NALMA by a single genus, *Notophthalmus*. *Notophthalmus viridescens* occurs in Maryland and West Virginia, and *Notophthalmus* sp. indet. occurs in southeastern Texas, within the range of two modern species of the genus.

Rancholabrean North American Land-mammal Age. The Rancholabrean is represented in North America by seven families of living salamanders: Sirenidae, Cryptobranchidae, Proteidae, Plethodontidae, Amphiumidae, Ambystomatidae, and Salamandridae. The Sirenidae is represented by both of its living genera, *Pseudobranchus* and *Siren*; the Cryptobranchidae by its single living genus, *Cryptobranchus*; the Proteidae by its single living North American genus, *Necturus*; the Plethodontidae by *Aneides, Desmognathus, Eurycea, Gyrinophilus, Hydromantes. Plethodon*, and *Pseudotriton*; the Amphiumidae by its single living genus, *Amphiuma*; the Ambystomatidae by its single North American genus, *Ambystoma*; and the Salamandridae by *Notophthalmus*. Among these families, at least 23 species are represented; only 1, **Pseudobranchus robustus*, is considered extinct (4.3% of the total).

The family Sirenidae in the Rancholabrean NALMA is represented by two species known from Florida, the living *Siren lacertina* and the extinct **Pseudobranchus robustus* Goin and Auffenberg, 1955. The family Cryptobranchidae is represented by *Cryptobranchus alleganiensis* from Alabama, Tennessee, and Virginia and by *Cryptobranchus* sp. indet. from Missouri. There is little doubt that the Missouri specimen is *C. alleganiensis*. The family Proteidae is represented by *Necturus maculosus* from Arkansas and Tennessee. In Florida, *Necturus* sp. indet. is recognized well south of its present range.

Among Rancholabrean NALMA Plethodontidae, *Aneides lugubris* is known from California; *Desmognathus fuscus*, from Virginia and West Virginia; *Desmognathus monticola*, from West Virginia; *Desmognathus ochrophaeus*, from Alabama, Virginia, and West Virginia; *Desmognathus* sp. indet., from Alabama, Georgia, Pennsylvania, Tennessee, and Virginia; *Eurycea cirrigera*, from Tennessee; *Eurycea lucifuga*, from Tennessee; *Eurycea* sp., from Alabama, Georgia, Tennessee, Pennsylvania, and Virginia; *Gyrinophilus porphyriticus*, from Tennessee, Virginia, and West Virginia; *Gyrinophilus* sp. indet., from Georgia; *Hydromantes* sp. indet., from California; the *Plethodon glutinosus* complex of species, from Florida, Georgia, Tennessee, Texas, Virginia, and West Virginia; *Plethodon* sp. indet., from Georgia, Pennsylvania, Tennessee, and Virginia; *Pseudotriton ruber*, from Georgia; and *Pseudotriton* cf. *Pseudotriton ruber*, from West Virginia.

The Rancholabrean NALMA Amphiumidae and Ambystomatidae are each represented by a single genus. *Amphiuma* represents the family Amphiumidae in the form of *Amphiuma means* from Florida and Texas. The family Ambystomatidae is represented by the genus *Ambystoma*, which is in turn represented by the following Rancholabrean NALMA species: the *Ambystoma laterale–Ambystoma jeffersonianum* complex, from Ohio and West Virginia; *Ambystoma maculatum*, from Alabama, Arkansas, Missouri, Tennessee, Virginia, and West Virginia; the *A. maculatum* group, which occurs in Georgia, Pennsylvania, Tennessee, and Virginia; *Ambystoma opacum*, from Missouri, Pennsylvania, Tennessee, and West Virginia; the *A. opacum* group, which occurs in Georgia, Tennessee, and Virginia; *Ambystoma texanum*, from Texas; and *Ambystoma tigrinum*, from Arkansas, Florida, Georgia, Kansas, Missouri, Nebraska, New Mexico, Tennessee, Texas, and Virginia. Parenthetically, *A. tigrinum* is the most widely distributed salamander in the Rancholabrean NALMA of North America.

Finally, the family Salamandridae in the Rancholabrean of North America is made up of *Notophthalmus viridescens*, from Georgia, Pennsylvania, Virginia, and West Virginia; and *N. viridescens* or *Notophthalmus perstriatus*, from northwest Florida (both species occur together in the area today).

Overview of Salamanders in the Pleistocene

Surprising to many observers, the fossil record clearly shows that amphibians (in contrast to many mammals and some birds) were remarkably resilient in the face of the rapid climatic changes, habitat loss, and other ecological and physiological stresses that affected the rest of the Pleistocene biota. This knowledge makes the present worldwide decline of frogs and salamanders even more difficult to accept. According to a recent nonspeculative and comprehensive statistical study, nearly 32% of the amphibian species in the world are threatened (IUCN et al., 2004). This represents an astounding total of 1856 species of amphibians.

The following pages summarize the current knowledge about the apparent evolutionary stasis ("holding pattern") of salamander species during the North American Pleistocene, their range adjustments, and their assumed pattern of recolonization of formerly glaciated areas. As mentioned in chapter 1, the Pleistocene was characterized by the periodic landscape-covering and -uncovering movements of massive ice sheets (Fig. 77) and their many associated changes. These changes profoundly affected the organization of biotic communities, especially in the northern hemisphere. Undoubtedly, these events must have put severe stress on

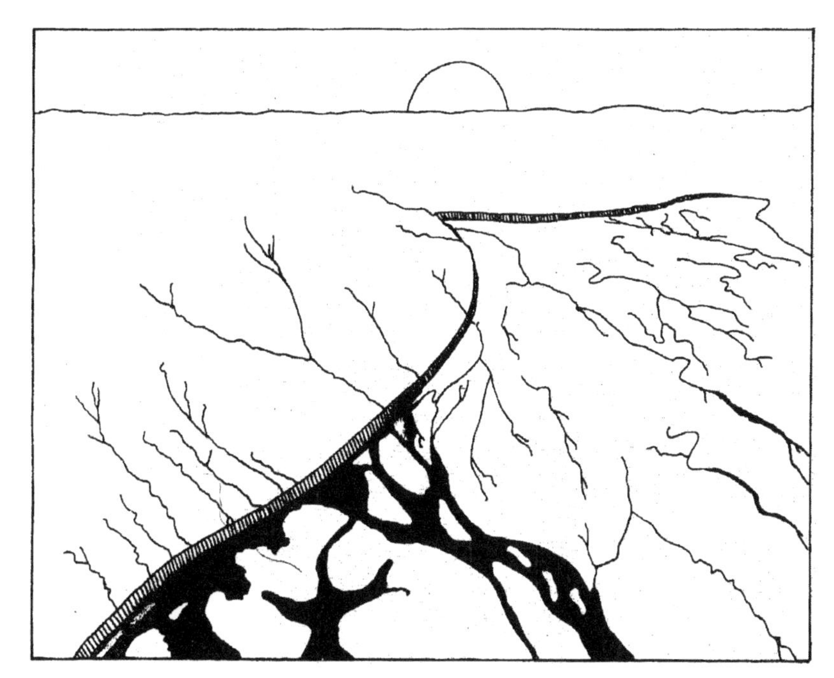

FIGURE 77. Schematic drawing of the final withdrawal of the ice sheet in Michigan at the end of the Pleistocene. The leading edge of the ice is seen as a sheer cliff. Meltwater floods the newly exposed land in various patterns.

salamanders, as well as other vertebrate groups. Complete obliteration of a community by ice up to 2 miles (about 3.2 km) thick is an event that is difficult to comprehend. In many ways it is similar to what would result if seawater a mile or two deep slowly advanced over the land in North America, stayed for hundreds or even many thousands of years, and then retreated again. In North America, during the late Wisconsinan glacial age, about 20 ka BP, the Laurentide Ice Sheet extended southward to similar latitudes in southern Illinois, Indiana, and Ohio (Fig. 78). There is no question that amphibian and reptile life was wiped out in most of Canada, all of Michigan, much of Wisconsin and Illinois, and the northern two-thirds of Indiana and Ohio. Succeeding ice advances occurred at about 1000-year intervals, with a final advance extending to northern Indiana and Ohio at about 15 ka BP. About 14 ka BP (Fig. 79), the ice began its final retreat in the northeastern United States, exposing the land for a re-invasion of salamanders, as well as other plant and animal life.

Each time that salamanders re-invaded former territory occupied by masses of ice, they entered topographically altered areas, as every time the ice sheets advanced or retreated, they etched new landforms into the landscape. The thickness of the ice varied from place to place, but it has been estimated that the average thickness (depth) was about 2 km and that the maximum thickness of some lobes may have been 3 km or more (Flint, 1971). Valleys and ridges, various types of small hills, moraines the size of small mountains, and lakes, streams, swamps, and bogs formed; were re-covered by ice; then formed again in different patterns, depending on the dynamics of the flowing ice.

Of the many changes associated with glacial cycles, climatic change probably has received the most scientific attention. We found out a few decades ago (see Graham, 1976; Lundelius et al., 1983; Graham and

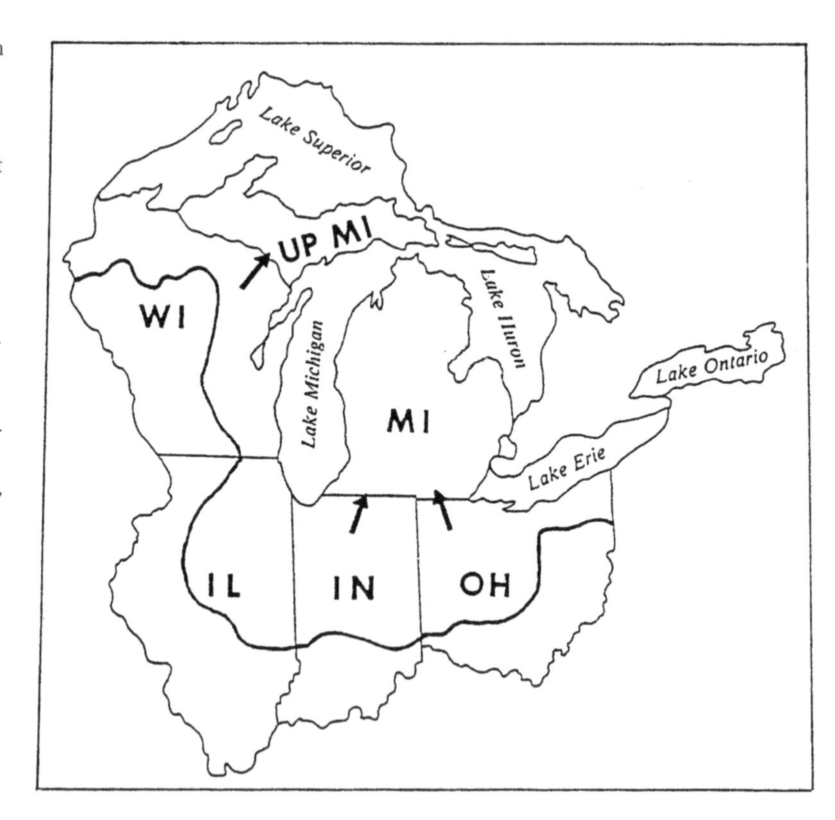

FIGURE 78. Maximum extent (dark line) of the Laurentide Ice Sheet in the Great Lakes region, where it reached southern Illinois, Indiana, and Ohio in the Wisconsin glacial age. The arrows indicate routes of re-entry of amphibians and reptiles into Michigan after the final glacial retreat. Abbreviations: IL, Illinois; IN, Indiana; MI, Michigan; OH, Ohio; UP, Upper Peninsula of Michigan; WI, Wisconsin.

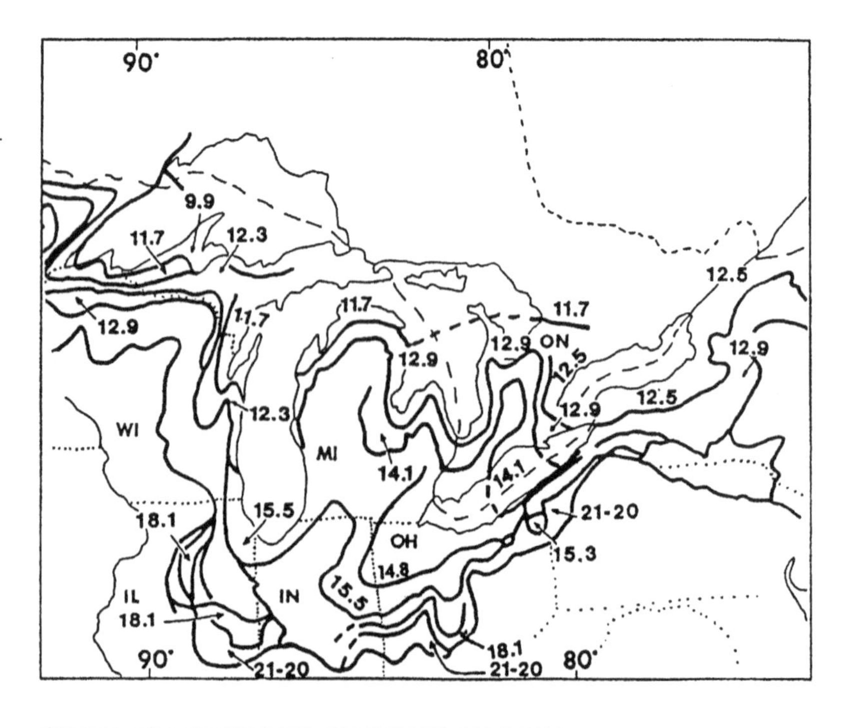

FIGURE 79. Pleistocene ice margins at various dates, in thousands of years before present (BP), in the Great Lakes region. Abbreviations: IL, Illinois, IN, Indiana; MI, Michigan; ON, Ontario; WI, Wisconsin.

Lundelius, 1984) that the classic idea of alternating cold glacial climates and warm interglacial climates was an oversimplification. Modern evidence shows that climates in proglacial areas, such as in southern Wisconsin and Michigan, were extremely cold but that in the central and southern United States, the climate was more equable than it is today, with cooler summers and warmer winters. This theory, often termed the Pleistocene climatic equability model, has been important in the consideration of mixtures of northern and southern herpetological species in Pleistocene faunas (Holman, 1995a).

Vegetational changes brought on by glacial and interglacial glacial ages undoubtedly affected biotic communities south of the glaciated areas. But the classical idea that major vegetation associations were pushed southward en masse (Fig. 80) by the advancing ice has been replaced by new ideas. Although tundra- or taiga-like vegetation occurred in proglacial areas near the ice, it is now generally believed that many vegetational communities in the central and southern United States existed as a mosaic of plants. Some of these plants consisted of species that "stayed put," and others represented northern invading species.

This notion has developed from the realization that both plants and animals reacted individually to the changes in the Pleistocene and that mosaic communities of plants and animals were able to exist in the equable climates south of the proglacial areas (Graham and Lundelius,

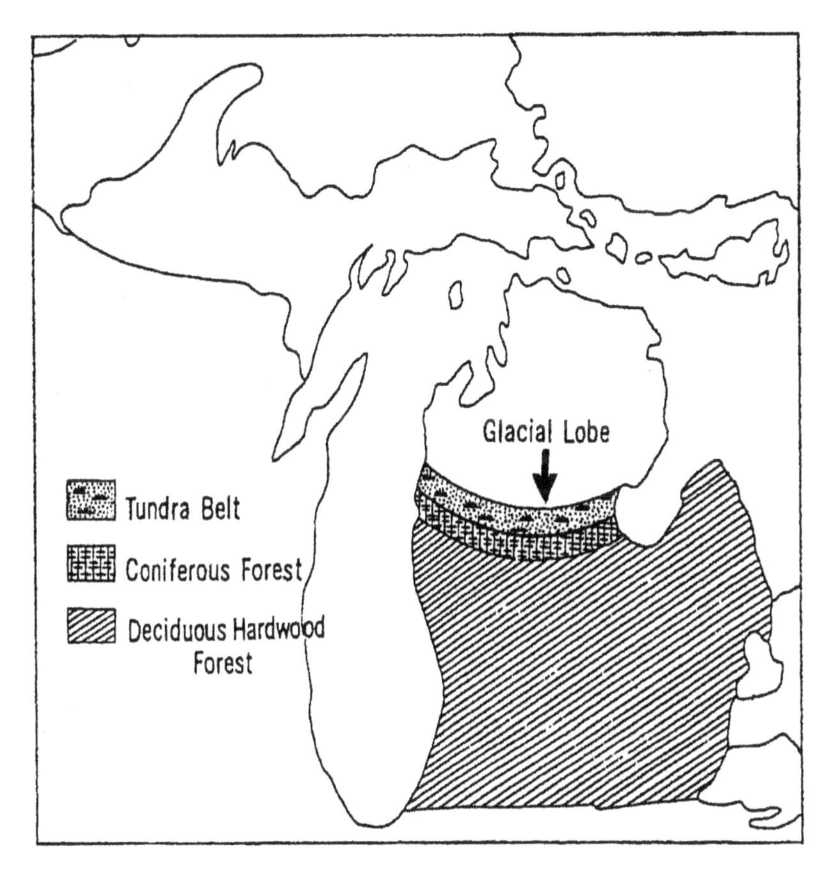

FIGURE 80. "Stripe hypothesis": biological communities in an orderly advance ahead of the ice sheet during the Pleistocene of Michigan.

1984). The classic concept has been referred to as the stripe hypothesis (Fig. 80) and the modern one as the plaid hypothesis (Holman, 1995a).

Salamander populations were not only disrupted by the obliteration of habitats by the ice sheet but also confronted with the problem of alternate flooding and re-emergence of the land caused by rising and falling lake, river, and sea levels. Much of the global water supply was bound up in the ice sheets during much of the Pleistocene. During full glacial periods, lake, river and sea levels dropped and seaways shrank. But during interglacial periods the opposite occurred, enhanced by the fact that isostatic rebound occurred in areas that had been overloaded with masses of ice. Florida, for instance, was so flooded by the sea during interglacial times that it was cut off from the mainland, with only its highest central part, the "Ocala Island," left above sea level (Cooke, 1945). In southeastern Canada the St. Lawrence River, which is a freshwater system today, was part of an inland marine system called the Champlain Sea (Fig. 81; also see Harington, 1988).

Of the temporal cycles within the Pleistocene, the glacial periods appear to have lasted much longer than the interglacial ones, at least in North America and Europe. This must have had a great impact on the salamander populations of the Pleistocene. In North America the last glacial stage of the Pleistocene Rancholabrean NALMA, the Wisconsinan, is considered to have lasted about 110 000 years, whereas the last interglacial stage, the Sangamonian, is thought to have lasted only about 10 000 years.

Unfortunately, very little information is available on the amount of time it took for stable communities to develop after the withdrawal of ice sheets in North America. Retreating ice sheets in the Pleistocene left

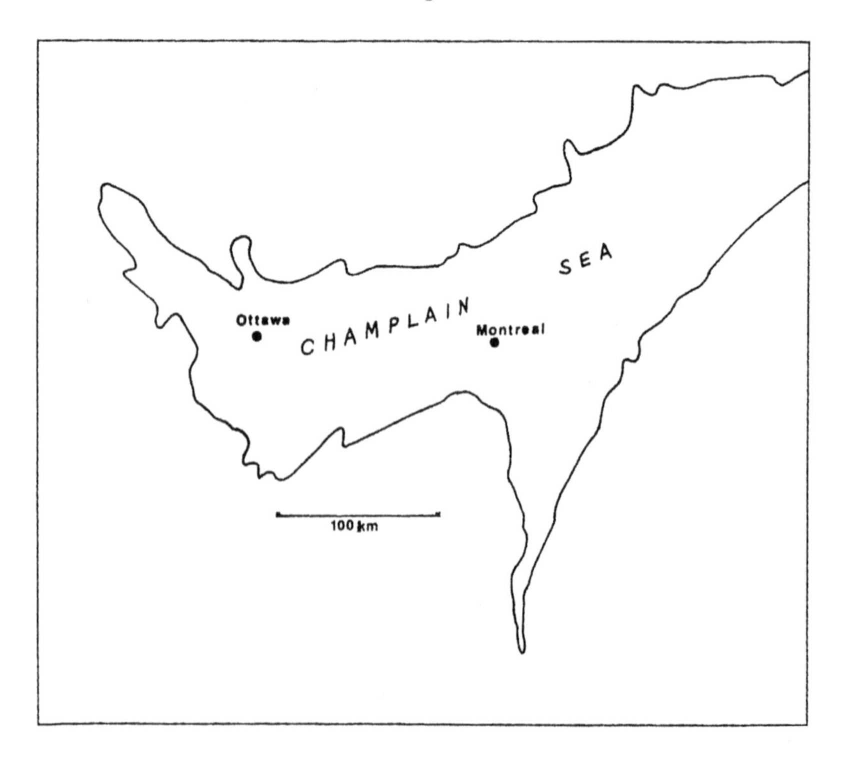

FIGURE 81. General outline of the Champlain Sea near its maximum in the late Wisconsinan glacial age.

mainly a sterile mass of mud, silt, sand, and gravel behind. Obviously, there must have been a series of floras and faunas that succeeded one another before stable communities were established. Some ambystomatid salamanders, for instance, would need rodent holes for shelter; many small plethodontids would need leaf litter and rotting logs for habitation. Aquatic salamanders would be unable to exist in the sterile ponds left by recently melted ice.

Apparent Evolutionary Stasis in North American Pleistocene Amphibians. Despite all the Pleistocene stresses that had to be coped with, North American salamanders appeared to exist in a state of evolutionary stasis during the epoch. Eight families, 46 genera, and about 191 species of mammals became extinct during the Pleistocene, and most of the large mammals that became extinct had dominated the ecosystem in which they lived. This was a veritable mass extinction, and it occurred during human times! Of the birds in North America, in the Late Pleistocene alone, 2 families, 19 genera, and a currently undetermined number of species became extinct. On the other hand, in North America, no families, genera, or unquestioned extinct species of salamanders are known from the Pleistocene; the three questionable species are **Cryptobranchus guildayi*, **Pseudobranchus robustus*, and **Ambystoma alamosensis*. The first two need more study; the third (*A. alamosensis*) is almost certainly an aberrant form of *Ambystoma tigrinum*. Virtually the same situation occurs in Britain and continental Europe (see Stuart, 1991; Holman, 1998; Sanchiz, 1998), where many higher and lower taxa of mammals became extinct during the Pleistocene.

Before we speculate too much about the causes of the apparent evolutionary stasis of amphibians through the Pleistocene, some possible modifications of the concept should be considered. Because of the loss of information that occurs during the death and preservation of most vertebrate fossils, especially amphibians because of their relatively fragile bones, we tend to find bits and pieces of salamander fossils, rather than complete skeletons. In fact, I am not aware of a single complete salamander fossil in North America. Thus, specific identifications of amphibians are made on the basis of individual elements, such as ilia in frogs, and as we have seen in this book, vertebrae in salamanders.

Obviously, one of the most serious problems is that extinct species, and possibly even genera, may have gone unrecognized. Cryptic salamander species are those that appear to be osteologically identical but have been separated on the basis of molecular or behavioral data or a combination of both. These taxa are impossible to identify to species on the basis of skeletal remains (e.g., species of the *Plethodon glutinosus* complex of salamanders). Obviously, many cryptic salamander species may have existed in the Pleistocene.

A daunting problem with salamanders (and anurans as well) is the fact that when they do turn up in fossil digs as relatively complete skeletons, they may be so embedded in matrix that they present a two-dimensional view only, often with the articular surfaces and muscle scars obscure and the individual skull bones indistinguishable. Often a complete salamander atlas or trunk vertebra, unbroken and free of matrix, is much more diagnostic.

An attitude that is difficult to resolve is the ingrained idea among many paleoherpetologists that all Pleistocene herpetological material (other than large tortoises or island taxa) must belong to modern species. This bit of reasoning probably emerged from the negative reaction of the "new" generation of paleoherpetologists of the 1950s to the overzealous naming of Pleistocene amphibians and reptile species by their predecessors. Too many of us have picked up a portion of a turtle shell in a museum tray and were able to fit it together perfectly with another portion of turtle shell with a different scientific name. Such occurrences have helped to foster an attitude that anyone who attempts to name an extinct Pleistocene amphibian species should be the recipient of intense scrutiny by other paleoherpetologists. Oddly, by contrast, one who identifies a modern amphibian species from the Pleistocene on the basis of fragmentary material is seldom questioned.

Probably the best way to resolve the above issue is a series of intense studies of individual variation in the skeletons of modern and fossil salamanders and frogs. But limitations do exit. One seldom gets an adequate sample of fossil amphibian species—an obvious exception being the hundreds of specimens of *Ambystoma tigrinum* that J. A. Tihen was able to study in Kansas. Also, because of the need to conserve living species, getting a comparative modern skeletal collection that is adequate for variation studies may become a legal, or at least an ethical, problem. Unfortunately, such skeletal collections are woefully lacking in most museum and other collections in the world.

Rancholabrean NALMA sites are much more numerous in North America than the earlier Irvingtonian NALMA ones (Holman, 1995a). Thus, it is possible that extinctions of Irvingtonian salamanders might have been undetected. All this having been said, the Pleistocene herpetofaunas, documented at many sites in continental North America, the British Isles, and western continental Europe, have been strikingly more stable than the mammalian and avian faunas (e.g., Holman, 1993, 1995a, 1998; Sanchiz, 1998).

The why of the apparent stasis of Pleistocene amphibians may be gleaned from additional studies of modern as well as fossil amphibians. It appears obvious that ectothermic (cold-blooded) animals, such as amphibians, that have very low metabolic rates and are able to hibernate (in winter) or estivate (in summer) during inclement conditions would have many advantages over endothermic (warm-blooded) birds and mammals during intervals of climatic fluctuations. The ability of many amphibians to freeze solid in the winter, sometimes several times, without dying (e.g., Pinder et al., 1992; Layne and Lee, 1995; Pough et al., 1998) certainly must have been adaptive during the glacial periods of the Pleistocene. In some species of amphibians glycogen from the liver is converted to glucose, which acts as an antifreeze. In others, such as the Old World salamander *Salamandrella keyserlingii*, converted glycerol is used. I would think that continued experimental studies of estivation and hibernation rhythms in salamanders could be important in establishing guidelines for estimating their Pleistocene adjustments.

Aquatic salamanders, such as *Necturus maculosus* and *Notophthalmus viridescens*, are able to remain moderately active under the ice and continue feeding in the winter. I have accidentally caught N. *maculosus* on live bait when ice fishing in Michigan in February and early March, as have other Michigan ice anglers I know. Although N. *viridescens* becomes rather sluggish in Massachusetts during the coldest winter months, about 20–30% of these salamanders collected from December to February had food in their stomachs (Morgan and Grierson, 1932). I studied the food habits of 23 south-central Indiana *Plethodon dorsalis* that were active under leaf litter from October through February of 1952–1953 (Holman, 1955). Their most active feeding period was on February 20, when their stomachs contained spiders, larval and adult beetles, red mites, other unidentifiable arthropod remains, and slugs.

The small size of many salamanders would appear to be advantageous to them in the face of Pleistocene habitat shrinkage, especially compared with larger endotherms, especially the megaherbivores (giant mammals such as mastodonts and mammoths) that need very large tracts of habitat (see Owen-Smith, 1987). In the early part of the 1900s, when salamanders were much more abundant than they are today, relatively large numbers of individuals of some species occurred in very small areas, such as under patches of litter on Lake Michigan beaches (e.g., Ruthven et al., 1928).

A distinct probability exists that few frogs and salamanders were directly dependent, either as predators, scavengers, or commensals, on the large endothermic species that became extinct at the end of the Pleistocene. At the well-studied Pleistocene site at Rancho La Brea in Los Angeles, California, the scavenging birds and the dung beetles that were dependent on the megaherbivores became extinct along with these giants (Harris and Jefferson, 1985). The amphibian and reptile species at Rancho La Brea, however, all survived to the present.

In contrast, in Australia a giant monitor lizard, *Megalania*, and a giant snake, *Wonambi* (Murray, 1984), were top predators. *Megalania* (at first thought to be a dinosaur) specialized on the megaherbivores of the region during the Pleistocene. Neither of these predators survived the Pleistocene, probably because so many large marsupial herbivores died out. Oddly, few suggestions have been made in Pleistocene studies about possible amphibian ecological relationships to the large extinct mammals. It is possible that these amphibians existed well outside the interactions that took place between the large endotherms and their predators and scavengers. Perhaps studies on how small amphibians tie into the food web in African or Southeast Asian areas that support large herbivorous mammals today would be interesting in light of the differential geographic patterns of extinctions during the Pleistocene. Of course, salamanders occur only peripherally on the African continent.

The act of parenting in the endothermic birds and mammals is obviously a part of the reproductive stress syndrome and would appear to be a considerable drain on their energy budgets during times of climatic change. Obviously, parenting, at least in the sense of birds and mammals, is lacking in amphibians. The higher reproductive potential of many small salamanders and frogs, especially compared with that of megaherbivores such as elephants and rhinos, would appear to favor the amphibian species in times of climatic change.

The hypothesis of human overkill as a dominating factor in the extinction of the large herbivorous mammals and their scavengers and predators has been put forward now for about four decades (e.g., Martin and Wright, 1967; Martin and Klein, 1984). Holman (1959a) compared the extinction percentages of amphibians and reptiles with those of mammals in a large Pleistocene fauna in Florida. He suggested the possibility that the rise of humans might have been an important factor in the much lower extinction percentages in the herpetological component of the fauna than in the mammalian component. This might reflect the unpalatability of most amphibians and reptiles (especially salamanders, with their noxious skin secretions), compared with large mammalian herbivores, as a food source for humans. On the other hand, the huge Florida tortoises (*Hesperotestudo*) that became extinct at the end of the Pleistocene must have been an easily obtained source of food for humans. Attesting to this, the Little Salt Spring Late Pleistocene site in Florida yielded the remains of a giant tortoise, apparently killed and eaten by humans (Clausen et al., 1979; Holman, 1981; Holman and Clausen, 1984).

In summary, further study of salamander physiology, cryobiology (effects of cold temperatures on organisms), and ecology may bear directly on the question of why there was a general stasis in these amphibians during the Pleistocene in the face of climatic oscillations and great environmental changes. Moreover, further study of salamanders (and frogs) in the food web of the Pleistocene, as well as in modern times, especially in megaherbivore-dominated communities, should provide added insights. Finally, studies of the late Wisconsinan glacial age in North America may eventually provide information about the importance of salamanders as a Pleistocene food resource for humans. In Michigan, at least, I am not aware of any salamander remains associated with Paleo-Indian sites (Holman, 2006).

Range Adjustments of Salamanders in the Pleistocene. Both amphibians and reptiles made far fewer range adjustments in the Pleistocene than did mammals (Holman, 1995a). In fact, it now seems that in many places in the world the tendency for both salamanders and frogs was to stay put during the Pleistocene, especially south of the areas near the ice margins (e.g., Fay, 1988; Holman, 1995a, 1998). Whether this was due to wide physiological and ecological tolerances already present in Pleistocene amphibians or to rapid evolution of new physiological and ecological tolerances in the face of climatic selective pressures is an arguable point. Actually, a lot of evolutionary selection could theoretically have taken place during the almost 2 million years of severe climatic fluctuations in the Pleistocene, especially in amphibians and reptiles living in areas marginal to the ice sheets.

From a study of more than 150 North American Pleistocene amphibian faunas (Holman, 1995a), it appears that most amphibian species in the unglaciated areas tended to remain in place. The situation is especially documented in the Late Pleistocene of the Appalachian region, where several large herpetofaunas were studied by Fay (1988). Here, oddly enough, the few invading species were shown to be from the south, rather than from the north. However, Fay pointed out that it would be difficult to detect any northern amphibian invaders, as several amphibian species in Appalachia today also occur much farther north as well.

Now, taking a look at salamander population movements in other regions, we find that in peninsular Florida *Ambystoma* sp. and *Necturus* sp. both occurred south of their present ranges in Florida (see specific accounts in chapter 2). The Ozark region in Missouri and Arkansas is today an isolated uplifted area with numerous caves and interesting salamander taxa. Most of the Pleistocene amphibian faunas collected from this region come from fissures and caves, all but one of which date from Late Pleistocene times. Here the amphibian situation is even more striking than in Appalachia, as all the fossil species are extant and currently living in the vicinity of their fossil sites (Holman, 1995a, 2003).

Turning to the High Plains area of the United States, we find that some interesting amphibian range adjustments occurred in Pleistocene extralimital (falling outside its present range by at least one degree of latitude) eastern and northern species. This appears to reflect the fact that Pleistocene herpetofaunas of earlier glacial ages (Kansan and Illinoian) were prevalent in the High Plains, whereas southern and eastern North America mainly had Late Pleistocene (Sangamonian interglacial age and late Wisconsinan glacial age) sites. The Albert Ahrens Local Fauna (Irvingtonian NALMA) in southeastern Nebraska (Ford, 1992) is considered to represent a full glacial period during Kansan times. Here the northern and northeastern extralimital *Ambystoma laterale–Ambystoma jeffersonianum* complex and *Rana sylvatica* (Wood Frog) occur with the eastern extralimital taxa *Bufo americanus* (American Toad), *Hyla versicolor* or *Hyla chrysoscelis* (Gray Treefrogs), and *Rana clamitans* (Green Frogs) and with the southern extralimital species *Pseudacris* cf. *Pseudacris clarkii* (Spotted Chorus Frog). The nearest area of sympatry (area where species co-exist) of most of the Albert Ahrens herpetological species (this includes reptiles) lies in extreme northwestern Missouri. Ford (1992) suggested that this herpetological assemblage lived in a more equable climate than occurs in the area today.

In north-central Texas, the Easley Ranch Local Fauna (Lynch, 1964), a Rancholabrean site in north-central Texas, yielded an extralimital southern frog *Eleutherodactylus marnockii* (Cliff Chirping Frog). Two other Late Pleistocene sites in north-central Texas, the Clear Creek and Howard Ranch (sometimes called Groesbeck Creek) sites yielded *Ambystoma texanum*, which at present live somewhat east of that area (Holman, 1963, 1964).

Patterns of Salamander Re-invasion in Previously Glaciated Areas. Pleistocene records in Britain and on the European continent show that amphibians were able to invade previously glaciated areas several times during the epoch (Holman, 1993, 1998; Sanchiz, 1998). Holman (1992) developed a simple model for the postglacial recolonization of Michigan, a state that was completely covered by the Laurentide Ice Sheet at least once in the Pleistocene (see Fig. 78). This model was based mainly on (1) the postglacial vegetational development documented by paleobotanical and palynological (fossil pollen) data; (2) the ecological tolerances of living amphibians and reptiles; and (3) the Michigan amphibian and reptile fossil record.

Routes of entry for glacially displaced amphibians and reptiles in Michigan have been suggested (see Fig. 78). The postglacial amphibian fauna of the Upper Peninsula was probably largely derived from a Wisconsin corridor, likely because of the limitations caused by the width and depth of the Straits of Mackinac. This invasion must have occurred some time after 9.9 ka BP, since at that time the Upper Peninsula had glacial ice along its northern edge (see Fig. 79). The Lower Peninsula probably derived its postglacial amphibian fauna through Indiana and Ohio.

Considering the paleobotanical record and the lag time necessary for stable plant communities to develop, I suggested (Holman, 1992, 2003) that the first frog species (perhaps the tundra-tolerant, freeze-tolerant species *Rana sylvatica*, Wood Frog) might have reached southern Michigan by about 13 ka BP. I here suggest that *Ambystoma laterale* might have been the first salamander to reach southern Michigan by about the same date (see Conant and Collins, 1998, map, p. 438).

The Holman model (1992, 2003) proposed three categories of invading herpetological species: (1) *primary invaders* whose ecological tolerances included coniferous forest and occasionally tundra; (2) *secondary invaders* adapted to areas of mixed coniferous–broadleaf forests; and (3) *tertiary invaders* adapted to broadleaf forests. The tertiary invaders are all restricted at present to the southern part of the Lower Peninsula of Michigan, where many of them exist as peripheral or isolated populations and are listed as threatened or endangered taxa.

As mentioned above, *Rana sylvatica*, which is able to exist in tundra conditions today and freezes during winter hibernation with no damaging effects, may well have been the first primary amphibian invader of postglacial Michigan as the great mass of sterile, glacially derived sand and silt gave way to tundra between about 14 and 13 ka BP. Other primary forms probably invaded southern Michigan about 12.5 ka BP as the first spruce forests developed and then later (about 11.8 ka BP) when jack pine and red pine communities began to develop in southern Michigan. These primary species probably did not reach the Upper Peninsula of Michigan until at least 9.9 ka BP, as glacial ice rimmed the northern border of the area at that time.

Salamanders considered primary species are *Ambystoma laterale*, *Ambystoma maculatum*, *Ambystoma tigrinum*, *Hemidactylium scutatum*, *Necturus maculosus*, *Notophthalmus viridescens*, and *Plethodon cinereus*.

Secondary amphibian invaders probably began to enter southern Michigan about 10.6 ka BP ago as mixed coniferous–broadleaf forests began to develop in the southern part of the state. No salamanders are considered secondary invaders.

The entrance of tertiary invaders probably began with the establishment of mixed hardwood forests in southern Michigan about 9.9 ka BP. Tertiary salamanders species include (1) *Siren intermedia*, currently confined to the extreme southwest portion of the Lower Peninsula; (2) *Ambystoma opacum*, also confined to this part of Michigan; and (3) *Ambystoma texanum*, currently confined to southeastern lower Michigan. All three of these species are endangered, especially *S. intermedia* and *A. opacum* which have not been seen for years.

Epilogue

The classification and phylogeny of salamanders and frogs are in an embryonic state because of an abysmal fossil record for these amphibians. In fact, we know much more about the classification and phylogeny of the dinosaurs in the Jurassic and Cretaceous than we know about the salamanders of the Pleistocene and Recent. Carroll (2000b, p. 1462) summed this up when he wrote, "Because of major gaps in knowledge of the fossil record, understanding of the interrelationships of the principal groups of amphibians remains incomplete. Any cladogram would give an erroneous impression of the precision of the comprehension of their phylogenetic history. All that is practical at the present time is an informal classification."

If anything, a perusal of this book will bear this out. (1) We include two families of salamander-like amphibians (Batrachosauroididae and Scapherpetontidae) with the salamanders, but we really do not know whether they relate to the living salamander groups. (2) We make cladograms depicting the relationships within the major groups of salamanders; but these trees are often mechanical and not really subject to scrutiny because of the overwhelming lack of rigorous studies dealing with individual variation. (3) Finally, we must deal with scores of modern species that are osteologically indistinguishable from one another (e.g., the *Plethodon glutinosus* complex of salamanders).

Perhaps the third problem will never be cleared up unless can we get fossil DNA from salamander bones. On the other hand, problems 1 and 2 can probably be significantly addressed if we send more people into the field to look for salamander fossils and build up comparative skeletal collections of presently living species.

REFERENCES

Adams, L., and Martin, H. 1929. A new urodele from the Pliocene of Kansas. American Journal of Science 217:504–520.

Albright, L. B. 1994. Lower vertebrates from an Arikareean (earliest Miocene) fauna near the Toledo Bend Dam, Newton County, Texas. Journal of Paleontology 68:1131–1145.

Arnold, E. N., and Ovenden, D. W. 2002. Reptiles and amphibians of Europe. Princeton: Princeton University Press.

Arnold, S. J. 1982. A quantitative approach to antipredator performance. Salamander defense against snake attack. Copeia 1982:247–253.

Auffenberg, W. 1959. The epaxial musculature of *Siren*, *Amphiuma*, and *Necturus*. Bulletin of the Florida State Museum 4:253–265.

Auffenberg, W. 1961. A new genus of fossil salamander from North America. American Midland Naturalist 66:456–465.

Bell, C. J., and Mead, J. I. 2000. Biochronology of North American microtine rodents. *In* J. S. Noller, J. M. Sowers, and W. R. Lettis, eds., Quaternary geochronology: methods and applications. Washington, DC: American Geophysical Union, AGU Reference Shelf 4, pp. 379–405.

Bolt, J. R. 1991. Lissamphibian origins. *In* H-P. Schultze and L. Trueb, eds., Origins of the higher groups of tetrapods: controversy and consensus. Ithaca: Cornell University Press, pp. 194–222.

Brame, A. H., Jr., and Murray, K. F. 1968. Three new slender salamanders (*Batrachoseps*) with a discussion of relationships and speciation within the genus. Bulletin of the Los Angeles County Museum of Natural History (Science) 4:1–35.

Brattstrom, B. H. 1953. Records of Pleistocene reptiles and amphibians from Florida. Florida Academy of Sciences, Quarterly Journal 16:243–248.

Brattstrom, B. H. 1955. Pliocene and Pleistocene amphibians and reptiles from southeastern Arizona. Journal of Paleontology 29:150–154.

Bryant, J. D. 1991. New early Barstovian (Middle Miocene) vertebrates from the upper Torreya Formation, eastern Florida Panhandle. Journal of Vertebrate Paleontology 11:472–489.

Burton, T. M. 1976. An analysis of the feeding ecology of the salamanders (Amphibians, Urodela) of the Hubbard Brook Experimental Forest, New Hampshire. Journal of Herpetology 10:187–204.

Bury, R. B. 1972. Small mammals and other prey in the diet of the Pacific giant salamander (*Dicamptodon ensatus*). American Midland Naturalist 87:524–526.

Cannatella, D. C., and Hillis, D. M., eds. 1993. Amphibian relationships: phylogenetic analysis of morphology and molecules. Herpetological Monographs 7:1–7.

Carpenter, K. 1979. Vertebrate fauna of the Laramie Formation (Maastrichtian), Weld County, Colorado. Contributions to Geology, University of Wyoming 17:37–49.

Carroll, R. L. 2000a. The fossil record and large-scale patterns of amphibian evolution. *In* H. Heatwole and R. L. Carroll, eds., Amphibian biology. Vol. 4: Palaeontology, the evolutionary history of amphibians. Chipping Norton, Australia: Surrey Beatty & Sons, pp. 973–978.

Carroll, R. L. 2000b. Introduction to Appendix 2. *In* H. Heatwole and R. L. Carroll, eds., Amphibian biology. Vol. 4: Palaeontology, the evolutionary history of amphibians. Chipping Norton, Australia: Surrey Beatty & Sons, p. 1462.

Chantell, C. J. 1971. Fossil amphibians from the Egelhoff Local Fauna of north-central Nebraska. Contributions from the Museum of Paleontology, University of Michigan 23:239–246.

Clark, J. M. 1985. Fossil plethodontid salamanders from the latest Miocene of California. Journal of Herpetology 19:41–47.

Clausen, C. J., Cohen, A. D., Emiliani, C., Holman, J. A., and Stipp, J. J. 1979. Little Salt Spring, Florida: a unique underwater site. Science 203:609–614.

Collins, J. T., and Taggart, T. W. 2002. Standard common and current scientific names for North American amphibians, turtles, reptiles, and crocodilians. 5th ed. Lawrence, Kans.: Center for North American Herpetology.

Conant, R., and Collins, J. T. 1998. Field guide to reptiles and amphibians, eastern and central North America. 3rd ed. Boston: Houghton Mifflin.

Cooke, C. M. 1945. Geology of Florida. Bulletin, Florida Geological Survey, No. 29, 339 pp.

Cope, E. D. 1889. Batrachia of North America. United States National Museum Bulletin 34:1–525.

Crother, B. I., ed. 2000. Scientific and standard English names of amphibians and reptiles of North America north of Mexico with comments regarding confidence in our understanding. Society for the Study of Amphibians and Reptiles, Herpetological Circular No. 29, 82 pp.

Crother, B. I., Boundy, J., Campbell, J. A., de Queiroz, K., Frost, D., Green, D. M., Highton, R., Iverson, J. B., McDiarmid, R. W., Meylan, P. A., Reeder, T. W., Seidel, M. E., Sites, J. W., Jr., Tilley, S. G., and Wake, D. B. 2003. Scientific and standard English names of amphibians and reptiles of North America north of Mexico: update. Herpetological Review 34:196–203.

Curtis, K., and Padian, K. 1999. An Early Jurassic microvertebrate fauna from the Kayenta Formation of northeastern Arizona: microfaunal change across the Triassic–Jurassic boundary. PaleoBios 19:19–37.

Davis, L. C. 1973. The herpetofauna of Peccary Cave, Arkansas. M.S. thesis, University of Arkansas, Fayetteville.

DeFauw, S. L. 1989. Temnospondyl amphibians: a new perspective on the last phase of evolution of the Labyrinthodontia. Michigan Academician 21:7–32.

Denton, R. K., Jr., and O'Neill, R. O. 1998. *Parrisia neocesariensis*, a new batrachosauroidid salamander and other amphibians from the Campanian of eastern North America. Journal of Vertebrate Paleontology 18:484–494.

Duellman, W. E., and Trueb, L. 1986. Biology of the amphibians. New York: McGraw-Hill.

Edwards, J. L. 1976. Spinal nerves and their bearing on salamander phylogeny. Journal of Morphology 148:305–328.

Eshelman, R. E. 1975. Geology and paleontology of the early Pleistocene (late Blancan) White Rock Fauna from north-central Kansas. Papers on Paleontology, No. 13, 60 pp.

Esteban, M., Castanet, J., and Sanchiz, B. 1995. Size inference based on skeletal fragments of the common European frog (*Rana temporaria* L.). Herpetological Journal 5:229–235.

Estes, R. 1963. Early Miocene salamanders and lizards from Florida. Florida Academy of Sciences, Quarterly Journal 26:234–256.

Estes, R. 1964. Fossil vertebrates from the Late Cretaceous Lance Formation, eastern Wyoming. University of California Publications in Geological Sciences, No. 49, 180 pp.

Estes, R. 1965a. Fossil salamanders and salamander origins. American Zoologist 5:319–334.

Estes, R. 1965b. A new fossil salamander from Montana and Wyoming. Copeia 1965:90–95.

Estes, R. 1969a. Prosirenidae, a new family of salamanders. Nature (London) 224: 87–88.

Estes, R. 1969b. Batrachosauroididae and Scapherpetontidae, late Cretaceous salamanders. Copeia 1969:225–234.

Estes, R. 1969c. The fossil record of amphiumid salamanders. Breviora, No. 322, 11 pp.

Estes, R. 1975. Lower vertebrates from the Fort Union Formation, Late Paleocene, Big Horn Basin, Wyoming. Herpetologica 31:365–385.

Estes, R. 1976. Middle Paleocene lower vertebrates from the Tongue River Formation, southeastern Montana. Journal of Paleontology 50:500–520.

Estes, R. 1981. Gymnophiona, Caudata. Handbuch der Paläoherpetologie, Part 2. Stuttgart: Gustav Fischer Verlag, 115 pp.

Estes, R. 1988. Lower vertebrates from the Golden Valley Formation, Early Eocene of North Dakota. Acta Zoologica Cracoviensia 31:541–562.

Estes, R., and Tihen, J. A. 1964. Lower vertebrates from the Valentine Formation of Nebraska. American Midland Naturalist 72:453–472.

Estes, R., Berberian, P., and Meszoely, C. A. M. 1969. Lower vertebrates from the Cretaceous Hell Creek Formation, McCone County, Montana. Breviora, No. 337, 33 pp.

Evans, S. E., and Milner, A. R. 1994. Middle Jurassic microvertebrate assemblages from the British Isles. *In* N. C. Frazer and H-D. Sues, eds., In the shadow of the dinosaurs: Early Miocene tetrapods. Cambridge: Cambridge University Press, pp. 303–321.

Evans, S. E., and Milner, A. R. 1995. Early Cretaceous salamanders (Amphibia: Caudata) from Las Hoyas, Spain. *In* Second International Symposium on Lithographic Limestones, Lleida, Cuencia (Spain), July 1995. Ediciones de la Universidad Autónoma de Madrid, Madrid, Extended Abstracts, pp. 63–65.

Evans, S. E., and Waldman, M. 1996. Small reptiles and amphibians from the Middle Jurassic of Skye, Scotland. *In* M. Morales, ed., The continental Jurassic. Museum of Northern Arizona Bulletin 60, pp. 219–226.

Evans, S. E., Milner, A. R., and Mussett, F. 1988. The earliest known salamander (Amphibia: Caudata): a record from the Middle Jurassic of England. Geobios 21:539–552.

Evans, S. E., Milner, A. R., and Werner, C. 1996. Sirenid salamanders and a gymnophionan amphibian from the Cretaceous of Sudan. Palaeontology 39: 77–95.

Evans, S. E., Lally, C., Chure, D. C., Elder, A., and Maisano, J. A. 2005. A Late Jurassic salamander (Amphibia: Caudata) from the Morrison Formation of North America. Zoological Journal of the Linnean Society 143:599–616.

Fay, L. P. 1984. Mid-Wisconsinan and mid-Holocene herpetofaunas of Eastern North America: a study in minimal contrast. Special Publication of Carnegie Museum of Natural History, No. 8, pp. 14–19.

Fay, L. P. 1988. Late Wisconsinan Appalachian herpetofaunas: relative stability in the midst of change. Annals of Carnegie Museum 57:189–220.

Fay, L. P. (n.d.) Hanover Quarry No. 1 fissure, Adams County, Pennsylvania, early (?) Irvingtonian NALMA. Letter to J. A. Holman, May 12, 1993.

Flint, R. F. 1971. Glacial and Quaternary geology. New York: John Wiley & Sons.

Ford, K. M., III. 1992. Herpetofauna of the Albert Ahrens Local Fauna (Pleistocene: Irvingtonian), Nebraska. M.S. thesis, Michigan State University, East Lansing.

Fox, R. C., and Naylor, B. G. 1982. A reconsideration of the relationships of the fossil amphibian *Albanerpeton*. Canadian Journal of Earth Sciences 19:118–128.

Francis, E. T. B. 1934. The anatomy of the salamander. Oxford: Oxford University Press.

Frost, D. R., ed. 1985. Amphibian species of the world, a taxonomic and geographical reference. Lawrence, Kans.: Allen Press and Association of Systematics Collections.

Gallagher, W. B., Parris, D. C., and Spamer, E. 1986. Paleontology, biostratigraphy, and depositional environments of the Cretaceous–Tertiary transition in the New Jersey Coastal Plain. Mosasaur 3:1–35.

Gardner, J. D. 1999a. Redescription of the geologically youngest albanerpetontid (?Lissamphibia): *Albanerpeton inexpectatum* Estes and Hoffstetter, 1976, from the Miocene of France. Annales de paléontologie 85:57–84.

Gardner, J. D. 1999b. The amphibian *Albanerpeton arthridion* and the Aptian–Albian biogeography of albanerpetontids. Palaeontology 4:529–544.

Gardner, J. D. 1999c. New albanerpetontid amphibians from the Albian to Coniacian of Utah, USA—bridging the gap. Journal of Vertebrate Paleontology 19:632–638.

Gardner, J. D. 2000a. Revised taxonomy of albanerpetontid amphibians. Acta Palaeontologica Polonica 45:55–70.

Gardner, J. D. 2000b. Albanerpetontid amphibians from the Upper Cretaceous (Campanian and Maastrichtian) of North America. Geodiversitas 22:349–388.

Gardner, J. D. 2000c. Comments on the anterior region of the skull in the Albanerpetontidae (Temnospondyli; Lissamphibia). Neues Jahrbuch für Geologie und Paläontologie, Monatshefte 2000:1–14.

Gardner, J. D. 2001. Monophyly and the affinities of albanerpetontid amphibians (Temnospondyli: Lissamphibia). Zoological Journal of the Linnaean Society 131:309–352.

Gardner, J. D. 2002. Monophyly and intra-generic relationships of *Albanerpeton* (Lissamphibia: Albanerpetontidae). Journal of Vertebrate Paleontology 22: 12–22.

Gardner, J. D. 2003a. Revision of *Habrosaurus* Gilmore (Caudata:Sirenidae) and relationships among sirenid salamanders. Palaeontology 46:1089–1122.

Gardner, J. D. 2003b. The fossil salamander *Proamphiuma cretacea* Estes (Caudata: Amphiumidae) and relationships within the Amphiumidae. Journal of Vertebrate Paleontology 23:769–782.

Gardner, J. D., and Averianov, A. O. 1998. Albanerpetontid amphibians from Middle Asia. Acta Palaeontologica Polonica 43:453–467.

Gilmore, C. W. 1928. Fossil lizards of North America. Memoirs of the National Academy of Sciences, 3rd series, No. 22, 201 pp.

Goin, C. J., and Auffenberg, W. 1955. The fossil salamanders of the family Sirenidae. Bulletin of the Museum of Comparative Zoology, Harvard University 113:497–514.

Goin, C. J., and Auffenberg, W. 1957. A new fossil salamander of the genus *Siren* from the Eocene of Wyoming. Copeia 1957:83–85.

Goin, C. J., and Auffenberg, W. 1958. New salamanders of the family Sirenidae from the Cretaceous of North America. Fieldiana: Geology 10:449–459.

Goin, C. J., and Goin, O. B. 1962. Introduction to herpetology. San Francisco and London: W. H. Freeman.

Goin, C., Goin, O., and Zug, G. 1978. Introduction to herpetology. San Francisco: Freeman and Co.
Gould, S. J. 1977. Ontogeny and phylogeny. Cambridge, Mass.: Harvard University Press.
Graham, R. W. 1976. Late Wisconsinan mammalian faunas and environmental gradients in the eastern United States. Paleobiology 2:343–350.
Graham, R. W., and Lundelius, E. L., Jr. 1984. Coevolutionary disequilibrium and Pleistocene extinctions. *In* P. S. Martin and R. G. Klein, eds., Quaternary extinctions: a prehistoric revolution. Tucson: University of Arizona Press, pp. 223–249.
Green, M., and Holman, J. A. 1977. A late Tertiary stream channel fauna from South Bijou Hill, South Dakota. Journal of Paleontology 51:543–547.
Green, N. B., and Pauley, T. K. 1987. Amphibians and reptiles in West Virginia. Pittsburgh: University of Pittsburgh Press.
Grinnell, J., and Storer, T. I. 1924. Animal life in the Yosemite. Berkeley: University of California Press.
Gut, H. J., and Ray, C. E. 1963. The Pleistocene vertebrate fauna of Reddick, Florida. Florida Academy of Sciences, Quarterly Journal 26:315–328.
Harington, C. R. 1988. Marine mammals of the Champlain Sea, and the problem of whales in Michigan. *In* N. R. Gadd, ed., The late Quaternary development of the Champlain Sea basin. Geological Association of Canada, Special Paper 35, pp. 225–240.
Harris, A. H. 1987. Reconstruction of mid-Wisconsinan environments in southern New Mexico. National Geographic Research 3:142–151.
Harris, J. M., and Jefferson, G. T., eds. 1985. Rancho La Brea: treasures of the tar pits. Los Angeles: Natural History Museum of Los Angeles County.
Hay, J. M., Ruvinsky, I., Hedges, S. B., and Maxson, L. R. 1995. Phylogenetic relationships of amphibian families inferred from DNA sequences of mitochondrial 12S and 16S ribosomal RNA genes. Molecular Biology and Evolution 12:928–937.
Hay, O. P. 1917. Vertebrates mostly from stratum No. 3 at Vero, Florida; together with description of a new species. Annual Report of the Florida Geological Survey 9:43–68.
Heatwole, H., and Carroll, R. L., eds., 2000. Amphibian biology. Vol. 4: Palaeontology, the evolutionary history of amphibians. Chipping Norton, Australia: Surrey Beatty & Sons.
Hibbard, C. W. 1940. A new Pleistocene fauna from Meade County, Kansas. Transactions of the Kansas Academy of Science 43:417–426.
Hibbard, C. W. 1949. Techniques of collecting microvertebrate fossils. Contributions from the Museum of Paleontology, University of Michigan 8:7–19.
Hibbard, C., and Dalquest, W. 1966. Fossils from the Seymour Formation of Knox and Baylor Counties, Texas, and their bearing on the late Kansan climate of the region. Contributions from the Museum of Paleontology, University of Michigan 21:1–66.
Hinderstein, B., and Boyce, J. 1977. The Miocene salamander *Batrachosauroides dissimulans* (Amphibia, Urodela) from east Texas. Journal of Herpetology 11: 369–372.
Holman, J. A. 1955. Fall and winter food of *Plethodon dorsalis* in Johnson County, Indiana. Copeia 1955:143.
Holman, J. A. 1958. The herpetofauna of Saber-tooth Cave, Citrus County, Florida. Copeia 1958:276–280.
Holman, J. A. 1959a. Amphibians and reptiles from the Pleistocene (Illinoian) of Williston, Florida. Copeia 1959:96–102.
Holman, J. A. 1959b. A Pleistocene herpetofauna near Orange Lake, Florida. Herpetologica 15:121–125.

Holman, J. A. 1961. Amphibians and reptiles of the Howard College Natural Area. Journal of the Alabama Academy of Science 32:77–87.
Holman, J. A. 1962. Additional records of Florida Pleistocene amphibians and reptiles. Herpetologica 18:115–119.
Holman, J. A. 1963. Late Pleistocene amphibians and reptiles of the Clear Creek and Ben Franklin local faunas of Texas. Journal of the Graduate Research Center 31:152–167.
Holman, J. A. 1964. Pleistocene amphibians and reptiles from Texas. Herpetologica 20:73–83.
Holman, J. A. 1965a. A small Pleistocene herpetofauna from Houston, Texas. Texas Journal of Science 27:418–423.
Holman, J. A. 1965b. A late Pleistocene herpetofauna from Missouri. Transactions of the Illinois Academy of Science 58:190–194.
Holman, J. A. 1966a. A small Miocene herpetofauna from Texas. Florida Academy of Sciences, Quarterly Journal 29:267–275.
Holman, J. A. 1966b. The Pleistocene herpetofauna of Miller's Cave, Texas. Texas Journal of Science 28:372–377.
Holman, J. A. 1967. A Pleistocene herpetofauna from Ladds, Georgia. Bulletin of the Georgia Academy of Science 25:154–166.
Holman, J. A. 1968a. A Pleistocene herpetofauna from Kendall County, Texas. Florida Academy of Sciences, Quarterly Journal 31:165–172.
Holman, J. A. 1968b. Lower Oligocene amphibians from Saskatchewan. Florida Academy of Sciences, Quarterly Journal 31:273–289.
Holman, J. A. 1969a. Pleistocene amphibians from a cave in Edwards County, Texas. Texas Journal of Science 31:63–67.
Holman, J. A. 1969b. Herpetofauna of the Pleistocene Slaton Local Fauna of Texas. Southwestern Naturalist 14:203–212.
Holman, J. A. 1969c. The Pleistocene amphibians and reptiles of Texas. Publications of the Museum, Michigan State University, Biological Series 4: 163–192.
Holman, J. A. 1971. Herpetofauna of the Sandahl Local Fauna (Pleistocene: Illinoian) of Kansas. Contributions from the Museum of Paleontology, University of Michigan 23:349–355.
Holman, J. A. 1972a. Amphibians and reptiles. *In* M. F. Skinner and C. W. Hibbard, eds., Early Pleistocene preglacial and glacial rocks and faunas of north-central Nebraska. Bulletin of the America Museum of Natural History 148: 55–71.
Holman, J. A. 1972b. Herpetofauna of the Calf Creek Local Fauna (Lower Oligocene: Cypress Hills Formation) of Saskatchewan. Canadian Journal of Earth Sciences 9:1612–1631.
Holman, J. A. 1973. New amphibians and reptiles from the Norden Bridge Fauna (Upper Miocene) of Nebraska. Michigan Academician 6:149–163.
Holman, J. A. 1974. A late Pleistocene herpetofauna from southwestern Missouri. Journal of Herpetology 8:343–346.
Holman, J. A. 1975a. Neotenic salamander remains. *In* F. Wendorf and J. Hester, eds., Late Pleistocene environments of the southern High Plains. Publications of the Fort Burgwin Research Center, No. 9, pp. 193–195.
Holman, J. A. 1975b. Herpetofauna of the WaKeeney Local Fauna (Lower Pliocene: Clarendonian) of Trego County, Kansas. *In* G. R. Smith and N. E. Friedland, eds., Studies on Cenozoic paleontology and stratigraphy in honor of Claude W. Hibbard. Papers on Paleontology, No. 12, pp. 49–66.
Holman, J. A. 1976a. The herpetofauna of the lower Valentine Formation, north-central Nebraska. Herpetologica 32:262–268.
Holman, J. A. 1976b. Owl predation on *Ambystoma tigrinum*. Herpetological Review 7:114.

Holman, J. A. 1976c. Cenozoic herpetofaunas of Saskatchewan. *In* C. S. Churcher, ed., Athlon: Essays in palaeontology in honour of Loris Shano Russell. Royal Ontario Museum, Life Sciences Miscellaneous Publication, pp. 80–92.

Holman, J. A. 1977a. Amphibians and reptiles of the Gulf Coast Miocene of Texas. Herpetologica 33:391–403.

Holman, J. A. 1977b. The Pleistocene (Kansan) herpetofauna of Cumberland Cave, Maryland. Annals of Carnegie Museum 46:157–172.

Holman, J. A. 1977c. America's northernmost Pleistocene herpetofauna (Java, north-central South Dakota). Copeia 1977:191–193.

Holman, J. A. 1978a. Herpetofauna of the Bijou Hills Local Fauna (Late Miocene: Barstovian) of South Dakota. Herpetologica 34:253–257.

Holman, J. A. 1978b. The late Pleistocene herpetofauna of Devil's Den Sinkhole, Levy County, Florida. Herpetologica 34:228–237.

Holman, J. A. 1979. Herpetofauna of the Nash Local Fauna (Pleistocene: Aftonian) of Kansas. Copeia 1979:747–749.

Holman, J. A. 1981. Florida sink hole archaeological site yields evidence of human life 12,000 years ago. Explorers Journal 59:114–116.

Holman, J. A. 1982a. The Pleistocene (Kansan) herpetofauna of Trout Cave, West Virginia. Annals of Carnegie Museum 51:391–404.

Holman, J. A. 1982b. New herpetological species and records from the Norden Bridge Fauna (Miocene: late Barstovian) of Nebraska. Transactions of the Nebraska Academy of Sciences 10:31–36.

Holman, J. A. 1984. Herpetofauna of the Duck Creek and Williams local faunas (Pleistocene: Illinoian) of Kansas. Special Publication of Carnegie Museum of Natural History, No. 8, pp. 20–38.

Holman, J. A. 1985a. Herpetofauna of Ladds Quarry. National Geographic Research 1:423–436.

Holman, J. A. 1985b. New evidence of the status of Ladds Quarry. National Geographic Research 1:569–570.

Holman, J. A. 1986. Butler Spring herpetofauna of Kansas (Pleistocene: Illinoian) and its climatic significance. Journal of Herpetology 20:568–570.

Holman, J. A. 1987. Herpetofauna of the Egelhoff site (Miocene: Barstovian) of north-central Nebraska. Journal of Vertebrate Paleontology 7:109–120.

Holman, J. A. 1992. Patterns of herpetological re-occupation of post-glacial Michigan: amphibians and reptiles come home. Michigan Academician 24:453–466.

Holman, J. A. 1993. British Quaternary herpetofaunas: a history of adaptations to Pleistocene disruptions. Herpetological Journal 3:1–7.

Holman, J. A. 1995a. Pleistocene amphibians and reptiles in North America. New York: Oxford University Press.

Holman, J. A. 1995b. Ancient life of the Great Lakes basin. Ann Arbor: University of Michigan Press.

Holman, J. A. 1996a. The large Pleistocene (Sangamonian) herpetofauna of the Williston IIIA site, north-central Florida. Herpetological Natural History 4: 35–47.

Holman, J. A. 1996b. Herpetofauna of the Trinity River Local Fauna (Miocene: early Barstovian), San Jacinto County, Texas, USA. Tertiary Research 17:5–10.

Holman, J. A. 1996c. Glad Tidings, a late Middle Miocene herpetofauna from northeastern Nebraska. Journal of Herpetology 30:430–432.

Holman, J. A. 1997. Amphibians and reptiles from the Pleistocene (late Wisconsinan) of Sheriden Pit Cave, northwestern Ohio. Michigan Academician 29: 1–20.

Holman, J. A. 1998. Pleistocene amphibians and reptiles in Britain and Europe. New York: Oxford University Press.

Holman, J. A. 2000a. Fossil snakes of North America: origin, evolution, distribution, paleoecology. Bloomington: Indiana University Press.

Holman, J. A. 2000b. Pleistocene Amphibia: evolutionary stasis, range adjustments, and recolonization patterns. *In* H. Heatwole and R. L. Carroll, eds., Amphibian biology. Vol. 4: Palaeontology, the evolutionary history of amphibians. Chipping Norton, Australia: Surrey Beatty & Sons, pp. 1445–1458.

Holman, J. A. 2001. In quest of Great Lakes Ice Age vertebrates. East Lansing: Michigan State University Press.

Holman, J. A. 2003. Fossil frogs and toads of North America. Bloomington: Indiana University Press.

Holman, J. A. 2006. Amphibians and reptiles of Michigan: a Tertiary and recent faunal adventure. Detroit: Wayne State University Press. (In press.)

Holman, J. A., and Clausen, C. J. 1984. Fossil vertebrates associated with paleo-Indian artifacts at Little Salt Spring, Florida. Journal of Vertebrate Paleontology 4:146–154.

Holman, J. A., and Grady, F. 1987. Herpetofauna of New Trout Cave. National Geographic Research 3:305–317.

Holman, J. A., and Grady, F. 1989. The fossil herpetofauna (Pleistocene: Irvingtonian) of Hamilton Cave, Pendleton County, West Virginia. National Speleological Society, Bulletin 51:34–41.

Holman, J. A., and Grady, F. 1994. A Pleistocene herpetofauna from Worm Hole Cave, Pendleton County, West Virginia. National Speleological Society, Bulletin 56:46–49.

Holman, J. A., and Harrison, D. L. 2002. A new *Thaumastosaurus* (Anura: familia incertae sedis) from the Late Eocene of England, with remarks on the taxonomic and zoogeographic relationships of the genus. Journal of Herpetology 36:621–626.

Holman, J. A., and Harrison, D. L. 2003. A new helmeted frog of the genus *Thaumastosaurus* from the Eocene of England. Acta Palaeontologica Polonica 48:157–160.

Holman, J. A., and McDonald, J. N. 1986. A late Quaternary herpetofauna from Saltville, Virginia. Brimleyana 12:85–100.

Holman, J. A., and Sullivan, R. M. 1981. A small herpetofauna from the type section of the Valentine Formation (Miocene: Barstovian), Cherry County, Nebraska. Journal of Paleontology 55:138–144.

Holman, J. A., and Voorhies, M. R. 1985. *Siren* (Caudata: Sirenidae) from the Barstovian Miocene of Nebraska. Copeia 1985:264–266.

Holman, J. A., and Winkler, A. J. 1987. A mid-Pleistocene (Irvingtonian) herpetofauna from a cave in southcentral Texas. Pearce-Sellards Series: An Occasional Publication of the Texas Memorial Museum, No. 44, 17 pp.

Holman, J. A., Bell, G., and Lamb, J. 1990. A late Pleistocene herpetofauna from Bell Cave, Alabama. Herpetological Journal 1:521–529.

Hudson, D. M., and Brattstrom, B. H. 1977. A small herpetofauna from the late Pleistocene of Newport Beach Mesa, Orange County, California. Bulletin, Southern California Academy of Sciences 76:16–20.

Huene, F. V. 1956. Paläontologie und Phylogenie der niederen Tetrapoden. Jena: Gustav Fischer.

Hulbert, R. C., Jr., ed. 2001. The fossil vertebrates of Florida. Gainesville: University Press of Florida.

Hulbert, R. C., Jr., and Morgan, G. S. 1989. Stratigraphy, paleoecology, and vertebrate fauna of the Leisey Shell Pit Local Fauna, early Pleistocene (Irvingtonian) of southwestern Florida. Papers in Florida Paleontology, No. 2, 19 pp.

Hulbert, R. C., Jr., and Pratt, A. E. 1998. New Pleistocene (Rancholabrean) vertebrate faunas from coastal Georgia. Journal of Vertebrate Paleontology 18: 412–429.

IUCN, Conservation International, and NatureServe. 2004. Global amphibian assessment. Internet: www.globalamphibians.org

Ivachnenko, M. F. 1978. Urodelans from the Triassic and Jurassic of Soviet central Asia. Paleontologicheskiy Zhurnal 1978:84–89. (In Russian.)

Johnson, E. 1987. Vertebrate remains. *In* E. Johnson, ed., Late Quaternary studies on the southwest High Plains. College Station: Texas A & M University Press, pp. 49–89.

Kasper, S., and Parmley, D. 1990. A late Pleistocene herpetofauna from the lower Texas Panhandle. Texas Journal of Science 42:289–294.

Klippel, W. E., and Parmalee, P. W. 1982. The paleontology of Cheek Bend Cave. Phase II report to the Tennessee Valley Authority [contract No. TVA-TV 49244A], 249 pp.

Krause, D. 1980. Early Tertiary amphibians from the Bighorn Basin, Wyoming. Papers on Paleontology, No. 24, pp. 69–71.

LaDuke, T. C. 1991. First record of salamander remains from Rancho La Brea. Annual meeting of the Southern California Academy of Sciences, May 1991. Abstract 7, pp. 10–11.

Larson, A. 1991. A molecular perspective on the evolutionary relationships of the salamander. Evolutionary Biology 25:211–277.

Larson, A., and Dimmick, W. W. 1993. Phylogenetic relationships of salamander families; an analysis of congruence among morphological and molecular characters. *In* D. Cannatella and D. Hillis, eds., Amphibian relationships: phylogenetic analysis of morphology and molecules. Herpetological Monographs 7:77–93.

Larson, A., Wake, D. B., Maxson, L. R., and Highton, R. 1981. Molecular phylogenetic perspective on the origins of morphological novelties in salamanders of the tribe Plethodontini (Amphibia: Plethodontidae). Evolution 35: 405–422.

Larson, J. H. 1963. The cranial osteology of neotenic and transformed salamanders and its bearing on interfamilial relationships. Ph.D. dissertation, University of Washington, Washington, DC.

Laurin, M. 1998. The importance of global parsimony and historical bias in understanding tetrapod evolution. Part I. Systematics, middle ear evolution and jaw suspension. Annales des sciences naturelles: Zoologie, 13e série 19: 1–42.

Laurin, M. 2002. Tetrapod phylogeny, amphibian origins, and the definition of the name tetrapods. Systematic Biology 51:364–369.

Laurin, M., and Reisz, R. R. 1997. A new perspective on tetrapod phylogeny. *In* S. S. Sumida and K. L. Martin, eds., Amniote origins. San Diego: Academic Press.

Layne, J. R., Jr., and Lee, R. E., Jr. 1995. Adaptions of frogs to freezing. Climatic Research 5:53–59.

Lincoln, R. J., Boxshall, G. A., and Clark, P. F. 1982. A dictionary of ecology, evolution, and systematics. Cambridge: Cambridge University Press.

Lindquist, S. B., and Bachmann, M. D. 1982. The role of visual and olfactory cues in the prey catching behavior of the tiger salamander, *Ambystoma tigrinum*. Copeia 1982:81–90.

Lundelius, E. L., Graham, R. W., Anderson, E., Guilday, J., Holman, J. A., Steadman, D. W., and Webb, S. D. 1983. Terrestrial vertebrate faunas. *In* H. E. Wright, ed., Late Quaternary environments of the United States. Vol. 1: The late Pleistocene, S. Porter, ed. Minneapolis: University of Minnesota Press, pp. 311–353.

Lynch, J. D. 1964. Additional hylid and leptodactylid remains from the Pleistocene of Texas and Florida. Herpetologica 20:141–142.

Lynch, J. D. 1965. The Pleistocene amphibians of Pit II, Arredondo, Florida. Copeia 1965:72–77.

Lynch, J. D. 1966. Additional treefrogs (Hylidae) from the North American Pleistocene. Annals of Carnegie Museum 38:265–271.

MacNamara, M. C. 1977. Food habits of terrestrial adult migrant and immature red efts of the red-spotted newt *Notophthalmus viridescens*. Herpetologica 33: 127–132.

Martin, P. S., and Klein, R. G., eds. 1984. Quaternary extinctions: a prehistoric revolution. Tucson: University of Arizona Press.

Martin, P. S., and Wright, H. E., eds. 1967. Pleistocene extinctions: the search for a cause. New Haven: Yale University Press.

Maxson, L. R., Highton, R., and Wake, D. B. 1979. Albumin evolution and its phylogenetic implications in the plethodontid salamander genera *Plethodon* and *Ensatina*. Copeia 1979:502–508.

McGowen, G., and Evans, S. E. 1995. Albanerpetontid amphibians from the Cretaceous of Spain. Nature (London) 373:143–145.

Mead, J. H., Sankey, J. T., and McDonald, H. G. 1998. Pliocene (Blancan) herpetofaunas from the Glenns Ferry Formation, southern Idaho. *In* W. A. Akersten, H. G. McDonald, D. J. Meldrum, and M. E. T. Flint, eds., And whereas . . . Papers on the vertebrate paleontology of Idaho honoring John A. White. Idaho Museum of Natural History, Occasional Paper 36, Vol. 1, pp. 94–109.

Mead, J. I., Van Devender, T. R., Cole, K. L., and Wake, D. B. 1985. Late Pleistocene vertebrates from a packrat midden in the south-central Sierra Nevada, California. Current Research in the Pleistocene 2:107–108.

Meszoely, C. 1966. North American fossil cryptobranchid salamanders. American Midland Naturalist 75:495–515.

Meszoely, C. 1967. A new cryptobranchid salamander from the Early Eocene of Wyoming. Copeia 1967:346–349.

Meylan, P. A. 1984. A history of fossil amphibians and reptiles in Florida. Florida State Museum, Plaster Jacket, No. 44, 28 pp.

Miller, L. 1944. Notes on the eggs and larvae of *Aneides lugubris*. Copeia 1944: 435–436.

Miller, M. D., Jr. 1992. Analysis of fossil salamanders from Cheek Bend Cave, Maury County, Tennessee. M.S. thesis, Appalachian State University, Boone, N.C.

Miller, W. E. 1971. Pleistocene vertebrates of the Los Angeles Basin and vicinity (exclusive of Rancho La Brea). Natural History Museum of Los Angeles County, Science Bulletin No. 10, 24 pp.

Milner, A. R. 1983. The biogeography of salamanders in the Mesozoic and early Caenozoic: a cladistic vicariance model. *In* R. W. Sims, J. H. Price, and P. E. S. Whalley, eds., Evolution, time and space: the emergence of the biosphere. New York: Academic Press; Systematics Association, Special Vol. 23, pp. 431–468.

Milner, A. R. 1988. The relationships and origin of living amphibians. *In* M. J. Benton, ed., The phylogeny and classification of tetrapods. Oxford: Clarendon Press; Systematics Association, Special Vol. 35A, pp. 59–102.

Milner, A. R. 1993. The Paleozoic relatives of the lissamphibians. Herpetological Monographs 7:8–27.

Milner, A. R. 1994. Late Triassic and Jurassic amphibians: fossil record and phylogeny. *In* N. C. Frazer and H-D. Sues, eds., In the shadow of the dinosaurs. Cambridge: Cambridge University Press, pp. 5–22.

Milner, A. R. 2000. Mesozoic and Tertiary Caudata and Albanerpetontidae. *In* H. Heatwole and R. L. Carroll, eds., Palaeontology, the evolutionary history of amphibians. Chipping Norton, Australia: Surrey Beatty & Sons, pp. 1412–1444.

Morgan, A. H., and Grierson, M. C. 1932. Winter habits and yearly food consumption of adult spotted salamander newts, *Triturus viridescens*. Ecology 13: 54–62.

Murray, P. 1984. Extinction down under: a bestiary of extinct Australian late Pleistocene monotremes and marsupials. *In* P. S. Martin and R. G. Klein, eds., Quaternary extinctions: a prehistoric revolution. Tucson: University of Arizona Press, pp. 600–628.

Naylor, B. 1978a. The frontosquamosal arch in newts as a defense against predators. Canadian Journal of Zoology 56:2211–2216.

Naylor, B. 1978b. The systematics of fossil and Recent salamanders (Amphibia: Caudata), with special reference to the vertebral column and trunk musculature. Ph.D. dissertation, University of Alberta, Edmonton.

Naylor, B. 1978c. The earliest known *Necturus* (Amphibia, Urodela), from the Palaeocene Ravenscrag Formation of Saskatchewan. Journal of Herpetology 12:565–569.

Naylor, B. 1979. A new species of *Taricha* (Caudata: Salamandridae), from the Oligocene John Day Formation of Oregon. Canadian Journal of Earth Sciences 16:970–973.

Naylor, B. 1981a. Cryptobranchid salamanders from the Paleocene and Miocene of Saskatchewan. Copeia 1981:76–86.

Naylor, B. 1981b. A new salamander from the family Batrachosauroididae from the Late Miocene of North America, with notes on other batrachosauroidids. PaleoBios 39:1–14.

Naylor, B., and Fox, R. C. 1993. A new ambystomatid salamander, *Dicamptodon antiquus* n. sp., from the Paleocene of Alberta, Canada. Canadian Journal of Earth Sciences 30:814–818.

Naylor, B., and Krause, D. 1981. *Piceoerpeton*, a giant early Tertiary salamander from western North America. Journal of Paleontology 55:505–523.

Nessov, L. A. 1988. Late Mesozoic amphibians and lizards of Soviet middle Asia. Acta Zoologica Cracoviensia 31:475–486.

Nessov, L. A., Fedorov, P. V., Potanov, D. O., and Golovyeva, L. S. 1996. The structure of the skull of caudate amphibians collected from the Jurassic of Kirgizstan and the Cretaceous of Uzbekistan. Vestnik Sankt-Petersburgskogo Universiteta, Seriia 7, Geologia 7:3–11. (In Russian.)

Owen-Smith, N. 1987. Pleistocene extinctions: the pivotal role of megaherbivores. Paleobiology 13:351–362.

Palmer, A. R., and Geissman, J., compilers. 1999. 1999 geologic time scale. Washington, DC: Geological Society of America.

Parmalee, P. W., Oesch, R. D., and Guilday, J. E. 1969. Pleistocene and Recent vertebrate faunas from Crankshaft Cave, Missouri. Reports of Investigations, Illinois State Museum, No. 14, 37 pp.

Parmley, D. 1984. Herpetofauna of the Coffee Ranch Local Fauna (Hemphillian land mammal age) of Texas. *In* M. Horner, ed., Festschrift for Walter W. Dalquest in honor of his sixty-sixth birthday. Wichita Falls, Tex.: Department of Biology, Midwestern State University, pp. 97–106.

Parmley, D. 1986. Herpetofauna of the Rancholabrean Schulze Cave Local Fauna of Texas. Journal of Herpetology 22:82–87.

Parmley, D. 1988. Additional Pleistocene amphibians and reptiles from the Seymour Formation, Texas. Journal of Herpetology 22:82–87.

Parsons, T. S., and Williams, E. E. 1963. The relationships of modern Amphibia: a re-examination. Quarterly Review of Biology 38:26–53.

Peabody, F. 1954. Trackways of an ambystomid salamander from the Paleocene of Montana. Journal of Paleontology 28:79–83.

Peabody, F. 1959. Trackways of living and fossil salamanders. University of California Publications in Zoology, No. 63, 72 pp.

Petranka, J. W. 1998. Salamanders of the United States and Canada. Washington and London: Smithsonian Institution Press.

Pinder, A. W., Storey, K. B., and Ultsch, G. R. 1992. Estivation and hibernation. *In* M. E. Feder and W. W. Burggren, eds., Environmental physiology of the amphibians. Chicago: University of Chicago Press, pp. 250–274.

Pough, F. H., Andrews, R. M., Cadle, J. E., Crump, M. L., Savitzky, A. H., and Wells, K. D. 1998. Herpetology. Upper Saddle River, N.J.: Prentice Hall.

Powers, J. H. 1907. Morphological variation and its causes in *Ambystoma tigrinum.* Studies of the University of Nebraska 7:197–273.

Preston, R. E. 1979. Late Pleistocene cold-blooded vertebrate faunas from the mid-continental United States, I. Reptilia: Testudines, Crocodilia. Papers on Paleontology, No. 19, 53 pp.

Prothero, D. R., and Emry, R. J., eds. 1996. The Eocene–Oligocene transition in North America. Cambridge: Cambridge University Press.

Rage, J-C., Marshall, L. G., and Gayet, M. 1993. Enigmatic Caudata (Amphibia) from the Upper Cretaceous of Gondwana. Geobios 26:515–519.

Raymond, L. R. 1991. Seasonal activity of *Siren intermedia* in northwestern Louisiana (Amphibia: Sirenidae). Southwestern Naturalist 36:144–147.

Rieppel, O., and Grande, L. 1998. A well-preserved fossil amphiumid (Lissamphibia: Caudata) from the Eocene Green River Formation of Wyoming. Journal of Vertebrate Paleontology 18:700–708.

Roček, Z., and Lamaud, P. 1995. *Thaumastosaurus bottae* De Stefano, 1903, an anuran with Gondwanan affinities from the Eocene of Europe. Journal of Vertebrate Paleontology 15:506–515.

Rogers, K. L. 1976. Herpetofauna of the Beck Ranch Local Fauna (Upper Pliocene: Blancan) of Texas. Paleontological Series (East Lansing) 1:167–200.

Rogers, K. L. 1982. Herpetofauna of the Courland Canal and Hall Ash local faunas (Pleistocene: early Kansan) of Jewell Co., Kansas. Journal of Herpetology 16:174–177.

Rogers, K. L. 1984. Herpetofaunas of the Big Springs and Hornet's Nest quarries (northeastern Nebraska, Pleistocene: late Blancan). Transactions of the Nebraska Academy of Sciences 12:81–94.

Rogers, K. L. 1987. Pleistocene high altitude amphibians and reptiles from Colorado (Alamosa Local Fauna: Pleistocene, Irvingtonian). Journal of Vertebrate Paleontology 7:82–95.

Rogers, K. L., Repenning, C. A., Luiszer, F. G., and Benson, R. D. 2000. Geology, history, stratigraphy, and paleontology of SAM Cave, north-central New Mexico. New Mexico Geology 22:89–117.

Ruthven, A. G., Thompson, C., and Gaige, H. T. 1928. The herpetology of Michigan. Ann Arbor: University of Michigan. Handbook Series, No. 3. 229 pp.

Sahni, A. 1972. The vertebrate fauna of the Judith River Formation, Montana. Bulletin of the American Museum of Natural History 147:321–412.

Salthe, S. 1967. Courtship patterns and phylogeny of the urodeles. Copeia 1967: 100–117.

Salthe, S. N. 1969. Reproductive modes and the number and sizes of ova in the urodeles. American Midland Naturalist 81:467–490.

Sanchiz, B. 1998. Handbuch der Paläoherpetologie. Part 4: Salientia. Munich: Verlag Dr. Friedrich Pfeil, 275 pp.

Sanchiz, B., Schleich, H. H., and Esteban, M. 1993. Water frogs (Ranidae) from the Oligocene of Germany. Journal of Herpetology 27:486–489.

Saunders, J. J. 1977. Late Pleistocene vertebrates of the western Ozark highlands. Reports of Investigation, Illinois State Museum, No. 33, 118 pp.

Schmidt, K. P. 1953. A check list of North American amphibians and reptiles 6th ed. Chicago: University of Chicago Press.

Schoch, R. R. 1998. Homology of cranial ossifications in urodeles: significance of developmental data for fossil basal tetrapods. Neues Jahrbuch für Geologie und Paläontologie, Monatshefte 1998:1–25.

Schoch, R. R., and Carroll, R. L. 2003. Ontogenetic evidence for the Paleozoic ancestry of salamanders. Evolution & Development 5:314–324.

Setoguchi, T. 1978. Paleontology and geology of the Badwater Creek area, central Wyoming. Part 16: The Cedar Ridge Local Fauna (Late Oligocene). Bulletin of Carnegie Museum of Natural History, No. 9, 61 pp.

Smith, H. T. U. 1940. Geological studies in southwestern Kansas. Kansas Geological Survey, Bulletin, No. 34, 212 pp.

Soler, E. I. 1950. On the status of the family Desmognathidae (Amphibia, Caudata). University of Kansas Science Bulletin 33:459–480.

Stuart, A. J. 1991. Mammalian extinctions in the late Pleistocene of northern Eurasia and North America. Biological Reviews of the Cambridge Philosophical Society 66:453–562.

Sullivan, R. M. 1991. Paleocene Caudata and Squamata from Gidley and Silberling quarries, Montana. Journal of Vertebrate Paleontology 11:293–301.

Tashjian, P. 1990. The sedimentology and stratigraphy of a fossiliferous layer in the Upper Cretaceous (Campanian) Englishtown/Marshalltown Formation near Ellisdale, N.J. M.S. thesis, Temple University, Philadelphia.

Taylor, E., and Hesse, C. 1943. A new salamander from the beds of San Jacinto County, Texas. American Journal of Science 241:185–193.

Tihen, J. A. 1942. A colony of fossil neotenic *Ambystoma tigrinum*. University of Kansas Science Bulletin 28:189–198.

Tihen, J. 1955. A new Pliocene species of *Ambystoma*, with remarks on other fossil ambystomatids. Contributions from the Museum of Paleontology, University of Michigan 12:229–244.

Tihen, J. A. 1958. Comments on the osteology and phylogeny of ambystomatid salamanders. Bulletin of the Florida State Museum, Biological Sciences 3: 1–50.

Tihen, J. A. 1974. Two new North American Miocene salamandrids. Journal of Herpetology 8:211–218.

Tihen, J. A., and Chantell, C. J. 1963. Urodele remains from the Valentine Formation of Nebraska. Copeia 1963:505–510.

Tihen, J. A., and Wake, D. B. 1981. Vertebrae of plethodontid salamanders from the Lower Miocene of Montana. Journal of Herpetology 15:35–40.

Trueb, L. 1993. Patterns of cranial diversity among the Lissamphibia. *In* J. Hanken and B. K. Hall, eds., The skull. Chicago: University of Chicago Press, pp. 255–343.

Trueb, L., and Cloutier, R. 1991. A phylogenetic investigation of the inter- and intrarelationships of the Lissamphibia (Amphibia: Temnospondyli). *In* H-P. Schultze and L. Trueb, eds., Origins of the higher groups of tetrapods: controversy and consensus. Ithaca, N.Y.: Cornell University Press, pp. 223–313.

Ultsch, G. R. 1973. Observations on the life history of *Siren lacertina*. Herpetologica 29:304–305.

Van Dam, G. H. 1978. Amphibians and reptiles (in the late Pleistocene Baker Bluff site, Tennessee). Bulletin of Carnegie Museum of Natural History, No. 11, pp. 19–25.

Van Devender, T. R., and Worthington, R. D. 1977. The herpetofauna of the Howell's Ridge Cave and the paleoecology of the northwestern Chihuahuan

Desert. *In* R. H. Wauer and D. H. Riskind, eds., Transactions of the Symposium on biological resources of the Chihuahuan Desert region, United States and Mexico. U.S. Department of the Interior, National Park Service Transactions and Proceedings Series, No. 3, pp. 85–106.

Van Frank, R. 1955. *Palaeotaricha oligocenica,* new genus and species, an Oligocene salamander from Oregon. Breviora, No. 45, 12 pp.

Van Valen, L., and Sloan, R. 1965. The earliest primates. Science 150:743–745.

Voorhies, M. R. 1990. Vertebrate paleontology of the proposed Norden Reservoir area, Brown, Cherry, and Keya Paha counties, Nebraska. Division of Archeological Research, Department of Anthropology, University of Nebraska, Lincoln, Technical Report 82-09, 731 pp.

Voorhies, M. R., Holman, J. A., and Xue X-X. 1987. The Hottell Ranch rhino quarries (basal Ogallala: medial Barstovian), Banner County, Nebraska. Part I: Geological setting, faunal lists, lower vertebrates. Contributions to Geology, University of Wyoming 25:55–69.

Wake, D. B. 1963. Comparative osteology of the plethodontid salamander genus *Aneides.* Journal of Morphology 113:77–118.

Wake, D. 1966. Comparative osteology and evolution of the lungless salamanders, family Plethodontidae. Memoirs of the Southern California Academy of Sciences 4:1–111.

Weigel, R. D. 1962. Fossil vertebrates of Vero, Florida. Special Publication, Florida Geological Survey, No. 10, 59 pp.

Wilson, M. V. H. 1980. Oldest known *Esox* (Pisces, Esocidae), part of a new Paleocene teleost fauna from western Canada. Canadian Journal of Earth Sciences 17:307–312.

Wilson, V. V. 1975. The systematics and paleoecology of two Pleistocene herpetofaunas of the southeastern United States. Ph.D. dissertation, Michigan State University, East Lansing.

Winkler, D. A., Murry, P. A., and Jacobs, L. L. 1990. Early Cretaceous (Comanchean) vertebrates of central Texas. Journal of Vertebrate Paleontology 10: 95–116.

Wood, H. E., II, Chaney, R. W., Clark, J., Colbert, E. H., Jepson, G. L., Reedside, J. B., Jr., and Stock, C. 1941. Nomenclature and correlation of the North American continental Tertiary. Geological Society of America Bulletin 52: 1–48.

Wood, J. T., and Goodwin, O. K. 1954. Observations on the abundance, food, and feeding behavior of the newt, *Notophthalmus viridescens viridescens* (Rafinesque) in Virginia. Journal of the Elisha Mitchell Science Society 70:27–30.

Woodburne, M. O., ed. 1987. Cenozoic mammals in North America: geochronology and biostratigraphy. Berkeley: University of California Press.

Worthington, R. D., and Wake, D. B. 1972. Patterns of regional variation in the vertebral column of terrestrial salamanders. Journal of Morphology 137:257–278.

Yatkola, D. A. 1976. Mid-Miocene lizards from western Nebraska. Copeia 1976: 645–654.

Zug, G. R. 1993. Herpetology: an introductory biology of amphibians and reptiles. San Diego: Academic Press.

FIGURE CREDITS

INDIVIDUALS

Rosemarie Attilio: 15, 31A–31D, 45, 57
Walter Auffenberg: 29, 30, 31G, 31H
Gene M. Christman: 43, 48
Merald Clark: 5, 38A, 56A, 57A, 58C, 58D
Sherri L. DeFauw: 2–4
James D. Gardner: 32, 49
Barbara Gudgeon: 80
Howard Hamman: 25–28, 64, 69, 70, 72
Diana Holingdale: 52
Donna R. Holman: 59 (in part), 60, 61F, 61G
J. Alan Holman: 1, 6–10, 12–14, 16–21, 33, 34D, 34E, 46, 47, 50, 51, 67, 68, 75–78, 81
Jane Kaminski: 38B–38D, 39, 56B–56E
Mark Marcuson: 35
Laszlo L. Meszoely: 34A–34C, 71
Bruce G. Naylor: 36, 40, 41
Teresa Petersen: 79
Lynn Sloan: 53 (in part)
Patricia Stinson: 31E, 31F
Robert Sullivan: 65
J. A. Tihen: 11, 54, 58A, 58B, 59 (in part), 61A–61C, 62A–62F
Alex Worm: 53
C. Vanderslice: 42, 44

JOURNALS AND INSTITUTIONS

Figures 2–4. From DeFauw (1989), courtesy of the *Michigan Academician.*
Figure 11. From Tihen (1958), courtesy of the *Bulletin of the Florida State Museum, Biological Sciences.*
Figure 22. From Holman (2003), courtesy of Indiana University Press.
Figure 23. From Holman (2003), courtesy of Indiana University Press.
Figure 24. From Holman (2003), courtesy of Indiana University Press.
Figure 25. From Estes (1964), courtesy of the University of California Press.
Figure 26. From Estes (1964), courtesy of the University of California Press.
Figure 27. From Estes (1964), courtesy of the University of California Press.
Figure 28. From Estes (1964), courtesy of the University of California Press.
Figures 29, 30, 31G, 31H. From Goin and Auffenberg (1955), courtesy of the Museum of Comparative Zoology, Harvard University.
Figure 32. From Gardner (2003a), courtesy of *Palaeontology.*
Figure 34A–34C. From Meszoely (1966), courtesy of the *American Midland Naturalist.*
Figure 36. From Naylor (1981a), courtesy of *Copeia.*

Figure 38. From Holman (1977b, 1982a), courtesy of *Annals of Carnegie Museum.*

Figure 39. From Holman (1982a), courtesy of *Annals of Carnegie Museum.*

Figures 40, 41. From Naylor (1978c), courtesy of the *Journal of Herpetology.*

Figure 42. From Clark (1985), courtesy of the *Journal of Herpetology.*

Figure 43. From Tihen and Wake (1981), courtesy of the *Journal of Herpetology.*

Figure 44. From Clark (1985), courtesy of the *Journal of Herpetology.*

Figure 48. From Tihen and Wake (1981), courtesy of the *Journal of Herpetology.*

Figure 49. From Gardner (2003b), courtesy of the *Journal of Vertebrate Paleontology.*

Figure 52. From Naylor and Fox (1993), courtesy of the *Canadian Journal of Earth Sciences.*

Figure 53. From Rogers (1987), courtesy of the *Journal of Vertebrate Paleontology.*

Figure 54. From Tihen (1955), courtesy of *Contributions from the Museum of Paleontology, University of Michigan.*

Figure 56A. From Holman (1975b), courtesy of *Papers on Paleontology.*

Figure 57A. From Rogers (1976), courtesy of *Publications of the Museum, Michigan State University, Paleontological Series (East Lansing).*

Figures 57B–57F. From Holman (1987), courtesy of the *Journal of Vertebrate Paleontology.*

Figure 58A, 58B. From Tihen (1958), courtesy of the *Bulletin of the Florida State Museum, Biological Series.*

Figure 59. From Tihen (1942), courtesy of the *University of Kansas Science Bulletin.*

Figure 60. From Holman (1968b), courtesy of the *Florida Academy of Sciences, Quarterly Journal.*

Figures 61A–61C. From Tihen (1974), courtesy of the *Journal of Herpetology.*

Figures 61D, 61E. From Estes (1963), courtesy of the *Florida Academy of Sciences, Quarterly Journal.*

Figure 61F, 61G. From Holman (1966a), courtesy of the *Florida Academy of Sciences, Quarterly Journal.*

Figure 62 (in part). From Naylor (1979), courtesy of the *Canadian Journal of Earth Sciences.*

Figure 62A–62F. From Tihen (1974), courtesy of the *Journal of Herpetology.*

Figure 63B. From Hinderstein and Boyce (1977), courtesy of the *Journal of Herpetology.*

Figure 64. From Estes (1969b), courtesy of *Copeia.*

Figure 65. From Sullivan (1991), courtesy of the *Journal of Vertebrate Paleontology.*

Figure 66. From Estes (1975), courtesy of *Herpetologica.*

Figure 69. From Estes (1969a), courtesy of *Nature (London).*

Figure 70, 71. From Estes (1976), courtesy of the *Journal of Paleontology.*

Figure 72. From Estes (1964), courtesy of the University of California Press.

Figure 73. From Fox and Naylor (1982), courtesy of the *Canadian Journal of Earth Sciences.*

Figure 78. From Holman (2003), courtesy of Indiana University Press.

Figure 79. From Holman (2003), courtesy of Indiana University Press.

Figure 80. From Holman (2001), courtesy of the Michigan State University Press.

Figure 81. From Holman (2001), courtesy of the Michigan State University Press.

GENERAL INDEX

Note: The general index covers chapters 1 and 3. The taxonomic and site indices to follow cover chapter 2 only.

TAXONOMIC INDEX

SITE INDEX

Fossil salamander sites are listed alphabetically by provinces and states, chronologically by time units from oldest and youngest, and then alphabetically by site names. Sites are abbreviated in this index because terms such as "fauna," "local fauna," "prospect," "locality," and "quarry" are often used inconsistently. Abbreviations: **JU** = Jurassic; **CR** = Cretaceous; **PA** = Paleocene; **EO** = Eocene; **OL** = Oligocene; **MI** = Miocene; **PLI** = Pliocene; and **PLE** = Pleistocene.

J. Alan Holman, Professor and Curator Emeritus of Vertebrate Paleontology at Michigan State University, has written seven books, including *Fossil Snakes of North America* (Indiana University Press, 2000) and *Fossil Frogs and Toads of North America* (Indiana University Press, 2003).

Milton Keynes UK
Ingram Content Group UK Ltd.
UKHW050829300723
425972UK00003B/73